Applied Environmental Science and Engineering for a Sustainable Future

Series editors

Applied Environmental Science and Engineering for a Sustainable Future (AESE) series covers a variety of environmental issues and how they could be solved through innovations in science and engineering. Our societies thrive on the advancements in science and technology which pave the way for better standard of living. The adverse effect of such improvements is the deterioration of the environment. Thus, better catchment management in order to sustainably manage all types of resources (including water, minerals and others) is of paramount importance. Water and wastewater treatment and reuse, solid and hazardous waste management, industrial waste minimisation, soil restoration and agriculture as well as myriad of other topics needs better understanding and application. This book series aims at fulfilling such a task in coming years.

More information about this series at http://www.springer.com/series/13085

Faisal I. Hai · Chettiyappan Visvanathan
Ramaraj Boopathy
Editors

Sustainable Aquaculture

Springer

Editors
Faisal I. Hai
Faculty of Engineering and Information Sciences
University of Wollongong
Wollongong, NSW
Australia

Ramaraj Boopathy
Nicholls State University
Thibodaux, LA
USA

Chettiyappan Visvanathan
Asian Institute of Technology
Klongluang Pathumthani
Thailand

Applied Environmental Science and Engineering for a Sustainable Future
ISBN 978-3-030-10341-5 ISBN 978-3-319-73257-2 (eBook)
https://doi.org/10.1007/978-3-319-73257-2

Printed on acid-free paper

This Springer imprint is published by Springer Nature
The registered company is Springer International Publishing AG
The registered company address is: Gewerbestrasse 11, 6330 Cham, Switzerland

Preface

Global human population will reach nine billion by 2050. The major food source to sustain such a large population is projected to be seafood. Aquaculture will be the prime source of seafood by 2030 according to the recent World Bank report. Fish farming can help provide livelihoods and feed the global population if practiced responsibly. For an aquaculture system to be truly sustainable, it must meet sustainability standard in three major areas including economic sustainability, environmental sustainability, and social and community sustainability.

We need to establish best aquaculture practice methods to avoid significant disruption to the ecosystem, the loss of biodiversity and substantial pollution impact to the environment. The system must be a viable business proposition with long-term prospects. Aquaculture system should contribute to community well-being and must be socially responsible.

Aquaculture can make an important contribution to global food security, but new investment is needed to meet the anticipated demand. Generally, small scale and organic growth of aquaculture has made it difficult to regulate and it contributes to the high level of risk to the new investors. The major risk factors in aquaculture are determined primarily by water management, production intensity, and closeness of fish farms to one another.

Life cycle assessment (LCA) defines sustainability in the broader sense by comparing food production systems in terms of impact on processes that govern global biogeochemical cycles. Thus concerted research should focus on development of a simplified biodiversity/water quality index of sustainability at the ecosystem level and adapt spatial planning to aquaculture zoning.

An effort has been made in this book to include important relevant recent research topics in sustainable aquaculture practices. This book contains chapters that cover socio-economic and environmental assessment for sustainable aquaculture production (Chaps. 1 and 2). Particularly, Chap. 8 presents an analysis of carbon footprint under an intensive aquaculture regime. Chapters 3 and 5 present sustainable fishing methods, while Chap. 4 critically assesses the aspect of sustainable aquaculture feed. Chapters 6 and 7 discuss aquaponics as a niche for sustainable modern aquaculture. The effect of use of pharmaceuticals to prevent fish disease on the surrounding marine environment is an emerging area of concern, and a critical discussion on this aspect is included in Chap. 9. The spread of organic waste and nutrients released by fish farms to natural water bodies has raised considerable concerns. Therefore, the methods to prevent their dispersion and removal (treatment) are the focus of Chap. 10. We believe that the current book will be very helpful to academician, researchers, and policy-makers in the area of aquaculture.

The editors of this book thankfully acknowledge the assistance of Muhammad Bilal Asif of the University of Wollongong in some of the baseline analyses. This book is a part of the book series "Applied Environmental Science & Engineering for a Sustainable Future (AESE)", and we gratefully acknowledge the cooperation of the series editors V. Jegatheesan, L. Shu, P. Lens, and C. Chiemchaisri.

Last but not the least, the editors are indebted to their family members for their wholehearted cooperation.

Wollongong, Australia — Faisal I. Hai
Klongluang Pathumthani, Thailand — Chettiyappan Visvanathan
Thibodaux, USA — Ramaraj Boopathy

Contents

Chapter 1
Aquaculture and the Environment: Towards Sustainability

Krishna R. Salin and Gabriel Arome Ataguba

Abstract The contribution of aquaculture to global fish production has increased in the last twenty years with the production level reaching 73.8 million tonnes in 2014, about 44% of total fish production. Asian and African aquaculture production accounts for a greater proportion of growth in aquaculture output. Aquaculture contributes to livelihoods as well as revenue in several countries even though the economic conditions have been inclement and environmental problems persist. Aquaculture will have to continue to grow to meet the increasing demand for fish. But growth would not be sustainable if the planning and management are not improved significantly. There is a need for local, national and international planning and management to cater for environmental, social, economic, health and animal welfare concerns. These form the core of best management practice in aquaculture. Aquaculture can impact on the environment negatively considering genetics, water quality, ecology, health and resource use while the environment affects aquaculture on three fronts: the cultured species, culture system and overall feasibility. These put together will demand some management effort in order to ensure sustainability of aquaculture depending on the application of site selection and carrying capacity assessment, aquaculture hazard and risk analysis, ecosystem-based approach to aquaculture, aquaculture governance and planning, and aquaculture certification and standards. These are discussed in this chapter.

Keywords Sustainable aquaculture · Intensification · Ecosystem approach
Risk analysis · Certification

K. R. Salin (✉) · G. Arome Ataguba
Aquaculture and Aquatic Resources Management, Asian Institute of Technology, Klong Luang, Pathumthani 12120, Thailand
e-mail: salinkr@ait.ac.th

G. Arome Ataguba
Department of Fisheries and Aquaculture, University of Agriculture, P.M.B. 2373, Makurdi, Nigeria

F. I. Hai et al. (eds.), *Sustainable Aquaculture*, Applied Environmental Science and Engineering for a Sustainable Future,
https://doi.org/10.1007/978-3-319-73257-2_1

1.1 Aquaculture Growth

Aquaculture production in the world (excluding aquatic plants) has grown by about 62.2% from the production level of 45.4 million tonnes in 2004 to 73.8 million tonnes in 2014, and today it contributes 44% to total fish production worldwide (FAO 2014, 2016). The increase in aquaculture production is expected to be sustained via increased production from Asia and Africa with the expansion of intermediate systems and small-scale pond aquaculture, which will be aided by sound nutrition (Hasan 2001).

Aquaculture contributes to livelihoods as well as revenue in several countries even though the economic conditions have been inclement and environmental problems persist. Without considering the secondary fisheries sector and other value chain stakeholders, the FAO (2016) estimates that as at 2014, there are about 57 million people engaged in the fisheries and aquaculture sector with aquaculture accounting for about 33% of this population and Asia alone has 96% of world fish farmers. The production of fish from wild fisheries has stagnated over the last decade (Fig. 1.1) while aquaculture production and per capita supply have increased, accompanied by increase in population. However, annual per capita consumption of fish is disproportionate across the regions of the globe with a projected level of 21 kg by 2022 following increased consumption in developed nations and greater growth in Asia and Oceania but with weak development in Africa (OECD/FAO 2015).

Fast growth of aquaculture without proper planning and management has raised increasing concern over its sustainability. Aquaculture will have to continue to grow to meet the increasing demand for fish. But growth would not be sustainable if the planning and management are not improved significantly. There is a need for

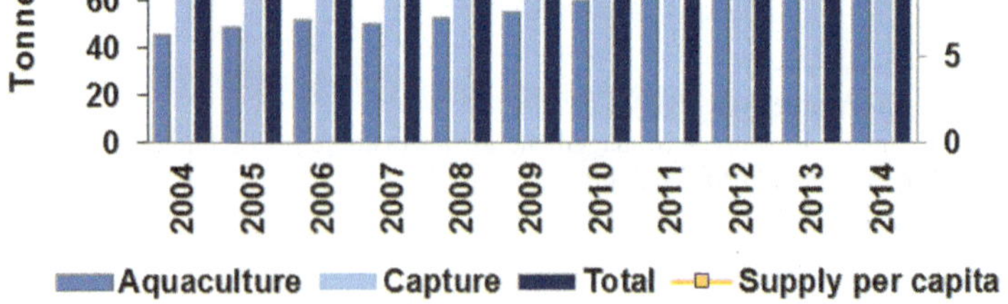

Fig. 1.1 World fish production and supply: 2004–2014. *Source* FAO (2007a, 2009, 2012, 2014, 2016)

local, national and international planning and management to cater for environmental, social, economic, health and animal welfare concerns. These form the core of best management practice as regards aquaculture.

Given the necessity to survive and maintain livelihoods among people who are employed in the aquaculture sector as well as those who are traditionally involved in fish farming both in Asia and in Sub-Saharan Africa, the exploitation of aquatic resources for aquaculture will continue. Diversity and flexibility in income generation as well as the assurance of food security are two important benefits derivable from these resources (Edwards 2002). Although aquaculture can be used as a means of eliminating unemployment and poverty, it will be totally unacceptable if the environmental impacts associated with it are sacrificed at the altar of poverty alleviation. Laxity in management and mishaps are two factors that bring about negative effects of aquaculture on the environment (Nugent 2009; Dominguez and Martín 2004).

The production of shrimps in many parts of the world have led to removal of mangrove vegetation while in some areas, inland aquaculture of marine species poses a threat to the fidelity of freshwater. In addition, the use of wild fish to feed cultured fish has become an issue of great concern. Tacon and Metian (2008) reported the empirical values for fish meal and fish oil use as against wet weight of fish gained in growth (Fish in-Fish out) for several species. Carnivorous species have the highest values in comparison with herbivorous and omnivorous fish. Filter feeding species such as oysters, clam, scallops and mussels are good nutrient absorbers and can utilize natural feed in water hence taking excessive nutrient load from the water. Integrated Multi-trophic Aquaculture uses waste products from other species as input for another species hence a combination of fed species and non-fed species in correct proportions alongside species that can synthesize food from inorganic sources such as seaweeds will result in a balanced feeding ecology bearing in mind site specificity, carrying capacity and food safety (Barrington et al. 2009).

Aquaculture makes use of land, water, wild fish and other natural resources in order to provide the right conditions for the cultured organisms to grow. Aquaculture covered about 18.8 million ha of land worldwide as at 2012 (Waite et al. 2014) while over 1.4 million ha of rice fields were used for aquaculture in China as at 2008 (FAO 2011a). Aquaculture tends to have a land utilization advantage over agriculture with the former utilizing only 0.5% of land (22.5 million ha) compared to agriculture (4.9 billion ha) in 2010 and yet producing more tonnage of fish per hectare (Boyd and McNevin 2014). Aquaculture modifies the environment, habitats, flora and fauna, scenery, proximal or in vivo water bodies as well as soil (Dosdat 2009). Continual consumption of these resources by aquaculture without a thought about sustainability will lead to depletion notwithstanding the competing claims on these resources by other sectors of productive economy. The culture system holds the key to ensuring sustainability and this according to Dalsgaard et al. (1995) can be achieved by focusing on the system and its ecology with a view to minimise the use of external inputs and to maximize the output in an integrated system. This is basically the core concept behind the ecosystem approach to aquaculture (FAO 2010) and it encompasses social dimensions (Staples and Funge-Smith 2009; Johnson 2007), governance (White and Diego-McGlone 2008) and climate change (Burrows et al. 2010).

1.1.1 Production Systems

Aquaculture production systems can be classified using several schemes. Production systems vary depending on several factors but the basic differences between them lie in water use and feeding (Emerson 1999).

1.1.1.1 Level of Intensity

The terms intensive, semi-intensive and extensive are commonly used to define culture methods. In practice, the distinction between them is often less than clear. They are, however, generally linked to the level of management input (Huntingford et al. 2012).

Aquaculture is classified according to the intensity of operations, in terms of nutrient inputs, areas used and stocking levels (Chuenpagdee et al. 2008; WRC-Report 2010). Today a lot of questions have been raised as to what constitutes each of these categories. Although Stevenson et al. (2007) were of the opinion that classifying aquaculture systems based on intensity was not easy, they maintained that the use of important variables such as stocking density, feeding rate and fertilizer application rate was necessary. The classification based on intensity as given here makes use of stocking density and feed/fertilizer application and management. However, Stevenson et al. (2007) believe that definition and measurement issues are necessary to classify production based on intensity and opined that an economic view be applied considering the fact that aquaculture production uses variable inputs (fry, feeds, fertilizer) in relation to land which is fixed.

Shang (1981), WRC-Report (2010) and Chuenpagdee et al. (2008) have given classifications based on the use of inputs as adapted above. However, as with economic measures of partial productivity, the inputs can be substituted for one another to some extent so that measuring one input cannot be totally satisfactory. With a multivariate approach to classification we can look at the particular sets of combinations of inputs that currently define production practices (Stevenson et al. 2007).

1.1.1.2 Culture Units

Earthen ponds, tanks, cages, pens and raceways are common culture units used to culture aquatic organisms (Photo 1.1). These units have to be constructed on land or in water and do carry with them an environmental impact. Cages and pens take

Photo 1.1 Intensive shrimp culture ponds are lined with HDPE liners, and well aerated with intensive feeding and regular water quality monitoring. High survival and good growth are obtained under this system. Effluents are regularly discharged from the pond by pumping from a central pit throughout the crop

advantage of the natural water resources available to produce fish without recourse to land. This is ideal for marine coastal areas and inland water bodies where current is low [<1.0 m s^{-1} Chen et al. (2008)].

1.1.1.3 Species Combination

The culture of single species (monoculture), two or more species (polyculture) and more recently a species integrated combination (Integrated Multi-Trophic Aquaculture; IMTA) are three examples of species combination as an aquaculture system. One of the fundamental concepts of IMTA is that animals and plants in the system must provide a benefit to the system and/or have significant economic value (Butterworth 2010). In this system, the by-products from one species are used as inputs (fertilizers, food and energy) for another such that fed aquaculture species (e.g. finfish/shrimp) are combined, in the appropriate proportions, with organic extractive aquaculture species (e.g. suspension feeders/deposit feeders/herbivorous fish) and inorganic extractive aquaculture species (e.g. seaweeds). This enables a balanced ecosystem management approach aimed at environmental sustainability through bio-mitigation, economic stability through product diversification and risk reduction, and social acceptability through better management practices (Barrington et al. 2009).

1.2 Threats from Aquaculture to the Environment

Aquaculture relies on water as a medium for holding the organisms under culture hence their survival depends largely on the quality of water as determined by parameters like dissolved oxygen, CO_2, carbonates, pH, NH_3, NO_2^- and NO_3^- among others. Water quality is impacted by aquaculture activities considering the use of feed, the release of waste by cultured organisms and the difficulty in separating waste from the water, which in turn has an effect on the organisms under culture. Culture systems vary and with each system, there is a unique environmental effect as a result of waste generated and management techniques. Dosdat (2009) classified waste of food origin to include: faeces, indigestible materials, un-ingested feed and ingested but undigested feed. Eutrophication is a form of organic pollution that results from the discharge of materials such as dissolved nutrients, un-ingested feed, faecal matter and deceased fish into water bodies either holding aquaculture cages or receiving aquaculture effluent. Aquaculture effluent (Photo 1.2) typically contains dissolved and suspended solids and nutrients including nitrogen and phosphorus that play a major role in eutrophication. Intensive culture systems with high stocking densities face problems resulting from bad water quality that can stress the cultured species predisposing them to disease. To counter the problems of

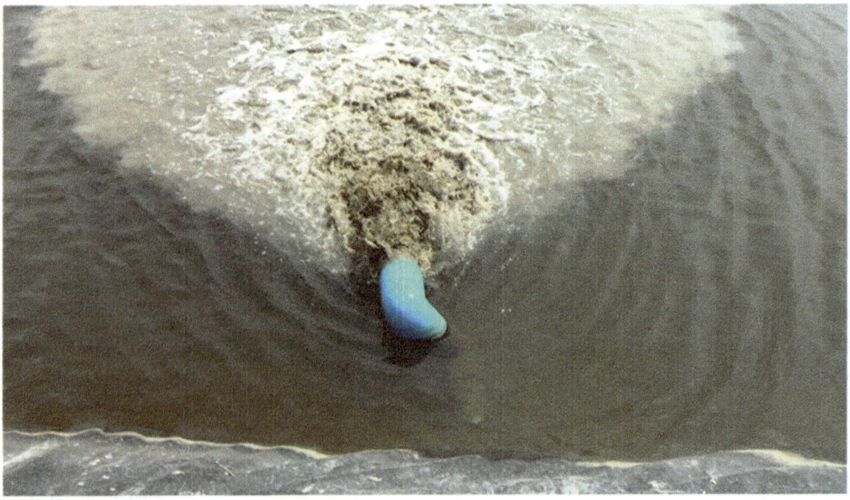

Photo 1.2 Effluent pumped from an intensive shrimp pond. This impacts receiving waters with increased turbidity as clearly seen here, unless directed to a water recirculating system

Table 1.1 Typical nutrient load from production of selected aquaculture species in ponds and tanks (kg/tonne of product)

Species	TSS	Total N	Total P	BOD_5	Carbon	References
Shrimp	476	15.9	1.46	63.3	730	Prapaiwong and Boyd (2012)
Trout	289–839	47–87	4.8–18.7	>944	101–565	Axler et al. (1997), Tekinay et al. (2009)
Salmon	191–606	20.3–39.3	9.1–10	410	226	Strain and Hargrave (2005), Hennessy et al. (1996)
Tilapia	382	44.95	14.26	10.4	145.6	Lin et al. (1997), Tabthipwon (2008), Neto and Ostrensky (2015)
Pangasius	2050	46–46.8	14.4–26.6	740	305.5	Anh et al. (2010), De Silva et al. (2010), Phanna (2011)
Channel catfish	353	83.6	12.7	25.6	713.5	Boyd et al. (2000)

bad water quality and disease, there is heavy reliance on chemicals and medication in the form of antibiotics (Ozbay et al. 2014).

The levels of effluent from various species under culture are presented in Table 1.1. These values were either quoted as given by the authors or derived from data presented. In cases where values were not expressed directly in kg/tonne, derivations were made based on harvest weight, concentration of variables (mg/L), volume of water and other factors relevant to the estimation. These values give a fairly good idea of the waste loading from various aquaculture species under cultivation.

Advances in aquaculture technology has created room for culture of hitherto uncultured species of high value in developed countries with increased demand for feed (Photo 1.3) and other inputs that impact on the environment. Intensification of aquaculture for production of export value species has been a point of focus by environmentalists in their quest for environment friendly food production,

Photo 1.3 Types of aqua-feeds: pellets for tilapia grow-out (left) and broodstock feed for groupers (right)

Table 1.2 Environmental threats from aquaculture (Emerson 1999; Kura et al. 2004; USAID 2013; Ozbay et al. 2014; Boyd and McNevin 2014)

Threat	Hazard	Risk
Genetic	Escapes	Fitness issues
	Exotics/GMO's	Genetic contamination; loss of biodiversity
	Wild broodstock	Introgression
	Stock enhancement	Extinction
Water quality	Effluent	Eutrophication; pollution
	Sediments	Habitat loss
Ecology	Land modification	Habitat alteration
	Salinization	Loss of freshwater
	GHG emission	Pollution; climate change
Health	Antibiotics	Resistance
	Chemicals	Pollution; bioaccumulation; toxicity
	Escapes	Disease
Resource use/Inputs	Fishmeal	Depletion of wild fish population
	Wild seedstock	
	Water extraction	Water shortage

GMO genetically modified organism, *GHG* green house gas

considering species like trout and salmon in Europe and the United States, and recently tilapia and pangasius in Asia (Bosma et al. 2011; Boyd and McNevin 2014).

Several attempts have been made to categorize the various threats from aquaculture to the environment. A concise classification of the threats posed by aquaculture to the environment is given in Table 1.2. Aquaculture facilities impact directly on water bodies that feed them water since the same water bodies receive effluents discharged. The effects are more pronounced in lakes and stagnant canals that serve as water sources, through changes in microbial communities as well as toxicity of discharged chemicals (Ozbay et al. 2014). While fertilizers can cause nutrient levels to rise leading to eutrophication, lime does not present any environmental threat; but the use of human waste is a potential hazard that raise food safety concerns (Boyd and Massaut 1999).

1.2.1 Genetic

Anthropogenic interventions in aquatic ecosystems do not stop at extraction alone since there are efforts to restore depleted feral fish populations through enhancement activities that come under three broad categories: Sea ranching, stock enhancement and of fish introduced in therestocking (Bell et al. 2006). Fisheries enhancement through hatchery produced fish has long been practiced as a means of recovery for depleted wild fish populations as well as conservation (Wada 1998; Antunes et al. 1999). Inland fisheries enhancements have utilized introductions and stocking as a means to improve fish populations (Cowx et al. 2012). The use of stock enhancement techniques in marine fisheries has come with mixed economic results ranging from failure: shrimp in Western Australia (Loneragan et al. 2006), Japan (Hamasaki and Kitada 2006), and China (Wang et al. 2006) to successes: scallop in Japan (Uki 2006) and New Zealand (Lorenzen 2008), and chum salmon in Japan (Hilborn 1998; Kitada 2014). In terms of biological impacts, stocking has been reported to have led to high mortality of feral salmon as a result of cannibalism by stocked fish (Pearsons and Fritts 1999).

Risks from introduction of exotic fish for culture depends on the probability of their establishment and that of occurrence of an adverse effect following their establishment (Miller et al. 2004). This is particularly true considering the fact that out of eleven species of fish introduced in the Hawaiian islands, only three species became established although factors such as number, duration before maturity, larval survival and water depth are key predictors (Johnston and Purkis 2016). Risk assessment of non-native species for introduction in Brazil revealed that all non-natives were unsuitable for use considering the high level of ecological risk associated with them (Britton and Orsi 2012).

Genetic manipulations and escapes of farmed species have the potential to initiate and establish losses in genetic diversity. Intra-specific diversity can be lost or degraded through genetic drift in bottlenecked populations, extinction and hybridization. Genetically modified organisms (GMO's) pose unknown and undetermined threats to natural populations. The use of risk assessment in determining effects of GMO's to natural populations is strongly advocated (Muir 2004).

Escapes of fish from aquaculture facilities can lead to fitness issues as observed in several studies (McGinnity et al. 2003; Weir and Grant 2005; Weir et al. 2005). Although Weir et al. (2005) reported differences between wild and farmed male salmon in terms of mate preference and reproductive success, Lehnert et al. (2012) reported that sperm fitness was greater in farmed than wild male chinook salmon. Susceptibility to predation of salmon is not related to size with equal probability to predation being reported (Solberg et al. 2015). On the whole, mathematical modelling suggests a strong non-linear relationship between impacts of escape, population of escapes and their reproductive viability while less adapted populations escaping at steady and low-levels can lead to proliferation of mal-adaptation in wild populations (Baskett et al. 2013). Poor broodstock management and breeding has led to production of highly inbred lines of giant river prawn, *Macrobrachium rosenbergii* in India (Nair and Salin 2012).

1.2.2 Water Quality

Aquaculture effluent contains both organic and inorganic materials that tend to increase the load in the environment where the effluent is released. In receiving waters, changes have been observed in the community structure of organisms with an increase in the number of organisms that depend on deposits from mussel aquaculture cages in South Africa (Stenton-Dozey et al. 1999). However, mussel cage aquaculture is reported not to affect the holding water body negatively in the Western Adriatic sea (Fabi et al. 2009) and South-eastern Brazil (da Costa and Nalesso 2006). Poor water quality is often dependent on other anthropogenic factors apart from aquaculture (Boyd and McNevin 2014). This is particularly true for mollusc aquaculture that is often used as a remediation for effluents than as a contributor. According to Rawson et al. (2002), bivalves and molluscs can effectively remove nutrients in moderately enriched waters as against heavily enriched waters. Effluent from ponds have less impact on receiving waters than domestic waste water, except for high total suspended solid concentration (Boyd and McNevin 2014). However, large scale aquaculture as well as clustered small holder farms tend to pose a problem to future aquaculture development within the same area they are located, and by extension world aquaculture due to eutrophication (Rawson et al. 2002). Nutrient enrichment has led to unsustainable economics of shrimp production in Krung Krabaen Bay and Welu wetlands in Thailand (Ataguba et al. 2014). In India, direct effluent discharge from *Macrobrachium* farms was responsible for eutrophication in the receiving waters of the densely populated state of Kerala compared to Andhra Pradesh (New et al. 2008).

Aquaculture can be carried out using either fed or non-fed species with the latter having a goal to reduce resource use in the form of feed while also ensuring environmental integrity (Photo 1.4). Feed use however tends to cause high levels of

Photo 1.4 Shrimp grown together with mussel in a pond in Thailand for environmental integrity

nutrient loading in receiving waters considering uneaten feed, faeces, and other biological waste produced through metabolism (Davis 2015). Depending on the prevailing circumstances in receiving waters as well as released concentration, discharge of nutrients as a result of bad feeding strategy as well as use of poor quality feed can either be assimilated or accumulated leading to pollution (White 2013). Clustering of farms tends to create a huge pool of feed that produce large quantities of waste (Craig 2002). The extent of pollution from nutrients of feed origin depends on the hydrodynamics of the water body, windswept, rate of loading, stocking density, FCR and spacing of farm units (White 2013). Modelling has been proposed as a means to create a balance between fed and no-fed aquaculture as well as human activities (Rawson et al. 2002).

1.2.3 Ecology

Shrimp farming has negatively impacted coastal ecosystems via mangrove loss and associated biodiversity changes (WorldBank 1998; Rajitha et al. 2010; Paez-Osuna 2001). Mangrove forests mitigate erosion, maintain coastal water quality, provide breeding grounds for aquatic organisms, and provide vital ecosystem services for people along the coasts (Valiela 2006; FAO 2007b). World mangrove forest cover has reduced from about 188,000 km^2 in 1980 to 152,300 km^2 in 2005 (Fig. 1.2).

Mangroves link terrestrial and marine ecosystems in about 124 countries with tropical or sub-tropical climate, are halophilic, evergreen, and thrive on sheltered coastlines, estuaries and deltas (FAO 2007b). Mangrove deforestation has occurred in all five continents of the world between 1980 and 2005 with a total of 36,000 km^2 lost within 25 years (Fig. 1.3).

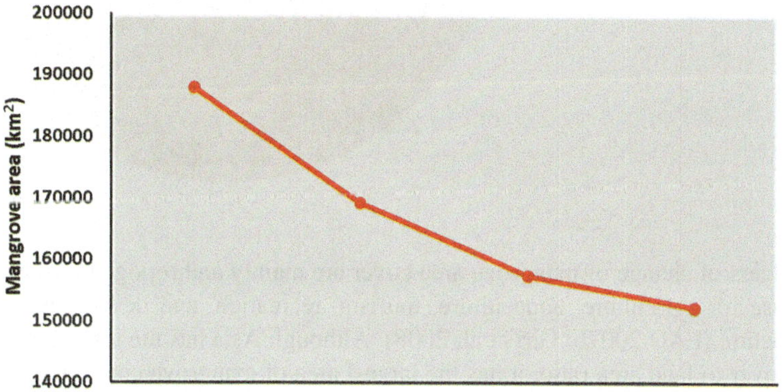

Fig. 1.2 Change in world mangrove area cover due to aquaculture and other human activities, 1980–2005. *Source* Spalding et al. (2010)

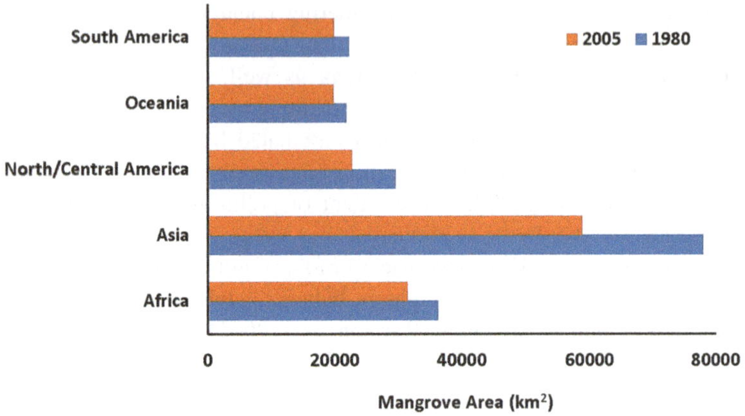

Fig. 1.3 Regional mangrove cover change, 1980–2005. *Source* FAO (2007b), Spalding et al. (2010)

Fig. 1.4 Regional percentage of world mangrove area lost, 1980–2005. *Source* FAO (2007b)

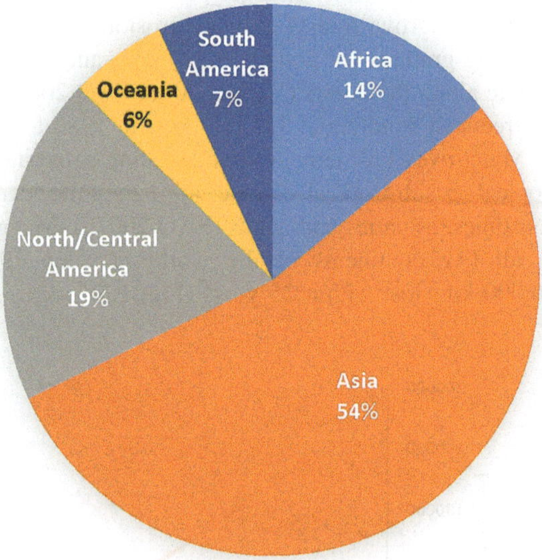

Drivers of change of mangrove area cover are mainly anthropogenic and include land use for agriculture, aquaculture, tourism, recreation, and development of infrastructure (FAO 2007b; Giri et al. 2008). Although Asia has the lowest mangrove area cover to land area ratio, it has the largest area of mangrove cover in the world, but the loss is also high. About 54% of total world mangrove forest area lost between 1980 and 2005 was from Asia (Fig. 1.4) and aquaculture contributed to this loss by 12% (Giri et al. 2008).

Construction of shrimp ponds entails mangrove destruction and from experience, excessive numbers of farms clustered along the shorelines lead to reduced productivity with attendant disease conditions and collapse of the ventures. A shift in livelihoods from aquaculture to fishing becomes impossible since breeding grounds and nursery areas have been destroyed hence recruitment altered and population of feral aquatic organisms must have migrated to favourable grounds or died out totally. This underscores the need to consider livelihood options in the context of sustainable aquaculture (Emerson 1999).

Coastal aquaculture of shrimps utilizes seawater or brackish water. Introduction of saline water for inland shrimp farming would increase soil and water salinities due to seawater and sediment discharge into inland water bodies (Tucker et al. 2008). Salinization of soil due to shrimp aquaculture has been reported in Bangladesh (Chowdhury et al. 2011) and Thailand (Teng 2008).

Energy use in aquaculture ponds for aeration is largely responsible for greenhouse gas emission associated with aquaculture (Pelletier and Tyedmers 2010). However, emission from aquaculture appears to be quite low (2.2%) compared to other food production sectors (Boyd and McNevin 2014) with tilapia production having an emission intensity that is quite lower than pork and beef production but comparable to broiler and Atlantic salmon (Pelletier and Tyedmers 2010). Removal of mangroves to pave way for shrimp aquaculture is detrimental to the environment since mangroves are very good storage units for carbon and their removal has the potential to contribute to climate change (Ahmed and Glaser 2016).

The persistence of organic matter in the environment depends on several factors including moisture, temperature and related microbial/chemical activity (Estrada and Soares 2017). There is a direct relationship between plant biomass and soil organic carbon with areas that have heavy vegetation being storage depots for emitted carbon (Alavaisha and Mangora 2016).

Carbon stocks vary across the globe with increasing concentration at the equator and it currently stands at a global average of 78 tonnes C ha^{-1} $year^{-1}$ while sequestration is at a rate of 2.9 tonnes C ha^{-1} $year^{-1}$ (Estrada and Soares 2017). Research has shown that above the ground plant biomass hold more carbon with reports of mangroves holding between 414 and 684 Mg C ha^{-1} in two areas of Tanzania (Alavaisha and Mangora 2016), 147 Mg C ha^{-1} in the Eastern coast of India (Sahu et al. 2016) and 853–1311 Mg C ha^{-1} in mangrove wetlands around Papua and East Indonesia (Taberima et al. 2014). The impact of mangrove removal for aquaculture as well as other anthropogenic needs on the ability of mangroves to sequester carbon from the environment is not readily quantifiable considering lack of information covering the extent of carbon sequestration by mangroves in wetlands and even below the earth surface (Donato et al. 2011).

1.2.4 Health

Water which is the medium of aquaculture is also the cradle of life since it supports numerous organisms both beneficial and harmful to human beings. The nutrient load that emanates from aquaculture gives a favourable environment to micro-organisms. Pathogenic organisms are ever present in the environment but their pathogenicity depends on the level of management and the aquaculture facility being used. Semi-closed systems present the greatest risk of pathogen transfer from farmed to wild fish (Huntington et al. 2006). Transmission of pathogens and parasites between farmed and wild fish and vice versa is however difficult to determine considering three critical points mentioned by Murray (2015), which include the presence of the pathogen in the fish at harvest, presence during processing and transmission from source to recipient. The latter can create unreliable outcome since the determination of points of outbreak and emergence of either a parasite or pathogen may not be accurate except where product traceability is complete. Sepúlveda et al. (2004), presented a report that is contrary to the wide-held thought that cultured fish can transmit pathogens and parasites to wild fish. They observed that wild fish in southern Chile that were in full interaction with cultured fish had greater parasite load than cultured salmon. Similarly, Sanil et al. (2010), reported that the intensity of protozoan (*Perkinsus olseni*) infections in wild pearl oyster (*Pinctada fucata*) was higher than cultured oysters. However, sea lice infection in farmed Atlantic salmon was found to be intermediate between two wild stocks (Glover et al. 2004).

The use of antibiotics in aquaculture is perceived to cause antibiotic resistance in unintended species in the environment. In Asia, a total of 36 antibiotics are used in aquaculture (Rico et al. 2012) and according to Anka et al. (2013), farmers use antibiotics without receiving advice and they use these at self-determined doses (Photo 1.5). In China, bacterial strains in shrimp hatcheries with known resistance to antibiotics were found to differ in resistance with pond water bacteria having a

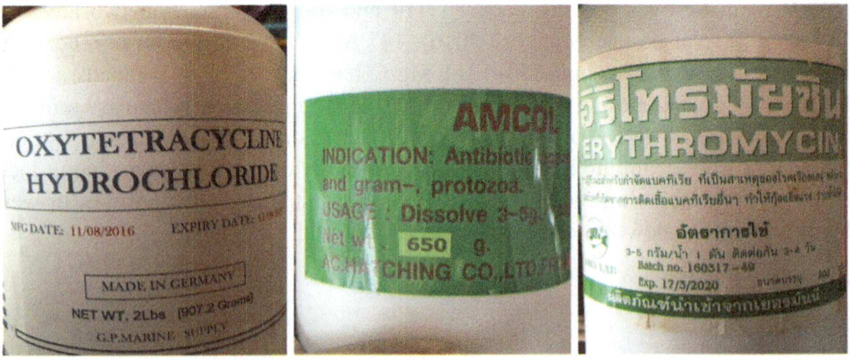

Photo 1.5 Some of the common antibiotics used in a commercial shrimp hatchery: oxytetracycline (left), product of unknown composition (centre) and erythromycin (right)

strong influence on the resistance capacity of sediment bacteria (Zhang et al. 2011). Le et al. (2005), also reported the incidence of antibiotic resistance in bacteria from shrimp farms in Vietnam especially *Bacillus* and *Vibrio*. The diversity of antibiotic resistant genes is also greatly enhanced by aquaculture (Harnisz et al. 2015).

The use of chemicals in aquaculture is generally believed to be safe except in circumstances where management is negligent. Chemicals used in aquaculture include lime, fertilizer, therapeutants, anaesthetics, hormones, oxidants, algicides, coagulants, feed additives, antifoulants, fuels and lubricants (Boyd and McNevin 2014). Treatments for parasites in fish often rely on chemicals such as $KMnO_4$, formalin, salt, iodine, organophosphates, hydrogen peroxide, chlorine and chloride compounds, and $CuSO_4$ among others (Rico et al. 2012; Ataguba et al. 2013; Boyd and McNevin 2014). Disinfectants such as $KMnO_4$, chlorine compounds, formalin and iodine can be highly toxic to macroinvertebrates and planktonic organisms but their persistence in the environment is low (Rico et al. 2012). Treatment of sea lice in salmon involves the use of various chemicals including organophosphates, hydrogen peroxide, pyrethroid and pyrethrins, which end up being toxic to non-target organisms such as insects, crustaceans and other macroinvertebrates (Page et al. 2005).

1.2.5 Resource Use/Inputs

Despite the declining wild stock, small pelagic fish are being caught and rendered as fishmeal and fish oil to produce aquaculture feeds. Excessive use of wild fish to manufacture aquaculture feed hurts the environment since sustainability is not possible. According to FAO (2014), about 15% of the total of 891 million tonnes of fish produced between 2007 and 2012 was used for feed and other non-food uses, but this quantity was reported to be in decline. This decline can be attributed to the increased use of ingredients of plant origin as well as other animal by-products in formulated feeds for aquaculture.

Although the collection of wild mullet fry for aquaculture had not shown any visible effects in Egypt between 1983 and 2008 (Saleh 2008), in India and Bangladesh, giant tiger shrimp fry collection had led to discards of up to 160 other fry per tiger shrimp (Naylor and Burke 2005), leading to a ban on its collection due to the obvious effect this had on the recruitment of bycatch species (Siriwardena 2007). However, wild fry collection in Bangladesh had shifted to the freshwater prawn (*M. rosenbergii*) with varying percentages of bycatch for either types of gear used (Ahmed and Troell 2010).

The agriculture sector is a major consumer of water. One of the major drivers for increasing world water demand is the abstraction for agriculture, including crops, livestock and aquaculture, apart from the rising demands for domestic and industrial use. A lion's share of this freshwater demand is for irrigation in crop production. Globally, agriculture uses up about 70% of the total freshwater abstraction, while this could be as high as 90% in most developing countries (FAO 2011a). In developed countries the use of freshwater for agriculture is quite low (up to 5% of global consumption), but more water is exploited for industries and energy production (15%). Domestic and municipal consumption account for 10% of the global extraction of freshwater (WWAP 2016).

Abstraction of water for aquaculture is inevitable since water is the most important factor in aquaculture. Extensive and semi-intensive pond aquaculture have low water demand compared to intensive flow-through systems (Beveridge and Phillips 1990; Boyd and McNevin 2014). The production of between 4 and 12 kg of rainbow trout annually in flow-through systems in Ontario utilized about 526,000 l of water (Moccia and Bevan 2005). Aquaculture consumes less than 1% of the world's freshwater that is renewable and reachable (Boyd and McNevin 2014). Considering this, water use by aquaculture may have little impact at the global environmental level.

1.3 Threats from the Environment to Aquaculture

Aquaculture is dependent on the environment hence there is a cyclic relationship between them such that the environment affects aquaculture while aquaculture also impacts the environment. Basically, environmental factors as they affect aquaculture can be managed through proper site selection but environmental hazards that occur as a result of weather and climate are often deleterious to aquaculture. Furthermore, anthropogenic activities also place some form of stress on aquaculture and the aquatic organisms under culture with concomitant effects on humans who in the first place caused the upset. These are considered as challenges to aquaculture (Mazur and Curtis 2006).

The environment limits the ability to culture certain species or adopt certain aquaculture practices. This is particularly true considering the fact that culture of cold water species like trout is difficult in the tropics and culture of marine species in land locked areas is also difficult. This notwithstanding, control of the environment for aquaculture of certain species can be achieved considering their plasticity.

According to Shang and Tisdell (1997), sustainable aquaculture as affected by the environment occurs at two fronts: the local or farm level and the wide society level. We categorize threats (Fig. 1.5) from the environment to aquaculture under three broad classes following Lawson (2013).

1.3.1 Threats to Cultured Species

1.3.1.1 Water Quality

Water supply for aquaculture is the most important criteria for siting and location of farm units. However, anthropological activities create a dynamic water quality in

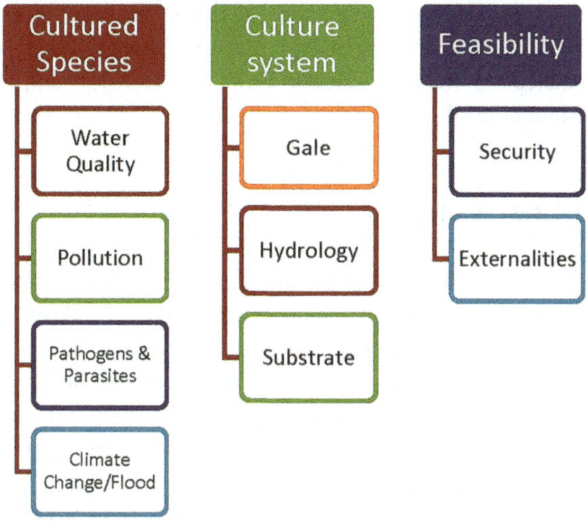

Fig. 1.5 Categories of environmental threats to aquaculture

water bodies that feed aquaculture facilities. Water for aquaculture must have suitable dissolved oxygen, temperature, pH, and low levels of nitrogenous compounds. Unfavourable levels of these parameters demand investment to correct the levels in intake water. Temperature and dissolved oxygen are the most difficult to regulate and would require greater variable cost to maintain.

Discharge of cooling water from power plants into water bodies that serve as water source for aquaculture can affect the temperature of water intake for aquaculture. Temperatures at discharge points can vary but typically range from 12 to 42 °C depending on the region of the world (Langford 1990). Deviations from optimal water temperatures for aquatic organisms lead to changes in respiration and metabolism with sub-optimal levels of water quality parameters causing stress and its related consequences (Lawson 2013). Temperature tolerance in fish runs from the inhibiting level through the loading level to the lethal level and finally the lethal threshold (Coutant 2013). When water temperatures in water bodies holding cages for aquaculture change drastically, supporting and beneficial organisms in the ecosystem may be eliminated in favour of organisms that may be detrimental to the stock under culture and vice versa (Echols et al. 2009). Lake stratification period during the summer has been found to be extended by the combined effects of thermal pollution and climate change (Kirillin et al. 2013) with nuclear and coal fed power plants contributing half the quantity of heat dumped into rivers and lakes (Raptis and Pfister 2016). Increased water temperatures lead to increased metabolic rates in aquatic organisms and this happens in an environment that is oxygen deficient since warm water contains less oxygen and places great stress on the organisms.

Domestic and industrial waste accumulation in coastal areas as well as highly populated settlements tend to make the environment unsuitable for aquaculture by serving as either direct drivers of toxicity or indirect drivers through the depletion of dissolved oxygen as a result of decomposition of organic matter (Gesamp 2001). Hypoxic conditions in the Gulf of Mexico have been linked to eutrophication caused by nutrient release through the Mississippi River (Rabalais et al. 2007; Bianchi et al. 2010).

1.3.1.2 Pollution

Pesticides, heavy metals and organic compounds that are released into the environment are a threat to aquaculture organisms given the fact that they can alter basic physiological functions, damage anatomical structures and can also lie latent and get passed to humans that consume fish. Echols et al. (2009) had rightly observed that aquatic pollution resulting from heavy metals, hydrocarbons, radioactivity and synthetic organic chemicals occur as a result of human negligence or inappropriate use and disposal of waste.

Tsangaris et al. (2010), reported a reduction in acetylcholinesterase (AchE) activity in mussels caged in sites close to agricultural land as well as sites impacted by urban and industrial waste suggesting some level of neurotoxicity. Similarly, Cappello et al. (2013), reported gill tissue damage in mussels raised in cages placed in water impacted by anthropogenic factors hence a form of environmental distress was affecting the mussels. Salmon feed has been indicted in the persistence of organochlorine compounds in salmon with fish flesh having the highest concentrations as against fish oil and fish feed suggesting bioaccumulation of these compounds in the fish (Jacobs et al. 2002). This is further strengthened by the report of Hites et al. (2004) that farmed salmon contained higher levels of organochlorine compounds than wild salmon. However, recent reports by Masci et al. (2014) and Nostbakken et al. (2015) have shown that levels of organochlorine compounds in farmed fish have declined but fish feed still remains a source of the pollutants getting into farmed fish.

1.3.1.3 Pathogens and Parasites

Disease and parasite infections occur as a result of interactions between the host, the pathogen/parasite and the environment. The prevalence of a pathogen in an environment does not necessarily translate to disease infection. Several predisposing factors relating to each of these factors that cause disease are discussed by St-Hilaire et al. (1998).

Marine invertebrates have been thought to be agents of bacteria transmission to aquaculture organisms (Olafsen 2001). Cryptobiosis is a parasitic infection that is transmitted to salmon in hatcheries through the leech as an intermediate host that is present in intake water (Guo and Woo 2009). Bryozoans have been reported as the carriers of the myxozoan parasite *Tetracapsuloides bryosalmonae* which causes proliferative kidney disease (PKD) in salmon and trout (McGurk et al. 2006). The vertical transmission of Myxosporea and Malacosporea in Bryozoans has been demonstrated by Morris and Adams (2006). Small and Pagenkopp (2011) gave an extensive review of the reserves and intermediate hosts of pathogens of crustaceans of economic importance and this included the water and sediments, other aquatic animals, algae and biofilm.

1.3.1.4 Climate Change/Flood

Climate change leads to several changes that are critical to both fisheries and aquaculture. These include drought, altered precipitation pattern, intensity of storms, changes in sea temperature, rising sea level, El Niño's and increased inland water temperatures (WorldFish 2007). Climate change has both positive and negative effects on aquaculture. According to Weinert et al. (2016), about 49 species of benthos have lost their habitats in the North sea between 2001 and 2009 while 11 species consolidated their habitats reach. Furthermore, the use of modelling effectively predicted gains in habitats coverage for an invasive mussel species against a native species in Europe while invasive crayfish would lose range in favour of the local species up to the year 2050 (Gallardo and Aldridge 2013). These corroborate the implication as proposed by WorldFish (2007) that sea surface temperature increase can increase aquaculture production but may be obscured by changes in number of species available for culture. Increased temperature of water has been found to affect fish muscle and mechanisms responsible for detoxication as exemplified in *Sparus aurata* (Madeira et al. 2016). Salt water intrusion will also create a shift from freshwater aquaculture species to brackish water species (Williams 2011).

1.3.2 Threats to the Culture System

1.3.2.1 Gale

Marine cage aquaculture is affected by adverse weather conditions which include violent storms, cold weather and strong winds. Storms and winds tend to affect cage structure and can destroy the cage entirely. Waves that build up as wind blows across the surface of the ocean can cause severe damage to cages that are constructed in open areas without wind breaks. Wave heights of 1–1.5 m can be detrimental to small cages (Lawson 2013). Climate change is expected to come with large waves and heavy storms in flood prone areas hence heavy precipitation that will lead to loss of aquaculture installations, increased capital expenditure on stronger cage moorings, pond dykes, reservoir walls and other farm facilities (WorldFish 2007). Computer aided modelling has been used to determine the mooring dynamics (Fredriksson et al. 2003; DeCew et al. 2010) and the movements as well as load impacting cages (Colbourne and Allen 2000).

Inland water based cage aquaculture (Photo 1.6) is also prone to effects of adverse weather conditions. Excessive as well as inadequate water flow has been

Photo 1.6 Tilapia cage farming in Chao Phraya River, Thailand

found to damage cage structures hence heavy economic losses in Northern Thailand (Lebel et al. 2015). Interestingly, there is a moderately positive correlation between wind velocity and concentration of 2-methylisoborneol (an off-flavour causing compound) in catfish pond waters in the United States (Hurlburt et al. 2009).

1.3.2.2 Hydrology

Water current tends to impact cages through increased loading on the cage and the supporting structures, and the effects of currents are far greater than waves (Huang et al. 2008). Although Lawson (2013) proposed that tidal currents in the range of 0.1–0.6 ms^{-1} was favourable for cages and values greater than 1 ms^{-1} (Huang et al. 2008) being totally destructive, Lader et al. (2008) reported that currents of 0.13 and 0.35 ms^{-1} caused cage volume reductions of 20 and 40%, respectively and that location influenced the volume reduction. Current distribution on the water surface is vital to stability of cage structures (DeCew et al. 2010) and several models have been estimated to determine design parameters that will ensure the stability of cages under strong currents (Zhao et al. 2007; Kim et al. 2014; Cui et al. 2013). The tested models have indicated that the tension on mooring lines is directly proportional to wave height and wave period with a recommendation on the use of square nets

(Cui et al. 2013) while (Zhao et al. 2007) posited that diamond shaped nets are better considering possibility of a reduction in structure size ratio.

Climate change is expected to cause changes in rainfall pattern, volume and hence affect water volume in water bodies leading to increase or decrease in water depth depending on the area affected. The use of cages and net pens in aquaculture will be affected since these structures depend on water level. Water exchange in floating cages is effective at depths with the cage bottoms off the floor at low tide (Lawson 2013).

1.3.2.3 Substrate

The choice of culture systems depends on available substrate. Changes in substrate type are expected under climate change scenario given increased precipitation, flooding and sediment flow into water bodies. Net pens are better constructed on muddy substrates while floating cages perform better on hard rocky surfaces (Lawson 2013). However, changes in substrate materials will tend to affect the culture system's performance as rightly observed by Tidwell and Coyle (2008). Boyd (2012), summarised the basic soil properties and the processes they affect in aquaculture ponds. Soil particle size and texture, organic matter content and sediment depth tend to affect the integrity of the pond as a holding facility.

Pond culture in areas with acid sulphate soils is affected by low pH levels due to oxidation of iron pyrite catalysed by *Thiobacillus* in the upper level of the soil when exposed to air (Boyd 2012; Yoo and Boyd 2012). Acid sulphate soils have been reported to affect aquaculture ponds in Thailand and the Philippines with pH values of 3.9–4.4 and 3.4–6.3 (Singh 1980), respectively. Harvesting operations that require pond drainage tend to expose the pond bottom soil to air hence oxidation leading to increased acidity and lower pH levels. Runoff from exposed dykes with oxidized iron pyrite into the pond will also lead to acidic condition in the pond (Mahmood and Saikat 1995).

The cost of remediation of acid sulphate soils for aquaculture is the basic problem posed by this environmental factor to aquaculture system construction (Hechanova 1984). Currently, the methods used for remediation and management include: Inundation/drainage cycles with seawater, liming, barricades and induced oxidation (Hechanova 1984; Sammut 1996). However, wise use of site selection criteria remains the best option if cost avoidance can be made else mitigation is needed when land propriety cannot be changed.

1.3.3 Threats to Aquaculture Feasibility

1.3.3.1 Security

Poaching and predation are two major issues that affect aquaculture facilities with origin from the environment. Cage aquaculture is prone to vandalism and poaching

(Ranchan 1984; Beveridge 2008; Parker 2011). This affects the profit margin of aquaculture ventures and is therefore a serious threat to aquaculture sustainability. Several measures have been proposed to curb poaching including row boats with barbed wire to pick up poaching nets, fencing, canine surveillance, and natural barriers in the form of trees. Poaching of oysters has recently been reported in Maryland in the USA (Lessner 2015; Calvert 2016). Predation can also be prevented by using nets to cover the pond tops to prevent entry by birds as well as around the sides to trap reptiles like monitor lizards and snakes.

1.3.3.2 Externalities

These are factors that are associated with economics and society with associated relevance to aquaculture itself. Support facilities for aquaculture such as feed manufacturing, economies of scale, foreign and local markets, policy and legal requirements, and allocation of space are among other factors.

Policy makers, aquaculture entrepreneurs and ancillary occupations seem to be disconnected hence aquaculture production is not at its peak or close to its potential (Krause et al. 2015). Aquaculture policy development must carry the people along and focus on national interest, rational use of resources and market driven development. Policies need to encourage the use of environmentally friendly production technologies (Olalo 2001). However, in cases where policy making does not involve the end users of the policy, a mismatch is created. Krause et al. (2015), argued that the divide between policy and end users tends to create inequity, mismatch between gains and needs leading to food insecurity and health issues. In China, the policy drive of the government focused on markets and information in the late 1990s (Huang et al. 2001). A policy framework and process that considers the environment, society and the economy with precautionary recognition of impacts within the social sphere, carrying the people along in policy making and re-evaluation process with due regards for production capacity and prevailing global demand, and feedback mechanisms that examine and depict the communal aspects of aquaculture in a multi-dimensional manner is ideal for sustainable aquaculture development (Krause et al. 2015). The policy interventions necessary for assimilation of aquaculture technology can be determined by using models (Nobre et al. 2009; Slater et al. 2013).

Legal issues relating to aquaculture development involve licensing, competing use of space and local regulations pertaining to use of land and water surfaces. Stakeholders in the marine finfish aquaculture industry in Europe advocate for greater representation and share in benefits and issues with ability to influence decisions within the social context (Ertor and Ortega-Cerda 2015). The tourism sector has been found to compete stiffly with coastal aquaculture in Europe (Hofherr et al. 2015).

The development of local and international markets in tandem is ideal for aquaculture development in order to encourage small holder aquaculture and ensure food security. However, a lopsided approach that considers foreign markets will harm national food security situation. The development of support industries that service the aquaculture sector is also vital in the quest for sustainable aquaculture (Lawson 2013).

1.4 Site Selection and Carrying Capacity Assessment of Aquaculture

Sustainability of aquaculture with consideration of ecological and environmental impacts will be achieved using several approaches. One important approach is proper siting of facilities so as to mitigate negative impacts. A combination of site selection, proper facility design, construction and management will form the bedrock for sustainable aquaculture. Policy is also vital in this regard as it ensures regulation of entry into the business of aquaculture and also creates specific locations that are suitable and capable of absorbing the effects of aquaculture on the environment. There is a paucity of regulatory framework for environmental quality in the aquaculture industry in most nations (USAID 2013).

The peak loading of aquaculture organisms that can be supported by the environment without deleterious effects on the stock, the culture system and the environment holding the system is what is referred to as carrying capacity (Stigebrandt 2011). Within an aquaculture system, the biomass under culture presents scenarios of an input-output relationship with inputs being additive and outputs being subtractive. According to Sowles (2003), there is a cultural dimension to carrying capacity hence it tends to have a dynamic nature. Carrying capacity of an environment for particular nutrients will be lowered with increasing levels of the nutrient but tends to increase with sustained removal by flora and fauna subject to their ability to reproduce and rebound from stressful conditions.

Acceptable levels of water quality within water bodies that surround an aquaculture site is the major determinant of quality of effluent being discharged by the aquaculture firms considering the stock under culture (Stigebrandt 2011). Viability and profitability of aquaculture ventures are linked to the level of environmental impact they have (Gegner and Rinehart 2009) hence production, livelihoods and competition for resource use must be in equilibrium.

1.4.1 Application of Site Selection and Carrying Capacity in Aquaculture

There is an inextricable link between policy and scientific evidence which in this case involves the determination of standards that are meant to regulate aquaculture through scientific processes to avoid exceeding carrying capacity. Determination of limits and standards using cutting edge ideas and contrivances that are dependable is the first step in the process of policy formulation that is closely followed by a political phase (Stigebrandt 2011). Aquaculture can be limited by space and competitive use of land (Hofherr et al. 2015) hence site selection must consider both carrying capacity and available space (Kapetsky and Aguilar-Manjarrez 2013).

According to Ross et al. (2013), the aim of carrying capacity assessment is achieved when there is no undesirable change in the ecosystem resulting from aquaculture while maximum set limits of production are executed with resultant social satisfaction considering competing interests within the given environment. Sustainability is therefore achieved since communal satisfaction is achieved through sufficient economic benefits using technologies that are eco-friendly while also providing room for other resource users. Aquaculture installations can be conveniently installed at capacities that are in tandem with the carrying capacity following its determination. Carrying capacity also serves as a tool for aquaculture planning, determination of suitable aquaculture zones and areas of appeal to aquaculture.

1.4.2 Theoretical Basis/Framework/Approach

Considering mollusc aquaculture, Inglis et al. (2000) gave four classes of carrying capacity: production carrying capacity, physical carrying capacity, social carrying capacity and ecological carrying capacity. Further explanation was provided by several authors considering the fact that the farm is a small unit within the larger environment that is subject to social and cultural norms (McKindsey et al. 2006; Byron et al. 2011; Ross et al. 2013).

There is both a temporal and spatial scope to site selection and carrying capacity in aquaculture planning and regulation. Activities pertaining to aquaculture planning and development progress spatially and temporally from identification of potential through sectoring or zoning and selection of the site to final implementation of aquaculture activities. Site selection is the final step in the process and it has the least scope with mapping of zones being intermediate and identification of potential having a broad spatial coverage.

Sustainability ought to drive aquaculture development with ecosystem stability being drawn from carrying capacity. This therefore means that all the categories of carrying capacity must be considered together within a given space over time to achieve ecosystem stability as proposed by Kapetsky and Aguilar-Manjarrez (2013). The FAO (2010) accentuated three basic principles of an ecosystem approach to aquaculture to include: use of all ecosystem functions and services without probability of becoming unsustainable, factor in livelihoods of all stakeholders with concern for improved wellbeing and finally reckon with activities, policies and targets of other productive sectors.

1.4.3 Carrying Capacity and Site Selection: When Is the Right Time to Apply These Tools?

Increased demand for aquaculture space is expected to occur in the future considering fast growth of the sector particularly in Asia and Africa with resultant effect being the use of virgin coastal and continental shelves hence altering the ecosystem services and products derivable. Local factors concerning the ecological integrity of a site and market volume are two conventional criteria that drive site selection for aquaculture. The need for proximity to infrastructure and associated services for aquaculture is a driver for the location of aquaculture facilities with impacts that will affect the environment, access to transportation and markets (Ross et al. 2013). The short term nature of this planning strategy will stifle aquaculture development even more so because it has a local spatial coverage with greater extent of effects on sustainability within the immediate environment.

Carrying capacity is considered all the way through the site selection process right from estimation of potential with emphasis at the point of actual siting. This follows temporal and spatial dimensions since the process must start at a given time and consider an entire gamut of available space before narrowing down to the actual spots that are suitable for aquaculture (Fig. 1.6). Advancements in aquaculture systems and introduction of novel aquaculture species is an on-going process hence carrying capacity estimation has to go hand in hand to suit the new system or species. In addition, expansion to new sites using old species still necessitates determination of carrying capacity.

Fig. 1.6 Spatial and temporal progression of aquaculture development activities

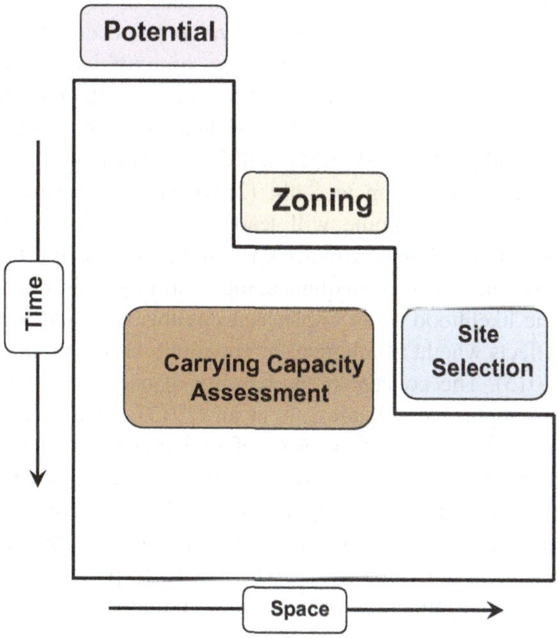

1.5 Aquaculture Hazard and Risk Analysis

1.5.1 Concept of Risk

Risk encompasses three basic concepts which according to Arthur (2008) includes uncertainty, probability and impact. Therefore, risk in aquaculture involves the possibility for an undesired outcome from production activities alongside damages that would occur. The presence of a factor, substance or object that causes or portends danger is referred to as a hazard (Johnson 1998). Biological nature of aquaculture systems is a hazard that brings about price uncertainty (Flaten et al. 2011).

Francis and Shotton (1998) considered the definitions of risk as covered under two points of view: the probability and the decision theory, the bottom-line being a consequence or effect of a situation and how large this can be. Surrounding the probability and scale of an undesirable event is the concept of uncertainty (Hargrave et al. 2005). A risk can be quantified if the probability of the undesirable event is known. Certain uncertainties can be reduced via accumulation of knowledge and data while others have a huge scale of uncertainty especially those related to natural occurrences (Gesamp 2001; Hargrave et al. 2005). Generally as issues move from the farm scale to a broader regional and national scale, predictability becomes difficult (Hargrave et al. 2005). McDaniels et al. (2006) have shown that decision making for risk management in the salmon aquaculture industry occurs at four different scales namely, local scale that deals with zoning and site selection, regional scale that deals with operating licences and monitoring of production, national scale that is involved in regulation, and finally the international scale that is quite different considering the absence of regulatory structures. While McDaniels et al. (2006) have proposed a value based decision making tool that is based on stakeholder opinion, the scientific community is divided as regards accepting public opinion in decision making (Young and Matthews 2011). This notwithstanding, the value based decision tool of McDaniels et al. (2006) encompasses five objectives that address sustainability and decision making.

Risk according to Sethi (2010) is the probability that a divergence from an anticipated outcome will lead to an undesirable effect with risk itself being a potential, while realised risk is risk that has been actually experienced. Considering magnitude and consequence, the World Organisation for Animal Health sees risk as the likelihood of an unpleasant circumstance and the magnitude of the unpleasant effects whether biological or economic on both human and animal wellbeing (OIE 2015). The concept of probability of occurrence and consequent losses was also highlighted by Olanrewaju et al. (2013). Although the presence of farmed salmon impacts wild trout in terms of infection with salmon lice, Taranger et al. (2011) opined that greater probability of occurrence of salmon lice in wild trout is not entirely explained by biomass of farmed salmon under culture. Brun (2013) sums up risk as an artefact with a subjective insight that is designed to help us grasp the meaning of danger and also be able to grapple with it considering the limitations of our environment, upbringing and culture (Olanrewaju et al. 2013).

Risk permeates fisheries management considering the presence of uncertainty, fluctuations, paucity and multiple goals (Sethi 2010; Olanrewaju et al. 2013). In aquaculture, uncertainties may include but not limited to diseases, natural disasters, poor outputs, breakdown of equipment, sediment accumulation, nutrient enrichment, exhaustion of dissolved oxygen, mortality, escapes, bloom of deleterious algae, conflicts on resource use, and bad water quality (Clark et al. 2010; Hargrave et al. 2005; McIntosh 2008; Tlusty 2002). Risk from aquaculture according to (Reantaso 2008; Ezekiel et al. 2011) include changes in habitat structure, pollution, climate change, genetic issues, food safety, and occupational risks. The scope of these risks is large since it affects the environment, the people and their wellbeing (Reantaso 2008). However, application of risk management is advocated as the best way to avoid loss and fritter (Secretan et al. 2007).

Ezekiel et al. (2011) delineated risk as either from aquaculture to the environment or from the environment to aquaculture hence effectively bringing in the perception of risk from different perspectives. Risks relating to production and marketing are vital for an aquaculturist while people who also rely on the same natural resources as the aquaculturist will view aquaculture as a risk within the environment. However way it is perceived, risk is associated with activities and outputs. Hazards affecting aquaculture development may come from biological, environmental or economic sources (see: Reantaso 2008; Ezekiel et al. 2011; Subasinghe et al. 2012; Swaminathan 2012).

1.5.2 Theoretical Framework of Risk Analysis in Aquaculture

The main goals of risk analysis in aquaculture are the identification and assessment of risk so that adequate mitigation can be directed at the risks in such a manner that benefits both the aquaculture enterprise as well as the society as a whole. To adequately mitigate risk, threat and vulnerability must be quantified so that controls can be applied to minimize risk. Risk management, assessment and communication are three key steps in risk analysis (Yoe 2012) with the addition of hazard identification being important considering aquatic animal health (OIE 2015), but it is important that this step considers the entire life cycle of the animal (Olanrewaju et al. 2013). A different risk analysis process is however proposed by Olanrewaju et al. (2013) to include assessment, screening, evaluation and management with assessment being a qualitative to quantitative process, screening being a specification step, evaluation being a stochastic process and the management step being mitigation. This approach however does not consider communication which is a very important step in the loop. For the purpose of this text, we shall consider four steps of risk analysis (Fig. 1.7).

Hazard identification brings to the fore all harmful objects, substances and conditions so that they can be assessed during the assessment step in order to

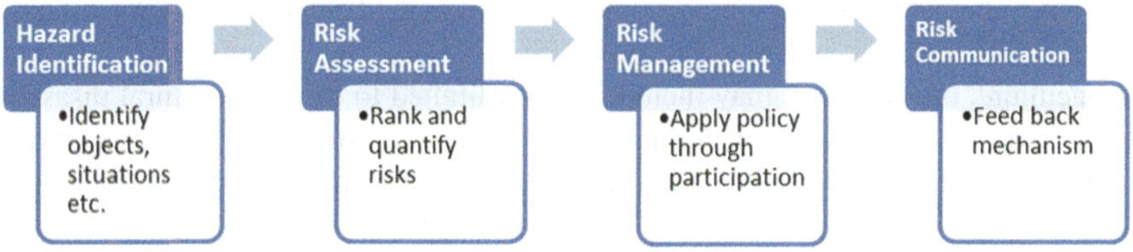

Fig. 1.7 Activities in a risk analysis process

determine qualitative and quantitative aspects of the risks posed by the hazards and therefore be able to make decisions and guide policy formulation accordingly. Hazard identification in aquaculture can be carried out using either robust techniques (Crawley et al. 2003; Wells 1996) or rapid tools (WHO 2003). Robust methods such as the fault tree (Hayes 2002a) and a variant of failure modes and effects analysis as applied to infection (Hayes 2002b) have been applied to estimate hazard from invasive species. The fault tree has been applied to investigate the hazard of introducing GIFT tilapia in Zambia (Lind et al. 2015). The risk management step is a participatory step that looks at all available and applicable policies by regulators and stakeholders with due consideration for social, economic and environmental wellbeing (Hargrave et al. 2005). Mis-matches in the system can be corrected through the risk communication step because it allows for re-evaluation of the process for re-adjustment. Yoe (2012) pointed out that the human mind can be quite good at analysing a situation to reach a decision but there could be pitfalls in human reasoning that necessitate the involvement of several people in the decision making process. It is therefore ideal that employees be involved in the risk management process at the enterprise level so that decisions can be taken through interactions between management and staff.

Risk management is a highly subjective step in the process since it relies on the perception of the risk manager. The risk of aquaculture to the environment and from the environment to aquaculture can be qualified using probability and magnitude. A risk matrix (Fig. 1.8) that is based on colour codes with progression of risk increase either through upward movement or movement to the right was presented by Brun (2013). Hazard identification and the determination of levels of impact and likelihood are vital and must be based on objectives such that the assessment process can delineate the effects of risks on the ecosystem, the environment and socio-economic wellbeing using available tools, conventions and procedures that effectively determine value (Olanrewaju et al. 2013; Aven 2012). The risk appetite of a farm will therefore be a yardstick to determine risk acceptance. The risk matrix will also guide regulatory agencies in the determination of the acceptability of aquaculture socially and environmentally. Magnitude is ranked below impact in the analysis involving the risk matrix hence impact is considered rather than magnitude.

Considering the spatial scope of risks, the risk matrix with its quantifiable probability of occurrence and impact will give us a vivid idea about the actual location of risks at all levels beginning from the farm to the environment, the social

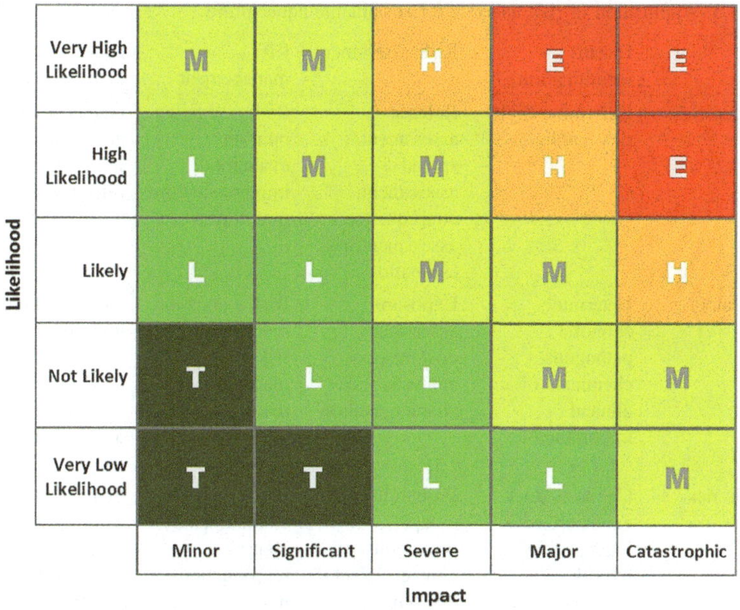

Fig. 1.8 Risk matrix. After Brun (2013): T = trivial Risk; L = low risk; M = moderate risk; H = high risk; E = extreme risk. Risk level is identified by the intersection of likelihood and consequence. Trivial risk will generally not require significant or specific resource use in its management while low risks can be managed using routine management with proper supervision. Moderate risk requires a timely higher level management intervention with the aid of an action plan and high risk requires ready to implement action plan. Extreme risk can be effectively managed using a detailed plan

and economic scopes. With this, potential effects of aquaculture on the environment such as erosion, loss of ecosystem services, pollution, nutrient loading and increased demand for water will easily be quantified. Application of risk analysis for pathogens, food safety, genetics, ecology, environment and social risks in aquaculture (Table 1.3) can ensure sustainability of aquaculture through a reduced impact of aquaculture to the environment as well as from the environment to aquaculture.

Pathogen risk analysis as applied to import of aquaculture products seeks to determine the presence of dangerous pathogens, the probability of transfer to importing nation, expected impacts of exposure of susceptible organisms, risk associated with each pathogen if products are allowed in, acceptability of risk of each pathogen and lastly, possible entry of aquaculture products with risks at acceptable level (Bondad-Reantaso and Arthur 2008). A concise analysis of risk associated with international trade in aquatic organisms in the Asia-Pacific region is presented by Diggles and Arthur (2010). Food safety risk analysis must first identify the hazards that are associated with the aquaculture product and determine the likely effects of exposure leading to identification of steps for risk management

Table 1.3 Application of risk analysis for sustainable aquaculture

	Hazard identification	Risk assessment	Risk management	Risk communication
Pathogen risk analysis	Determine high risk pathogens	Release assessment, spread assessment, consequence assessment, risk estimation	Risk evaluation, opinion evaluation, implementation/ monitoring	Gather knowledge and ideas, analyse information, pass information
Food safety	Determine zoonotic pathogens, chemicals of clinical importance	Exposure assessment, dose-response analysis, risk characterization	Risk evaluation, risk options, risk management/ review	Documentation of identified food borne risks for producers consumers and aquaculture value chain
Genetic risk analysis	Cultured stock, exotics, GMO, interspecific hybrids, non-selectively bred organisms	Probability of escape, probability of contact, effect assessment	Site location, containment, control reproduction, manage human activity/access	Science-industry meetings, extension service, publication of reports
Ecological risk analysis	Escapes, habitat modification	Qualitative assessments, quantitative assessments	Standards, inspection, prohibitions, permits, cost and benefits	Stakeholder participation, open communication
Environmental risk analysis	Suspended solids, nutrient enrichment, mangrove destruction, water abstraction	Release assessment, exposure assessment, effects assessment	Mitigation, monitoring, compliance	Quantitative aspects of risk, social dimensions, identify differences among stakeholders
Social risk analysis	Failure: farm level, community level, obstruction	Consequence, magnitudes, documentation	Hedging, aversion, attenuation, subsistence	Corporate social responsibility, continuous planning, implementation, observation and action

and subsequent communication. This principle has been utilized in AquaFRAM, a spreadsheet tool designed to estimate food safety and disease risk in aquacultured salmon (Soon and Baines 2012). Genetic risk in aquaculture is connected with introgression between cultured fish and wild stocks considering the fact that cultured fish have either been selectively bred for certain traits or genetically modified. Management of genetic risk will involve two basic approaches: reducing the possibility of escape as well as the ability to reproduce (Waples et al. 2012). In addition, the location of aquaculture facilities, effective containment and human access limits will go a long way to manage risks associated with escapes. A qualitative ecological risk analysis for the introduction of *Litopenaeus stylirostris* into Fiji from Brunei Darussalam favoured the introduction of the species and a critical evaluation of life cycle and habitat also did not portend any danger but competition, pathogen transfer and hybridization were contentious (Bondad-Reantaso et al. 2005).

Aquaculture makes use of resources from the environment and also discharges waste into the same environment. As such it is pertinent to determine if site preparation, waste discharge, and water use are hazards that pose a risk. In a risk assessment of the effect of shellfish aquaculture on the environment in Tasmania, the risk of organic nutrient loading was scored with a minor consequence rating with an unlikely rating in terms of likelihood and a low rating in terms of level of risk (Crawford 2001). Social risks resulting from aquaculture are based on hazard of failure of investment as well as obstruction of other commercial activities. The social risk of aquaculture can be assessed considering effects, scope and available records. Loss of investment in aquaculture is a risk that cannot be totally prevented considering the fact that aquaculture deals with biological subjects and the environment. Heavy losses have been incurred by households in poor communities in Vietnam (Luttrell et al. 2004). However, loss can be managed via insurance options with an example provided for the clam aquaculture industry in the United States (Beach and Viator 2008). Aside from insurance, the establishment of subsistence aquaculture facilities provides social benefits (Pillay 1997). Good site selection and planning will attenuate the social risks of aquaculture (Pillay 1997).

1.6 Ecosystem-Based Approach to Aquaculture (EAA)

A strategy that is aimed at blending aquaculture in the wider ecosystem for sustainability, equity and resilience was proposed by the FAO (2010) and dubbed the Ecosystem Approach to Aquaculture (EAA). EAA has created a way to ensure compliance with the Code of Conduct for Responsible Fisheries spatially and at various supervisory levels with regards for national policies, while providing an avenue for regulators and industry operators to work together to ensure sustainability with adequate consideration of the environment, the socio-economic well-being of culturists and goals of regulatory bodies (CAMFA II Policy Brief 2014). The EAA is a shift from conventional management to a method (Fig. 1.9) that puts emphasis on methodology of action with stakeholders having a voice (FAO 2010).

The FAO (2010), prescribed three principles on which EAA must operate that include the use of ecosystem services in planning for aquaculture development and management for sustainability, improved livelihoods of all people in the aquaculture value chain and finally a consideration of other sectors, current policies and expected results.

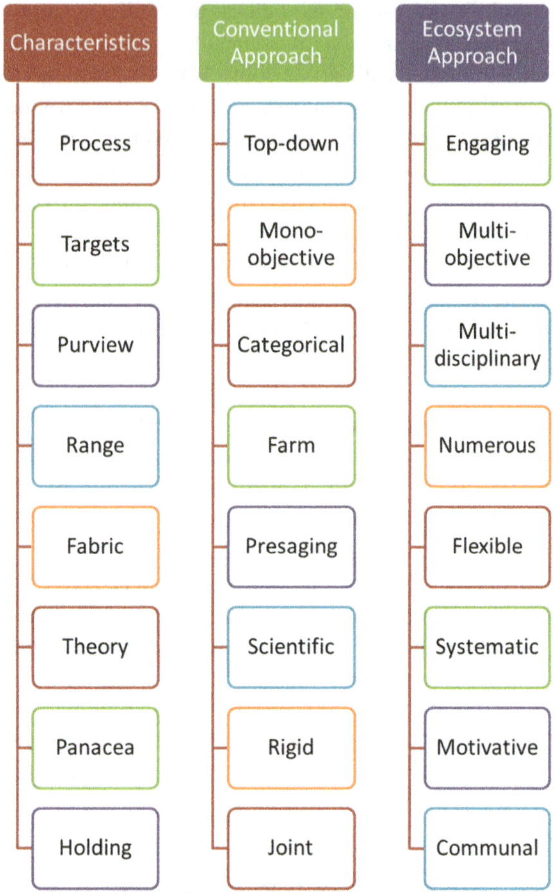

Fig. 1.9 Distinction between the conventional approach and ecosystem approach (Staples and Funge-Smith 2009)

1.6.1 EAA Planning and Implementation Framework

Planning for EAA (Table 1.4) involves five basic steps:

1. Scoping
2. Identification and prioritization of issues
3. Developing a management plan
4. Implementation
5. Enforcement.

The coverage of implementation according to Staples and Funge-Smith (2009) can be determined by the highest level policy goal in view which could cover national or regional aquaculture, provincial/state aquaculture or just one culture system.

Table 1.4 Framework for planning and implementation of EAA

Step	Function	Input	Outcome
Scoping	Extent of implementation	System boundaries, stakeholders	Coverage
Identification of issues	Expounds the need for intervention	Ecological wellbeing, socio-economic wellbeing, ability to achieve	Impacts and mitigation
Ranking of issues	Prioritization of issues for implementing solutions	Risk analysis	Hierarchy of issues to be tackled chronologically
Define objectives	To develop a working plan for implementation	Information on: use of sites, escapes, pollution, diseases and parasites, access to feeds	Better management
Formulate and implement management plan	Aquaculture development, logical management, adequate surveillance	Legal framework, institutional arrangements, competing livelihood options	Developed human capacity, purposeful research and dissemination, controlled internal and external threats
Implementation of EAA	Optimized feeding, better management, biosecurity, effluent management, environmental impact assessment	Quantitative data	Balance ecosystem with aquaculture
Monitoring and evaluation	Makes review and adaptation possible	Environmental indicators, socio-economic indicators	Reference points of indicators

1.6.2 Tools for Ecosystem Approach to Aquaculture

Sustainable aquaculture has a spatial scope therefore, spatial planning tools can effectively separate economic, social and environmental spheres of aquaculture with boundaries clearly defined even though interactions will occur (Aguilar-Manjarrez et al. 2010). Mathematical models are increasingly being used to model carrying capacities. These have the capacity to create replicas of real life situations and give efficient solutions to scenarios which are hitherto not ideal to determine in reality due to cost implications (Ross et al. 2013).

1.6.2.1 Environmental Models

Using baseline and monitoring data, mathematical algorithms can effectively predict environmental response to aquaculture stress. A total of 16 environmental models were considered by the European commission's Ecosystem Approach to Sustainable Aquaculture (ECASA) with each having a scale and a scope that covers species, culture system and motive (ECASA 2008). Carrying capacity, fluxes of matter and energy as well as environmental and ecological indicators serve as means of modelling.

Environmental models (Table 1.5) can be based on either just a single indicator within the environment or by using multiple indicators that include, particulate organic matter, soluble organic matter and chemicals involved in the production process (Southall et al. 2004). Simple linear regression using time series data can be used to estimate environmental capacity while more complex models are useful if there is a lack of data that predates present times hence the use of several of such tools in arriving at an environmental capacity.

Table 1.5 Environmental carrying capacity models

Model	Criteria/input	References
MOM (modelling ongrowing fish farms-monitoring)	Waste dispersion, waste dilution	Stigebrandt et al. (2004), (Stigebrandt 2011)
TRISULA/DELWAQ	Heat dispersion, sediment transport, water quality, heavy metals	Southall et al. (2004)
CORMIX	Near mixing	
CORMIX + WASP + QUALBAVI	BOD, DO, faecal coliform, N, P	
Ecopath + Ecosim (EWE)	Mass balance of trophic relationships	Christensen and Walters (2004)
Fuzzy models + GIS	Physical environment, pollutants	Navas et al. (2011)
Shellsim	Clearance rate, particle retention efficiency, filtration rate, absorption rate, rejection rate, ingestion rate, absorption efficiency	Ferreira et al. (2008)
EcoWin2000	One, two and three dimensional scaling, water exchanges, hydrodynamics (not appropriate for farm scale modelling)	
BEAST (benthic environmental assessment sediment tool)	Sediments, particle size, erosion	Walker et al. (2014)

1.6.2.2 Spatial Models

Determination of the worth of land and subsequent allocation for productive activity are very important steps in resource use and development. However, sustainability of productive and extractive use can only be determined using procedures and applications that will adequately guide decision as well as predict outcomes based on usage scenarios and available physical data (Table 1.6). This according to Kapetsky et al. (2007) can be achieved using GIS (Geographic Information System) tools. With the application of geo-spatial tools, sustainability friendly policies are now being made (Morgan and LaFary 2009).

The combination of planning and site selection is sine-qua-non to sustainable aquaculture. Information is critical to identifying suitable sites that have the capacity to hold aquaculture facilities. Comprehensive understanding of the environment, social, economic and political factors forms the bedrock of analysis of sites for aquaculture suitability. Unsustainable production would be the bane of ill-selected sites with impacts on the environment, livelihoods and the organisms under culture (Naylor and Burke 2005). Site selection that involves the use of GIS models are effectively done using logical analysis of spatial data to guide assessment of available resources and their management (Ragbirsingh and De Souza 2005; Longdill 2007).

Multi-Criteria Evaluation (MCE) is typically a multivariate analysis method that involves the use of various criteria and variables to arrive at the best combination of variables to achieve the best results desired. The combination of GIS and MCE was proposed by Carver (1991) and has been applied in the determination of potentials and suitable sites for carp culture (Salam et al. 2005), Japanese scallop (Radiarta et al. 2008), mussel (*Perna canaliculus*) (Longdill et al. 2008), giant river prawn (Hossain and Das 2010), Pacific oyster (Silva et al. 2011) and for general carrying capacity evaluation and site selection (Hossain et al. 2009; Dapueto et al. 2015).

Table 1.6 Steps for a spatial model estimation

Step	Description	Activity category
Goal identification	Quantifiable, practicable, germane, precise, temporal limit	Non-GIS activity
Identify criteria (factors and constraints)	Quantifiable else use proximates Details required determine criteria	
Align factors and criteria	Create common scale for factors Use fuzzy function to align criteria scores Decide functions for each criterion	
Assign weights to factors	Ranks, rates, pairwise comparison	
Map the criteria	Weighted linear combination	GIS activity
Confirm results	Determines reliability of results – Sensitivity analysis – Ground verification	

A combination of regulatory and social barriers, MCE and the Farm Aquaculture Resource Management (FARM) model was used by Silva et al. (2011) for site selection in areas with poor data for shellfish aquaculture.

GIS models are supported by decision making software that elucidates better judgement and decision making considering choices. Examples of software currently available include IDRISI, CommonGIS and ArcGIS (Nyerges and Jankowski 2010; Fisher 2006). ArcGIS does not have the robust capacity for MCE analysis but according to Nyerges and Jankowski (2010) weighted summation overlays give it the capability, while (Eldrandaly 2013) utilized component object model (COM) technology in tandem with ArcGIS to achieve MCE for site selection.

1.6.2.3 Socio-economic Models

There is no fully developed model for determining the social carrying capacity of aquaculture. Byron et al. (2014), utilized Ecopath, a mass balance model and IMPLAN, an input-output economic modelling software to integrate carrying capacity of bivalve aquaculture with social and ecological aspects to provide direction for policy that will ensure sustainability.

Models that have an economic and social approach to agricultural activity location dates back to 1826 with the von Thunen model which was premised on some assumptions that included isolation and independence, a surrounding of empty land around occupied space, flat terrain without obstructions, consistent climate and soil quality, self-distribution of produce without roads, and a goal for profit maximization by farmers (Chorley and Haggett 2013). Modifications to this model have been presented by various authors but according to Wilson and Birkin (1987), two critical shortcomings need to be addressed: the constraint that all products are sold at the nearest market and secondly the static nature of the model. Furthermore, Chorley and Haggett (2013) argued that the Thunen model required modification to remove its partial equilibrium status, account for non-economic factors with expanded scope and finally consider the varying scales of the central town.

Input-output models and spatial equilibrium models are a group of geographical location models that consider factors of production, consumers and the producers along the chain with competitive advantage differentiating regions. Chorley and Haggett (2013), advocated the use of the spatial equilibrium models as the most suitable for locating suitable sites for agricultural production provided there is adequate data.

Application of partial equilibrium models in terrestrial agriculture is widely reported. The need to improve these models using innovative additions and modifications has resulted in various versions of models. One common partial equilibrium model is the Common Agricultural Policy Regionalised Impact System (CAPRI). Espinosa et al. (2016), introduced structural changes in specialization, region and size of the economy into this model at the farm level and discovered that structural changes affected the area under cultivation, number of animals reared and the size of farms distributed considering their economic size hence the need to consider farm structural changes in policy decisions.

Land use change is driven by factors that are inherent in the society as well as the environment and it has a spatial outlook. The local changes in land use can be insignificant but regional and global land use changes are significant (Heistermann et al. 2006). Land use patterns have an effect on sustainability and are determined to a large extent by societal culture, livelihoods and appetite for commodities that require large expanses of land. This underscores the need for models that will help us understand the consequences of changes in natural and socioeconomic factors on production, exchange of goods, demand and prices of agricultural markets (von Lompe 2003). Economic models that deal with land use aim to elucidate the demand-supply pattern of production sectors with high demand for land and they are basically equilibrium models (Heistermann et al. 2006). Detailed descriptions of economic models are provided by Balkhausen and Banse (2005) and Heistermann et al. (2006). Economic models are classified under two categories: partial equilibrium models (PE) and computable general equilibrium models (CGE). The PE models are dynamic with a global scope, and consider other markets that are not agricultural as exogenous, hence there is homogeneity of goods traded. However, CGE models differ in the consideration of non-agricultural markets as endogenous, have a static approach, and consider trade as a bilateral interaction (Balkhausen and Banse 2005; Heistermann et al. 2006). The Global Trade Analysis Project (GTAP) model (a CGE Model) was combined with a biophysical model by Van Meijl et al. (2006) to predict the availability of land for agricultural activity in the EU without shortages between 2006 and 2036.

Merging geographic and economic models may be ideal for aquaculture judging from the insights gained in the application to terrestrial agriculture. A combination of CAPRI and CLUE (Conversion of Land Use and its Effects) have been reported to create robust results that link economics and policy (from CLUE) to agricultural production, income and price structure (from CAPRI) under the influence of land use dynamics (Overmars et al. 2013; Britz et al. 2011). A downscaling model that relied on the census of farms as well as land use type in Switzerland produced a reasonable measure of the spatial motif of use of land for agricultural purpose (Gärtner et al. 2013).

1.6.3 Aquaculture Governance, Planning and Management Practices

Policies, laws and their enforcement must be transparent with accountability hence in the absence of corruption, there will be development through a well-controlled economic environment aided by support from citizens with the will to grow. The difference between the developed and developing countries according to Hishamunda (2010) lies in good governance that creates conducive environment for wealth creation. This is also true for economic activities within the same productive sector (Hishamunda et al. 2014). Poverty is detrimental to the environment because it causes people to rely heavily on ecosystem services to the detriment of the environment hence good governance of aquaculture will prevent competition for resources, pollution and bad product quality while upholding human and social security (Hughes and Rose 2011).

Weak aquaculture governance is a threat to the application of the ecosystem approach to aquaculture because its absence creates room for bio-insecurity, poor quality of products, competition and conflict over aquaculture sites, lack of adherence to international standards and little or no trust from the communities (Hishamunda et al. 2012). Governments must ensure that there is conducive environment for aquaculture in the midst of market failure using aquaculture policies that create secure rights to property, encouraging market environments that have contract enforcement and small scale aquaculture ventures at its core and all driven by economic stability, effective research and extension and political security (Hishamunda et al. 2012). These according to Lent et al. (2008) will ensure a stable polity considering economic and social wellbeing. Gender inclusive policies that allow women equal participation in economic activities is important in the drive towards sustainable development (UNWomen 2014). The emerging role of women in the productive sector of the economy in non-OECD countries has been identified especially in export oriented business (OECD 2008). A gender sensitive aquaculture governance is also critical to sustainable aquaculture since the views of rights and their definitions differ between the genders (WorldFish 2011).

According to Hirst (2000), governance is the driving force behind activities, steering them in the direction they ought to go with appropriate controls so as to achieve set deliverables. Governance in the view of the Canadian Institute of Governance (CIG) is the determinant of possession of power to make decisions, voice concerns and show accountability (CIG 2016). From the foregoing, governance is adaptable to sections of society as well as economic groups within the society considering the people involved as stakeholders. The ecosystem approach to aquaculture is an all-inclusive approach that considers the stakeholders hence it is linked to governance and would create a sense of belonging and ownership among them that will lead to maximum impact from total participation.

The planning, management and control of aquaculture, using democratic principles and full participation by stakeholders in order to achieve sustainable livelihoods, social and environmental wellbeing can be referred to as aquaculture governance. However, issues such as aquaculture site availability, supporting inputs, increasing investments in marine aquaculture sector, effects of aquaculture on the environment, technology adaptation and access to finance and credit must be addressed clearly if aquaculture governance is expected to make an impact (Lent et al. 2008).

1.6.3.1 Making Aquaculture Governance Work

Bevir (2012) identified the overlapping and conflicting nature of aquaculture governance and has advocated a Deming's wheel approach to its implementation. According to the Asian Development Bank (ADB), there are four basic principles behind good aquaculture governance: accountability, inclusive participation, regularity and openness (ADB 1995). There are six indicators of governance as presented by the WorldBank (2015). These indicators presented by the ADB and the World Bank were harmonized by Hishamunda et al. (2014) as: accountability for actions, effectiveness/efficiency in dealing with issues, equity between people and predictability of actions (Table 1.7). Aquaculture governance must consider the role of gender in aquaculture as well as changes in livelihoods. Aquaculture has been reported to have a negative effect on the livelihoods of women in the fishing communities of West Africa due to excess work load and loss of livelihoods (Trottier 1987). For aquaculture to be sustainable, its governance must strive to close gender gaps in activities such that there is equity, increased access to resources by women and elimination of gender effects in policy implementation (Scott and Wilde 2006). Policy planning must be participative (FAO 2008) and address gender constraints relating to access to land, resources and capital (Trottier 1987) hence aquaculture governance must be participative, have representation from all relevant stakeholders, and be responsive to their needs.

1.7 Aquaculture Certification and Standards

A third party nonaligned assessment of quality standards that is aimed at confirming claims by a firm of meeting standards is referred to as certification. Standards are actually sets of collectively recognized keys that are beneficial to the aquaculture enterprise, the environment and the society. The setting of standards, endorsement and application of certification, designation of activities and outputs are the crux of certification. The major components in an aquaculture certification would include: food safety, social responsibility, animal welfare and environmental sustainability (Lee 2009). The credibility of certification depends on transparency and application of the scientific approach through integrity and accountability for big and small aquaculture enterprises. Traceability is also an important factor for certified products hence records of product movement is important so as to ensure responsible trading.

1.7.1 Challenges to Sustainable Aquaculture—The Quality Certification Perspective

Small holder aquaculture is at risk of exclusion from the globalization of aquaculture trade due to several reasons as highlighted by Subasinghe and Phillips (2007)

Table 1.7 Applying the principles of aquaculture governance with gender sensitivity

Principle	Outcome	Poor practice	Meeting and improving requirements
Accountability	• Certainty • Reduced graft	• Decisions by officials alone using dubious guidelines • Appeal of decisions is impossible • Scientific information is hardly valid	• Greater openness by regulators and industry • More criteria for transparency • Revelation of perks and outlay of aquaculture
Effectiveness/ efficiency	• Worthwhile service delivery	• Excessive governance • Opposing regulations • Extended red tape in license issuance • Mono-directional contexts of decision making • Low amplitude to monitor and enforce regulations • Insensitivity of communities and stakeholders	• Achievement based management system • Subject regulations to cost-benefit analysis • Expanded participation • Shape capacity
Predictability	• Curtailed risk and transaction costs • Inter-convertible property rights easing loan access	• Enigmatic property rights • Short-term licensure • Decision on site selection is based on rent seeking	• Make property law aquaculture friendly • Open and pellucid criteria and procedures
Equity	• Gender inclusive income and regional distribution	• Short term leases and associated short term objectives • Greater influence by large farms over small farms	• Spur female license bids • Uniform access to credit across genders • Tractability as panacea for license renewal
Participation	• Upsurge in acceptance and conformity • Cut down enforcement costs • Foment trust for aquaculture • Upsurge in demand and consumption • Use of aboriginal knowledge	• Reclining attitude of government officials • Exploiting decision communication process	• Effective communication between stakeholders and officials • Stimulate interest of women • Pass information on advantages of aquaculture and aquaculture products

(continued)

Table 1.7 (continued)

Principle	Outcome	Poor practice	Meeting and improving requirements
Gender representation	• Consider needs of women	• Stakeholders devoid of women • No female representation in regulatory agency	• Policies that fight discrimination and creates equal opportunities in public service • Affirmative action is supported in policies and legislation
Gender responsiveness	• Policies should have ranked needs of women	• Policies and activities in aquaculture that are not gender friendly	• Standardized and transparent methodology considering issues concerning women in aquaculture • Potent legal framework for women

Adapted from Scott and Wilde (2006), Hishamunda et al. (2014)

that include requirements for market access, risks and costs involved in meeting quality standards in the production process. The suitability of certification for small scale aquaculture is in doubt as pointed out by Marschke and Wilkings (2014) hence a deviation from the recommendation that certification schemes should be pro-small scale aquaculture (FAO 2011b). Standards are mostly for high value species that are popular in the western markets hence a very low share of market from developing countries (Jonell et al. 2013) with a lot of questions bothering on fair trade begging for answers (Lee 2009). In addition, the upsurge in eco-friendly certification schemes has been dubbed as modern day extra-territorial conquest (Vandergeest and Unno 2012). However, Global GAP has reported the certification of small holder Pangasius farmers in Vietnam, which was achieved as a result of collective effort (GlobalGap 2014). These challenges can be categorized (Table 1.8) as social, environmental/ethical, and financial challenges.

1.7.2 Certification Criteria for Sustainable Aquaculture

Sustainability is in itself not a quantifiable variable and therefore depends on indicators to identify its direction (Lee 2009). At both national and international levels, the aquaculture value chain is guided by regulations that are particularly targeted at food safety, disease control and conservation with a strong bias for international trade in processed aquatic organisms hence factors such as environmental and socio-economic wellbeing have become open to independent certification for compliance and responsible management (FAO 2011b). A summary of the rationales and relevant criteria for meeting food safety, animal welfare, social

Table 1.8 Challenges of certification and quality standards on small holder aquaculture

Challenge type	Effects
Social challenges	Social exclusion that can engender poverty
	Loss of confidence from consumers due to failure and difficulty in regaining same
	Less incentive for quality products in Least developed countries
	Questions as regards the openness of certification approach
	Level of commitment and involvement from stakeholders
Environmental and ethical challenges	Loss of equity and subsequently an unsustainable practice
	Divergent global perspectives and lack of support for indigenous technology
	Compliance may be evasive
Financial challenges	Huge costs of certification
	Small scale farmers lose access due to competitive disadvantage
	Inadvertent exclusion due to inability to obtain new markets

responsibility and environmental wellbeing by aquaculture is given in Table 1.9. The difficulty and high costs associated with data collection can be a serious setback in the process but with voluntary certification, there can be a huge database for environmental data that serve as indicators of the environmental sustainability of aquaculture (Lee 2009).

Table 1.9 Criteria for mitigating concerns of aquaculture in relation to its certification for sustainability

Concerns	Rationale	Criteria
Food safety	• Meet food safety regulations of the Codex Allimentarius • Be in line with all international food standards	• Use approved therapeutants, adhere to withdrawal periods • Use contaminant free feed with approved ingredients • Farm location is pollution free, suitable water quality free of microbial impact • Stock must be pathogen free • Traceability of inputs • HACCP hygiene conditions
Social welfare	• Equal market access for all producers • Worker's welfare	• Small holder, youth and gender inclusive post-certification market access • Involve all stakeholders with competing claims to resources • Adherence to labour and wage regulations • Clustering of small holder farmers for cost effective certification
Animal welfare	• Access to water and food, suitable environment, disease and injury control, functional improvement, behavioural/interactive freedom and mental and physical comfort	• Use international standards in aquatic health management • Quarantine must comply with FAO code of conduct for responsible fisheries • Consider specific pathogen free organisms • Stringent management of antibiotics and their discharge • Clean and hygienic culture environment for pathogen and parasite control • Limit harm to animals in production process with limited suffering at slaughter • Farm workers gain knowledge on animal welfare and health management

(continued)

Table 1.9 (continued)

Concerns	Rationale	Criteria
Environmental wellbeing	• Proper planning and determination of environmental impacts	• Adopt environmental impact assessment for site selection with support to manage impacts • Consider the environment in National aquaculture policy and plan • Inclusion of environmental quality and impact control in aquaculture governance • Water use with bias for conservation • Control fish escapes and import of exotics with restricted use of wild stocks in aquaculture • Censurable use of chemicals and drugs

1.7.3 Certification Schemes

Certification schemes are based on standards or codes of good practice that are documented by certifying organisations with the goal of sustainable aquaculture. These schemes are mostly species or region specific hence a thorough assessment and compliance requirements. However, the specific nature inhibits far-reaching acceptance and global coverage (Lee 2009).

According to Corsin et al. (2007), there are about 30 certification schemes in aquaculture with a bias for sustainability operating via different approaches. The number of schemes increased to a total of 40 schemes in 2009 (Global Trust Certification 2009). In Vietnam, Marschke and Wilkings (2014), reported that there are 4 certification schemes that have certified a total of 20 producers with certification still ongoing. The development of certification schemes is guided by the FAO (2011b) with guidelines for setting the standards, performing an accreditation and final certification.

Certification schemes can be categorized using several criteria. Lee (2009) used a classification scheme that divided the schemes as either organic or non-organic. Global Trust Certification (2009) also used a two category classification that includes trade and eco-friendly schemes while Corsin et al. (2007) adopted an eight category classification of the schemes using the advocate, trade and environment as anchors. We will attempt to also classify the schemes (Table 1.10) based on the two categories used by Lee (2009).

The scope of most of these certification schemes covers the environment but social welfare is excluded in organic schemes for aquaculture producers in developed nations (Lee 2009). The safe quality food (SQF) standard focuses entirely on food safety and traceability. Each scheme has advocates that prepare the standards and present the same to target producers with credibility being either inherent for state-owned agencies or derived from third party auditors especially for non-organic

Table 1.10 Aquaculture certification schemes

Scheme	Promoter	Coverage	Labelling	Species	Website
Non-organic					
Best Aquaculture Practice	Global Aquaculture Alliance	Global	Eco-label	Shrimp, tilapia, pangasius, salmon channel catfish	www.gaalliance.org www.aquaculturecertification.org
SEASAIP	Asia Seafood Improvement Collaborative	Asia	Seafood watch yellow	Shrimps: *Litopenaeus. vannamei* and *Penaeus monodon*	www.asicollaborative.org
Global GAP	Global GAP	Global	GGN	Pangasius, salmon, penaeid shrimp	www.globalgap.org
Aquaculture Stewardship Council	World Wildlife Fund	Global	Farmed responsibly	Currently 12 species, but envisages covering all major species	www.asc-aqua.org
Friend of the Sea	NGO: Friend of the Sea	Global	Eco-label	Multi-species (Salmon, seabass, shrimps etc.)	www.friendofthesea.org
Thai Quality Shrimp	Thai Dept. of Fisheries	National	Eco-label	Penaeid shrimp and Freshwater prawn	www.thaiqualityshrimp.com
ISO 9001 and ISO 14001	NGO: International Standards Organisation	Global	Quality	All species	www.iso.org
SQF Code 7	Safe Quality Food Institute	Global	Food safety	All species	www.sqfi.com www.fmi.org
Carrefour Filiere Qualite	Carrefour	Global	Eco-label	Oyster, shrimp, salmon	www.carrefour.com
Label Rouge	French Ministry of Agriculture	France, Madagascar, Scotland	Eco-label	Salmon, penaeid shrimp, seabass, oysters turbot	www.label-rouge.org
Malaysian Aquaculture Farm Certification	Malaysian Government	National	Nil	Penaeid and freshwater shrimp, fish, molluscs and ornamental fish	http://www.dof.gov.my/

(continued)

Table 1.10 (continued)

Scheme	Promoter	Coverage	Labelling	Species	Website
Scottish Salmon Producers' Organisation code of good conduct	Scottish Salmon Producers	National	COGP	Salmon	www.scottishsalmon.co.uk
SIGES-Salmon Chile	Fundacion Chile/CBPA	National	Nil	Salmon	www.salmonchile.cl/en
Organic					
Bioland	Bioland eV	Germany	Eco-label	Freshwater fish	www.bioland.de
International Federation of Organic Agriculture Movements	IFOAM Organics International	Global	Eco-label	Seaweeds, bivalves, etc.	www.ifoam.org
Bio-gro	New Zealand Biological Producers and Consumers Society	New Zealand	Eco-label	Fish, shellfish, crustaceans	www.biogro.co.nz
Bio-Suisse	NGO: Federation of Swiss organic farmers	Switzerland	Eco-label	Carp, char, trout, perch	www.bio-suisse.ch
Naturland	NGO	Global	Eco-label	Salmon, milkfish, penaeid shrimp, mussels, arapaima	www.naturland.org
Organic Food Federation	NGO	EU	Eco-label	Cod, salmon	www.orgfoodfed.com

schemes. Organic schemes must conform to guidelines specified by organic farming promoting organisations such as IFOAM (Germany), IMO (Switzerland), ACO (Australia) and a host of others. It is often a challenge for organic aquaculture certifying bodies to ensure that the recommended farm practices adhere well to the product requirements and to the rearing conditions for a particular species such that the farm outputs and market demands are reasonably balanced (Nair et al. 2014). The organic certifying bodies are audited to ensure that standards are met.

Two generic quality management standards that can be employed by any organisation are the ISO 9001 and ISO 14001 standards. Internal quality management in aquaculture firms as well as aquaculture product processing firms can be achieved by applying the ISO 9001 quality management standard, while environmental management can be made to conform with acceptable standards by the use of ISO 14001 management standards. In terms of specifics, the SQF code 7 covers aspects of food and feed safety in the aquaculture processing and production industry. The Aquaculture Certification Council (ACC) of the Global Aquaculture Alliance (GAA) seeks to promote Best Aquaculture Practices (BAP) for responsible aquaculture covering animal welfare, social and environmental wellbeing, food safety and traceability. The ACC standards target specific aquaculture species and include steps for risk analysis using guidelines embedded in the standards, and hence are more relevant to the aquaculture industry compared to the generic ISO 9001 and 14001 that both demand self-risk assessment from the farmer.

Another key international certification scheme is the GlobalG.A.P. (Global Good Agricultural Practice). GlobalGAP does not use eco-labelling but instead is a business to business certification scheme. It also covers food safety, social and environmental wellbeing, animal welfare and wellbeing of the workforce. Guidelines cover all areas of the production process from spawning through stocking to harvesting and processing.

Aquaculture dialogues have been initiated by the World Wildlife Fund (WWF) and management of the standards developed was transferred to the Aquaculture Stewardship Council (ASC). A total of 8 aquaculture standards that cover 12 aquaculture species have been developed. The ASC seeks to use efficient markets to reduce negative impact of aquaculture production considering social and environmental wellbeing through standards and compliance.

The seafood retail chain has also taken the responsibility of ensuring sustainable production seriously with the promotion of their own eco-labels and standards. These are embedded in internal business chain guidelines, ethics and codes of practice. The UK's Tesco adopts guidelines for sustainable fish products using the FAO Code of Conduct for Responsible Fisheries while sustainable aquaculture of salmon is being advocated by the development of standards for salmon products on offer by the giant retail outlet. Sainsbury's in the UK favours the use of Global GAP and organically certified farmed fish for its cultured fish supplies (Global Trust Certification 2009). In North America, the Food Market Institute (FMI) has engaged experts and franchises to develop proactive sustainable seafood standards with representation from Canada's Loblaw and Sobeys supply chains. Carrefour group in France has developed its own standards for the farmed fish that it sources, considering minimal environmental damage from their production hence a bias for tilapia, pangasius and other species with limited need for fishmeal based diets. The French Label Rouge is also used to certify quality of some aquaculture products including oysters, seabass, Scottish salmon and shrimp imported from Madagascar.

National certification schemes also exist in several countries in the south. In Thailand, there is the Thai Quality Shrimp certification scheme. Other schemes with national focus include the Vietnamese GAP, Hong Kong's Accredited Fish Farm Scheme, Chilean Code of Good Environmental Practices for salmon, and the Malaysian Aquaculture Farm Certification scheme.

1.8 Concluding Remarks

Aquaculture is a rapidly growing food production sector with level of intensification and number of people involved in the industry also increasing. In the wake of its expansion, there has been negligence on the part of operators that created environmental and ecological problems. Aquaculture can be a threat to genetic diversity, water quality, general ecology, health and natural resources. This notwithstanding, the environment can also impact negatively on aquaculture.

Sustainable aquaculture holds the key to mitigating the threats posed by aquaculture while also ensuring that this important food producing sector remains very productive. Several approaches must be used in order to ensure that aquaculture does not impact negatively on the environment. The determination of suitable sites with capacity to absorb the biomass from aquaculture as well as waste is the first step that needs to be taken to achieve the goal of sustainability. Site selection and carrying capacity estimation must follow each step of aquaculture from planning to actual culture, and it must be an ongoing process. In assessing sites for aquaculture, it is important to also understand the risks involved in operating aquaculture facilities in an area considering the introduction and transfer of pathogens and other risks. Risk assessment in aquaculture must be carried out in conjunction with local and indigenous communities so that knowledge of the area is built into the process. The use of stakeholder opinion is a bone of contention for the scientific community. However, the knowledge from community members can be very useful. Sustainable aquaculture can also be achieved with the use of the ecosystem approach to aquaculture when regulatory agencies are planning for aquaculture. This approach has scientific depth because it relies on modelling and seeks to ensure sustainable aquaculture through management, improvement of livelihoods as well as consideration of other resource users.

Furthermore, effective regulation of aquaculture through governance is expected to assist the enforcement of the ecosystem approach to aquaculture. The application of governance without corrupt practices will ensure good product quality, sound environmental management and general social wellbeing. Gender sensitivity is also crucial in governance hence gender representation and responsiveness are two key factors that must be considered in the process of governance. The emergence of aquaculture certification and standards has created room for operators to ensure they remain competitive in the market given the consumer demand for products that are produced sustainably and are safe for consumption. Transparency, integrity and accountability are the key drivers of certification. Certification of small scale aquaculture facilities is affected by social, environmental/ethical and financial challenges. Certification schemes are quite numerous with each scheme presenting unique standards that are used as reference in auditing aquaculture ventures. These standards can be categorized using several criteria. However, two basic categories included are the organic and non-organic standards.

Conclusively, since aquaculture can have adverse effects on the environment, it is highly desirable to understand and predict these impacts so that remedial actions can be taken to keep these consequences in check and within allowable thresholds. Continuous monitoring of aquaculture sites and zones by regulators is also important since it will lead to appropriate interventions when they are needed so as to foster environmental wellbeing, ecological integrity, desirable food quality attributes, and social security.

References

ADB (1995) Governance: sound development management. Asian Development Bank, Manila

Aguilar-Manjarrez J, Kapetsky J, Soto D (2010) Status and potential of spatial planning tools, decision-making and modelling in implementing the ecosystem approach to aquaculture. FAO, Rome

Ahmed N, Glaser M (2016) Coastal aquaculture, mangrove deforestation and blue carbon emissions: is REDD+ a solution? Mar Policy 66:58–66. https://doi.org/10.1016/j.marpol.2016.01.011

Ahmed N, Troell M (2010) Fishing for prawn larvae in Bangladesh: an important coastal livelihood causing negative effects on the environment. Ambio 39(1):20–29. https://doi.org/10.1007/s13280-009-0002-y

Alavaisha E, Mangora MM (2016) Carbon stocks in the small estuarine mangroves of Geza and Mtimbwani, Tanga, Tanzania. Int J For Res 2016:11. https://doi.org/10.1155/2016/2068283

Anh PT, Kroeze C, Bush SR, Mol APJ (2010) Water pollution by Pangasius production in the Mekong Delta, Vietnam: causes and options for control. Aquac Res 42(1):108–128. https://doi.org/10.1111/j.1365-2109.2010.02578.x

Anka IZ, Faruk MAR, Hasan MM, Azad MAK (2013) Environmental issues of emerging pangas (*Pangasianodon hypophthalmus*) farming in Bangladesh. Prog Agric 24(1–2):159–170

Antunes A, Apostolidis A, Berrebi P, Duguid A, Ferguson A, Garcia-Marin JL, Guyomard R, Hansen MM, Hindar K, Largiader C, Martinez P, Nielsen EE, Palm S, Ruzzante D, Ryman N, Triantaphyllidis C (1999) Conservation genetic managment of brown trout (*Salmo trutta*) in Europe. Report by the concerted action on identification, managment and exploitation of genetic resources in the Brown Trout (*Salmo trutta*) («TROUT CONCERT»; EU FAIR CT97–3882)

Arthur JR (2008) General principles of the risk analysis process and its application to aquaculture. In: Bondad-Reantaso MG, Arthur JR, Subasinghe RP (eds) FAO Fisher ed, Rome, pp 3–8

Ataguba GA, Kamble MT, Okayi GR, Chavan BR (2014) An overview and assessment of two wetlands in Eastern Thailand: Kung krabaen bay and welu wetlands in chanthaburi province. Int J Agric Environ Biotechnol 8(1):205–213. https://doi.org/10.5958/2230-732x.2015.00026.1

Ataguba GA, Okomoda VT, Unde ES (2013) Efficacy of copper sulphate as a prophylactic agent for fungal infection on egg, and its effect on hatching and early growth of *Clarias gariepinus* (Burchell 1822). Asian Fish Sci 26:156–166

Aven T (2012) Foundations of risk analysis. Wiley, West Sussex

Axler RP, Tikkanen C, Henneck J, Schuldt J, McDonald ME (1997) Characteristics of effluent and sludge from two commercial rainbow trout farms in Minnesota. Prog Fish Cultur 2:161–172. https://doi.org/10.1577/1548-8640(1997)059

Balkhausen O, Banse M (2005) Modelling of Land use and land markets in partial and general equilibrium models: the current state

Barrington K, Chopin T, Robinson S (2009) Integrated multi-trophic aquaculture (IMTA) in marine temperate waters. In: Soto D (ed) FAO Fisheries and Aquaculture technical paper. No. 529, Rome, pp 7–46

Baskett ML, Burgess SC, Waples RS (2013) Assessing strategies to minimize unintended fitness consequences of aquaculture on wild populations. Evol Appl 6(7):1090–1108. https://doi.org/10.1111/eva.12089

Beach RH, Viator CL (2008) The economics of aquaculture insurance: an overview of the U.S. pilot insurance program for cultivated clams. Aquac Econom Manag 12(1):25–38. https://doi.org/10.1080/13657300801959613

Bell JD, Bartley DM, Lorenzen K, Loneragan NR (2006) Restocking and stock enhancement of coastal fisheries: potential, problems and progress. Fish Res 80(1):1–8. https://doi.org/10.1016/j.fishres.2006.03.008

Beveridge M (2008) Cage aquaculture. In: Fishing news books. Wiley, NJ

Beveridge MCM, Phillips MJ (1990) Environmental impact of tropical inland aquaculture. In: Environment and third world aquaculture development, Bellagio

Bevir M (2012) Governance: a very short introduction. In: Very short introductions. OUP Oxford

Bianchi TS, DiMarco SF, Cowan JH, Hetland RD, Chapman P, Day JW, Allison MA (2010) The science of hypoxia in the northern Gulf of Mexico: a review. Sci Total Environ 408(7):1471–1484. https://doi.org/10.1016/j.scitotenv.2009.11.047

Bondad-Reantaso MG, Arthur JR (2008) Pathogen risk analysis for aquaculture production. In: Bondad-Reantaso MG, Arthur JR, Subasinghe RP (eds) Understanding and applying risk analysis in aquaculture. FAO Fisheries and Aquaculture technical paper, No. 519. FAO, Rome, pp 27–46

Bondad-Reantaso MG, Lovell ER, Arthur JR, Hurwood D, Mather PB (2005) Pathogen and ecological risk analysis for the introduction of Blue Shrimp, *Litopenaeus stylirostris*, from Brunei Darussalam to Fiji. Secretariat of the Pacific Community, Noumea, New Caledonia

Bosma R, Anh PT, Potting J (2011) Life cycle assessment of intensive striped catfish farming in the Mekong Delta for screening hotspots as input to environmental policy and research agenda. Int J Life Cycle Assess 16(9):903–915. https://doi.org/10.1007/s11367-011-0324-4

Boyd CE (2012) Bottom soils, sediment, and pond aquaculture. Springer, US

Boyd CE, Massaut L (1999) Risks associated with the use of chemicals in pond aquaculture. Aquacult Eng 20(2):113–132. https://doi.org/10.1016/S0144-8609(99)00010-2

Boyd CE, McNevin A (2014) Aquaculture, resource use, and the environment. Willey, New Jersey

Boyd CE, Queiroz J, Lee J, Rowan M, Whitis GN, Gross A (2000) Environmental assessment of channel catfish *Ictalurus punctatus* farming in Alabama. J World Aquac Soc 31(4):511–544

Britton JR, Orsi ML (2012) Non-native fish in aquaculture and sport fishing in Brazil: economic benefits versus risks to fish diversity in the upper River Paraná Basin. Rev Fish Biol Fisheries 22(3):555–565. https://doi.org/10.1007/s11160-012-9254-x

Britz W, Verburg PH, Leip A (2011) Modelling of land cover and agricultural change in Europe: combining the CLUE and CAPRI-spat approaches. Agr Ecosyst Environ 142(1–2):40–50. https://doi.org/10.1016/j.agee.2010.03.008

Brun E (2013) Risk analysis in aquaculture. OIE

Burrows D, Murray G, McGreer E (2010) Aquaculture and the ecosystem—based approach (EBA): concepts, case studies and lessons learned, pp 1–46

Butterworth A (2010) Integrated multi-trophic aquaculture systems incorporating abalone and seaweeds, vol 0914

Byron C, Link J, Costa-Pierce B, Bengtson D (2011) Calculating ecological carrying capacity of shellfish aquaculture using mass-balance modeling: Narragansett Bay, Rhode Island. Ecol Model 222(10):1743–1755. https://doi.org/10.1016/j.ecolmodel.2011.03.010

Byron CJ, Jin D, Dalton TM (2014) A social-ecological system (SES) that integrates carrying capacity for the sustainable management of bivalve aquaculture, pp 5–6

Calvert S (2016) Maryland marine police crack down on oyster poachers. Wall Street J. Dow Jones & Company

Cappello T, Mauceri A, Corsaro C, Maisano M, Parrino V, Lo Paro G, Messina G, Fasulo S (2013) Impact of environmental pollution on caged mussels *Mytilus galloprovincialis* using NMR-based metabolomics. Mar Pollut Bull 77(1–2):132–139. https://doi.org/10.1016/j.marpolbul.2013.10.019

Carver SJ (1991) Integrating multi-criteria evaluation with geographical information systems. Int J Geogr Inf Syst 5(3):321–339. https://doi.org/10.1080/02693799108927858

Chen J, Huang C, Xu H, Chen Z, Xu P, Yan X, Wang Y, Liu J (2008) Marine fish cage culture in China. In: Lovatelli A, Phillips MJ, Arthur JR, Yamamoto K (eds), vol 11. FAO, Rome, pp 285–299

Chorley RJ, Haggett P (2013) Socio-economic models in geography (Routledge revivals). Routledge, Routledge Revivals, London

Chowdhury MA, Khairun Y, Salequzzaman M, Rahman MM (2011) Effect of combined shrimp and rice farming on water and soil quality in Bangladesh. Aquacult Int 19(6):1193–1206. https://doi.org/10.1007/s10499-011-9433-0

Christensen V, Walters CJ (2004) Ecopath with ecosim: methods, capabilities and limitations. Ecol Model 172(2–4):109–139. https://doi.org/10.1016/j.ecolmodel.2003.09.003

Chuenpagdee R, Kooiman J, Pullin RSV (2008) Assessing governability in capture fisheries, aquaculture and coastal zones. J Transdiscipl Environ Stud 7(1):1–20

CIG (2016) Defining governance. http://iog.ca/defining-governance/. Accessed 24 May 2016

Clark JL, Weldon RN, Adams CM, Wirth FF (2010) Risk assessment of a shrimp aquaculture investment in Florida. Aquac Econom Manag 14(4):332–357. https://doi.org/10.1080/13657305.2010.526023

Colbourne DB, Allen JH (2000) Observations on motions and loads in aquaculture cages from full scale and model scale measurements. Aquacult Eng 24(2):129–148. https://doi.org/10.1016/S0144-8609(00)00069-8

Corsin F, Funge-Smith S, Jesper C (2007) A qualitative assessment of standards and certification schemes applicable to aquaculture in the Asia–Pacific region. FAO, Bangkok

Coutant CC (2013) Thermal Effects on fish ecology. Taylor & Francis, UK, pp 1167–1172 SE. https://doi.org/10.1081/E-EESE6-120048567

Cowx IG, Nunn AD, Harvey JP, Noble RAA (2012) Guidelines for stocking of fish within designated natural heritage sites. Scottish Natural Heritage Commissioned Report No. 513

Craig RK (2002) The other side of sustainable aquaculture: mariculture and nonpoint source pollution. Washington Univ J Law Policy 9(January):163–207

Crawford C (2001) Environmental risk assessment of shellfish farming in Tasmania. TAFI Marine Research Laboratories, Taroona, Tasmania

Crawley F, Tyler B, Centre EPS (2003) Hazard identification methods. Institution of Chemical Engineers

Cui Y, Guan CT, Wan R, Huang B, Li J (2013) Numerical simulation of a flatfish cage system in waves and currents. Aquacult Eng 56:26–33. https://doi.org/10.1016/j.aquaeng.2013.04.001

da Costa KGa, Nalesso RC (2006) Effects of mussel farming on macrobenthic community structure in Southeastern Brazil. Aquaculture 258(1–4):655–663. https://doi.org/10.1016/j.aquaculture.2006.04.023

Dalsgaard JPT, Lightfoot C, Christensen V (1995) Towards quantification of ecological sustainability in farming systems analysis. Ecol Eng 4(3):181–189. https://doi.org/10.1016/0925-8574(94)00057-C

Dapueto G, Massa F, Costa S, Cimoli L, Olivari E, Chiantore M, Federici B, Povero P (2015) A spatial multi-criteria evaluation for site selection of offshore marine fish farm in the Ligurian Sea, Italy. Ocean Coast Manag 116:64–77. https://doi.org/10.1016/j.ocecoaman.2015.06.030
Davis DA (2015) Feed and feeding practices in aquaculture. Woodhead Publishing Series in Food Science, Technology and Nutrition. Elsevier Science
De Silva SS, Ingram BA, Nguyen PT, Bui TM, Gooley GJ, Turchini GM (2010) Estimation of nitrogen and phosphorus in effluent from the striped catfish farming sector in the Mekong Delta, Vietnam. Ambio 39(7):504–514. https://doi.org/10.1007/s13280-010-0072-x
DeCew J, Tsukrov I, Risso A, Swift MR, Celikkol B (2010) Modeling of dynamic behavior of a single-point moored submersible fish cage under currents. Aquacult Eng 43(2):38–45. https://doi.org/10.1016/j.aquaeng.2010.05.002
Diggles BK, Arthur JR (2010) Pathogen risk analysis for aquatic animals: experiences from nine case studies. In: Bondad-Reantaso MG, Jones JB, Corsin F, Aoki T (eds) The seventh symposium on diseases in asian aquaculture, Taipei, Taiwan 2008. Fish Health Section, Asian Fisheries Society, Selangor, Malaysia, pp 271–290
Dominguez LM, Martín JMV (2004) Aquaculture environmental impact assessment. In: Popov V, Itoh H, Brebbia CA, Kungolos S (eds). WIT Press, pp 321–333
Donato DC, Kauffman JB, Murdiyarso D, Kurnianto S, Stidham M, Kanninen M (2011) Mangroves among the most carbon-rich forests in the tropics. Nature Geosci 4 (5):293–297. http://www.nature.com/ngeo/journal/v4/n5/abs/ngeo1123.html#supplementary-information
Dosdat A (2009) Environment impact of aquaculture, Fisheries ed. EOLSS Publishers
ECASA (2008) Ecosystem approach for sustainable aquaculture executive summary final activity report STREP. In: Sixth framework programme priority: integrating and strengthening the European research area. Scottish Association for Marine Science
Echols KR, Meadows JC, Orazio CE (2009) Pollution of aquatic ecosystems II: hydrocarbons, synthetic organics, radionuclides, heavy metals, acids, and thermal pollution. In: Encyclopedia inland waters, pp 120–128. https://doi.org/10.1016/B978-012370626-3.00223-4
Edwards P (2002) Aquaculture for poverty alleviation and food security. Aquacult Asia VII(2):53–56
Eldrandaly K (2013) Developing a GIS-based MCE site selection tool in arcGIS using COM technology. Int Arab J Inf Technol 10 (3)
Emerson C (1999) Aquaculture impacts on the environment. Cambridge Scientific Abstracts
Ertor I, Ortega-Cerda M (2015) Political lessons from early warnings: marine finfish aquaculture conflicts in Europe. Mar Policy 51:202–210. https://doi.org/10.1016/j.marpol.2014.07.018
Espinosa M, Gocht A, Heckelei T, Paloma SGY (2016) Incorporating farm structural change in models assessing the common agricultural policy: an application in the CAPRI farm type model. J Policy Model. https://doi.org/10.1016/j.jpolmod.2016.03.005
Estrada GCD, Soares MLG (2017) Global patterns of aboveground carbon stock and sequestration in mangroves. An Acad Bras Ciênc 89:973–989
Ezekiel EN, Abowei JFN, Ezekiel EF (2011) Hazard and risk analysis in culture fisheries. Res J Appl Sci Eng Technol 3(10):1108–1117
Fabi G, Manoukian S, Spagnolo A (2009) Impact of an open-sea suspended mussel culture on macrobenthic community (Western Adriatic Sea). Aquaculture 289(1–2):54–63. https://doi.org/10.1016/j.aquaculture.2008.12.026
FAO (2007a) The state of world fisheries and aquaculture 2006. Fisheries and Aquaculture Department, FAO, Rome
FAO (2007b) The world's mangroves 1980–2005, vol 153. FAO, Rome. ISBN: 978-92-5-105856-5
FAO (2008) Report of the expert consultation on improving planning and policy
FAO (2009) The state of world fisheries and aquaculture 2008. Fisheries and Aquaculture Department, FAO, Rome
FAO (2010) Aquaculture development 4. In: Ecosystem approach to aquaculture, vol 5, suppl 4. Rome. https://doi.org/10.1017/cbo9781107415324.004

FAO (2011a) The state of the world's land and water resources for food and agriculture: managing systems at risk. FAO/Earthscan, Rome

FAO (2011b) Technical guidelines on aquaculture certification. Food and Agriculture Organisation of the United Nations

FAO (2012) The state of world fisheries and aquaculture 2012. Fisheries and Aquaculture Department, FAO, Rome

FAO (2014) The state of world fisheries and aquaculture: opportunities and challenges. FAO, Rome

FAO (2016) The state of world fisheries and aquaculture: contributing to food security and nutrition for all. Fisheries and Aquaculture Department, FAO, Rome

Ferreira JG, Hawkins AJS, Monteiro P, Moore H, Service M, Pascoe PL, Ramos L, Sequeira A (2008) Integrated assessment of ecosystem-scale carrying capacity in shellfish growing areas. Aquaculture 275(1–4):138–151. https://doi.org/10.1016/j.aquaculture.2007.12.018

Fisher P (2006) Classics from IJGIS: twenty years of the international journal of geographical information science and systems. CRC Press, Boca Raton

Flaten O, Lien G, Tveteras R (2011) A comparative study of risk exposure in agriculture and aquaculture. Food Econ Acta Agricult Scand Sect C 8:20–34

Francis RICC, Shotton R (1998) Can J Fish Aquat Sci 1997 54(8):1699–1715. https://doi.org/10.1139/f97-100

Fredriksson DW, Swift MR, Irish JD, Tsukrov I, Celikkol B (2003) Fish cage and mooring system dynamics using physical and numerical models with field measurements. Aquacult Eng 27:117–146. https://doi.org/10.1016/S0144-8609(02)00043-2

Gallardo B, Aldridge DC (2013) Evaluating the combined threat of climate change and biological invasions on endangered species. Biol Cons 160:225–233. https://doi.org/10.1016/j.biocon.2013.02.001

Gärtner D, Keller A, Schulin R (2013) A simple regional downscaling approach for spatially distributing land use types for agricultural land. Agric Syst 120:10–19. https://doi.org/10.1016/j.agsy.2013.04.006

Gegner L, Rinehart L (2009) Aquaculture enterprises: considerations and strategies. ATTRA—National sustainable agriculture information service

Gesamp (2001) Planning and management for sustainable coastal aquaculture development. In: GESAMP reports and studies. FAO, Rome

Giri C, Zhu Z, Tieszen LL, Singh A, Gillette S, Kelmelis JA (2008) Mangrove forest distributions and dynamics (19752005) of the tsunami-affected region of Asia. J Biogeogr 35(3):519–528. https://doi.org/10.1111/j.1365-2699.2007.01806.x

GlobalGap (2014) First small-scale Pangasius farmers achieve GLOBALG.A.P. Aquaculture group certificate. http://www.globalgap.org/es/news/First-Small-Scale-Pangasius-Farmers-Achieve-GLOBALG.A.P.-Aquaculture-Group-Certificate/. Accessed 16 July 2016

Global Trust Certification (2009) Global market report

Glover KA, Hamre LA, Skaala, Nilsen F (2004) A comparison of sea louse (*Lepeophtheirus salmonis*) infection levels in farmed and wild Atlantic salmon (*Salmo salar* L.) stocks. Aquaculture 232(1–4):41–52. https://doi.org/10.1016/S0044-8486(03)00454-X

Guo FC, Woo PTK (2009) Selected parasitosis in cultured and wild fish. Vet Parasitol 163(3):207–216. https://doi.org/10.1016/j.vetpar.2009.06.016

Hamasaki K, Kitada S (2006) A review of kuruma prawn Penaeus japonicus stock enhancement in Japan. Fish Res 80(1):80–90. https://doi.org/10.1016/j.fishres.2006.03.018

Hargrave BT, Silvert W, Keizer PD (2005) Assessing and managing environmental risks associated with marine finfish aquaculture. In: Hargrave BT (ed) Environmental effects of marine finfish aquaculture. Springer Berlin Heidelberg, Berlin, Heidelberg, pp 433–461. https://doi.org/10.1007/b136021

Harnisz M, Korzeniewska E, Gołas I (2015) Chemosphere the impact of a freshwater fish farm on the community of tetracycline-resistant bacteria and the structure of tetracycline resistance genes in river water 128:134–141. https://doi.org/10.1016/j.chemosphere.2015.01.035

Hasan MR (2001) Nutrition and feeding for sustainable aquaculture development in the third millennium. In: Subasinghe RP, Bueno P, Phillips MJ, Hough C, McGladdery SE, Arthur JE (eds), Rome, 2001. NACA/FAO, pp 471–471

Hayes KR (2002a) Identifying hazards in complex ecological systems. Part 1: fault-tree analysis for biological invasions. Biol Invasions 4(3):235–249. https://doi.org/10.1023/A:1020979914453

Hayes KR (2002b) Identifying hazards in complex ecological systems. Part 2: infection modes and effects analysis for biological invasions. Biol Invasions 4(3):251–261. https://doi.org/10.1023/A:1020943231291

Hechanova RG (1984) Problems on pond construction and maintenance on cat clay soils, UNDP/FAO, Rome

Heistermann M, Muller C, Ronneberger K (2006) Land in sight? Achievements, deficits and potentials of continental to global scale land-use modeling. Agr Ecosyst Environ 114 (2–4):141–158. https://doi.org/10.1016/j.agee.2005.11.015

Hennessy MM, Wilson L, Struthers W, Kelly LA (1996) Waste loadings from two freshwater Atlantic salmon juvenile farms in Scotland. Water Air Soil Pollut 86(1–4):235–249. https://doi.org/10.1007/BF00279159

Hilborn R (1998) The economic performance of marine stock enhancement projects. Bull Mar Sci 62(2):661–674

Hirst P (2000) Democracy and governance. In: Pierre J (ed). Oxford University Press, pp 266–266

Hishamunda N (2010) Aquaculture governance: why does it matter? FAO Aquaculture Newsletter, Rome

Hishamunda N, Ridler N, Martone E (2014) Policy and governance in aquaculture lessons learned and way forward. FAO, Rome

Hishamunda N, Ridler NB, Bueno P, Satia B, Kuemlangan B, Percy D, Gooley G, Brugere C, Sen S (2012) Improving aquaculture governance: what is the status and options? In: Subasinghe RP, Richard Arthur J, Bartley DM, De Silva SS, Halwart M, N Hishamunda, Mohan CV, Sorgeloos P (eds), vol 1. pp 233–264

Hites RA, Foran JA, Carpenter DO, Hamilton MC, Knuth BA, Schwager SJ (2004) Global assessment of organic contaminants in farmed salmon. Science 303(5655):226–229. https://doi.org/10.1126/science.1091447

Hofherr J, Natale F, Trujillo P (2015) Is lack of space a limiting factor for the development of aquaculture in EU coastal areas? Ocean Coast Manag 116:27–36. https://doi.org/10.1016/j.ocecoaman.2015.06.010

Hossain MS, Chowdhury SR, Das NG, Sharifuzzaman SM, Sultana A (2009) Integration of GIS and multicriteria decision analysis for urban aquaculture development in Bangladesh. Landsc Urban Plan 90(3–4):119–133. https://doi.org/10.1016/j.landurbplan.2008.10.020

Hossain MS, Das NG (2010) GIS-based multi-criteria evaluation to land suitability modelling for giant prawn (*Macrobrachium rosenbergii*) farming in Companigonj Upazila of Noakhali, Bangladesh. Comput Electron Agricult 70(1):172–186. https://doi.org/10.1016/j.compag.2009.10.003

Huang CC, Tang HJ, Liu JY (2008) Effects of waves and currents on gravity-type cages in the open sea. Aquacult Eng 38(2):105–116. https://doi.org/10.1016/j.aquaeng.2008.01.003

Huang JJ, Xu J, Qiao F (2001) Production, accessibility and consumption patterns of aquaculture products in China, Rome, pp 275–275

Hughes S, Rose JB (2011) Governing aquaculture for human security. In: Taylor WW, Lynch AJ, Schechter MG (eds). American Fisheries Society, pp 377–377

Huntingford F, Jobling M, Kadri S (2012) Aquaculture and behavior. Wiley, NJ

Huntington TC, Roberts H, Cousins N, Pitta V, Marchesi N, Sanmamed A, Hunter-Rowe T, Fernandes TF, Tett P, McCue J, Brockie N (2006) Some aspects of the environmental impact of aquaculture in sensitive areas

Hurlburt BK, Brashear SS, Lloyd SW, Grimm CC, Thomson JL, Zimba PV (2009) Impact of weather on off-flavour episodes at a Louisiana commercial catfish farm. Aquac Res 40(5):566–574. https://doi.org/10.1111/j.1365-2109.2008.02132.x

Inglis GJ, Hayden BJ, Ross AH (2000) An overview of factors affecting the carrying capacity of coastal embayments for mussel culture. National Institute of Water & Atmospheric Research, New Zealand

Jacobs MN, Covaci A, Schepens P (2002) Investigation of selected persistent organic pollutants in farmed Atlantic salmon (*Salmo salar*), salmon aquaculture feed, and fish oil components of the feed. Environ Sci Technol 36(13):2797–2805. https://doi.org/10.1021/es011287i

Johnson R (2007) The ecosystem approach to aquaculture: an analysis of adaptive governance and management of small-scale coastal shrimp aquaculture in Thailand. Stockholm University, Stockholm

Johnson RW (1998) Risk management by risk magnitudes. Chem Heal Safe 5(5):1 – 2. (September/October 1998, Unwin Company Integrated Risk Management)

Johnston MW, Purkis SJ (2016) Forecasting the success of invasive marine species; lessons learned from purposeful reef fish releases in the Hawaiian Islands. Fish Res 174:190–200. https://doi.org/10.1016/j.fishres.2015.10.011

Jonell M, Phillips M, Ronnback P, Troell M (2013) Eco-certification of farmed seafood: will it make a difference? Ambio 42(6):659–674. https://doi.org/10.1007/s13280-013-0409-3

Kapetsky JM, Aguilar-Manjarrez J (2013) From estimating global potential for aquaculture to selecting farm sites: perspectives on spatial approaches and trends. In: Ross LG, Telfer TC, Falconer L, Soto D, Aguilar-Manjarrez J (eds) FAO/Institute of Aquaculture, University of Stirling, Expert Workshop, Stirling, the United Kingdom of Great Britain and Northern Ireland, 6–8 Dec 2010. FAO Fisheries and Aquaculture, pp 129–146

Kapetsky JM, Aguilar-Manjarrez J, Fao (2007) Geographic information systems, remote sensing and mapping for the development and management of marine aquaculture. FAO fisheries technical paper. Food and Agriculture Organization of the United Nations

Kim T, Lee J, Fredriksson DW, DeCew J, Drach A, Moon K (2014) Engineering analysis of a submersible abalone aquaculture cage system for deployment in exposed marine environments. Aquacult Eng 63:72–88. https://doi.org/10.1016/j.aquaeng.2014.10.006

Kirillin G, Shatwell T, Kasprzak P (2013) Consequences of thermal pollution from a nuclear plant on lake temperature and mixing regime. J Hydrol 496:47–56. https://doi.org/10.1016/j.jhydrol.2013.05.023

Kitada S (2014) Japanese chum salmon stock enhancement: current perspective and future challenges. Fish Sci 80(2):237–249. https://doi.org/10.1007/s12562-013-0692-8

Krause G, Brugere C, Diedrich A, Ebeling MW, Ferse SCA, Mikkelsen E, Pérez Agúndez JA, Stead SM, Stybel N, Troell M (2015) A revolution without people? Closing the people-policy gap in aquaculture development. Aquaculture 447:44–55. https://doi.org/10.1016/j.aquaculture.2015.02.009

Kura Y, Revenga C, Hoshino E, Mock G (2004) Fishing for answers making sense of the global fish crisis. World Resources Institute, Washington DC

Lader P, Dempster T, Fredheim A, Jensen Ø (2008) Current induced net deformations in full-scale sea-cages for Atlantic salmon (*Salmo salar*). Aquacult Eng 38(1):52–65. https://doi.org/10.1016/j.aquaeng.2007.11.001

Langford T (1990) Ecological effects of thermal discharges. Pollution monitoring series. Springer, Netherlands

Lawson T (2013) Fundamentals of aquacultural engineering, Illustrate edn. Springer, US

Le TX, Munekage Y, Kato SI (2005) Antibiotic resistance in bacteria from shrimp farming in mangrove areas. Sci Total Environ 349(1–3):95–105. https://doi.org/10.1016/j.scitotenv.2005.01.006

Lebel P, Whangchai N, Chitmanat C, Lebel L (2015) Risk of impacts from extreme weather and climate in river-based tilapia cage culture in Northern Thailand. Int J Glob Warm 8 (4):534–554. https://doi.org/10.1504/IJGW.2015.073054

Lee D (2009) Understanding aquaculture certification. Revista Colombiana de Ciencias Pecuarias 22(3):319–329

Lehnert SJ, Heath DD, Pitcher TE (2012) Sperm trait differences between wild and farmed Chinook salmon (*Oncorhynchus tshawytscha*). Aquaculture 344–349:242–247. https://doi.org/10.1016/j.aquaculture.2012.03.007

Lent R, English E, Wulff R, de Séligny JP, Mahon R, Cox A, Balton D, Tompkins C, Tinkham S, Metzner R, Terry J, McGoodwin JR, Mellano G, Bexten A, Moore K (2008) Policy brief: fisheries and aquaculture—sustainability and governance. In: 4th Global conference on oceans, coasts and islands. Working Group on Fisheries and Aquaculture: Sustainability and Governance

Lessner R (2015) New law aims to protect oyster farms from poachers. http://marylandreporter.com/2015/05/26/new-law-aims-to-protect-oyster-farms-from-poachers/. Accessed 16 July 2016

Lin CK, Shrestha MK, Thakurand DP, Diana JS (1997) Management to minimize the environmental impacts of pond effluent: harvest draining techniques and effluent quality. Aquacult Eng 25(2):125–135

Lind CE, Dana GV, Perera RP, Phillips MJ (2015) Risk analysis in aquaculture: a step-by-step introduction with worked examples, vol 2015–08. WorldFish, Penang, Malaysia

Loneragan NR, Ye Y, Kenyon RA, Haywood MDE (2006) New directions for research in prawn (shrimp) stock enhancement and the use of models in providing directions for research. Fish Res 80(1):91–100. https://doi.org/10.1016/j.fishres.2006.03.014

Longdill PC (2007) Environmentally sustainable aquaculture: an eco-physical perspective. University of Waikato, Waikato

Longdill PC, Healy TR, Black KP (2008) An integrated GIS approach for sustainable aquaculture management area site selection. Ocean Coast Manag 51(8–9):612–624. https://doi.org/10.1016/j.ocecoaman.2008.06.010

Lorenzen K (2008) Understanding and managing enhancement fisheries systems. Rev Fish Sci 16 (1–3):10–23. https://doi.org/10.1080/10641260701790291

Luttrell C, Son HV, Thuan HL, Viet CT, Lan N, Xiem VD, Hieu DTL (2004) Sustainable livelihood opportunities and resource management in coastline communes facing special difficulties. Ministry of Planning and Investment

Madeira D, Vinagre C, Diniz MS (2016) Are fish in hot water? Effects of warming on oxidative stress metabolism in the commercial species *Sparus aurata*. Ecol Ind 63:324–331. https://doi.org/10.1016/j.ecolind.2015.12.008

Mahmood N, Saikat SQ (1995) On acid sulfate soils of the coastal aquaculture. Pak J Mar Sci 4 (1):39–43

Marschke M, Wilkings A (2014) Is certification a viable option for small producer fish farmers in the global south? Insights from Vietnam. Mar Policy 50:197–206. https://doi.org/10.1016/j.marpol.2014.06.010

Masci M, Orban E, Nevigato T (2014) Organochlorine pesticide residues: an extensive monitoring of Italian fishery and aquaculture. Chemosphere 94:190–198. https://doi.org/10.1016/j.chemosphere.2013.10.016

Mazur NA, Curtis AL (2006) Risk perceptions, aquaculture, and issues of trust: lessons from Australia. Soc Nat Resour 19(9):791–808. https://doi.org/10.1080/08941920600835551

McDaniels T, Longstaff H, Dowlatabadi H (2006) A value-based framework for risk management decisions involving multiple scales: a salmon aquaculture example. Environ Sci Policy 9 (5):423–438. https://doi.org/10.1016/j.envsci.2006.03.005

McGinnity P, Prodöhl P, Ferguson A, Hynes R, Maoiléidigh NO, Baker N, Cotter D, O'Hea B, Cooke D, Rogan G, Taggart J, Cross T (2003) Fitness reduction and potential extinction of wild populations of Atlantic salmon, Salmo salar, as a result of interactions with escaped farm salmon. Proc Biol Sci R Soc 270(1532):2443–2450. https://doi.org/10.1098/rspb.2003.2520

McGurk C, Morris DJ, Auchinachie NA, Adams A (2006) Development of *Tetracapsuloides bryosalmonae* (Myxozoa: Malacosporea) in bryozoan hosts (as examined by light microscopy) and quantitation of infective dose to rainbow trout (*Oncorhynchus mykiss*). Vet Parasitol 135 (3–4):249–257. https://doi.org/10.1016/j.vetpar.2005.07.022

McIntosh D (2008) Aquaculture risk management. Northeastern Regional Aquaculture Center

McKindsey CW, Thetmeyer H, Landry T, Silvert W (2006) Review of recent carrying capacity models for bivalve culture and recommendations for research and management. Aquaculture 261(2):451–462. https://doi.org/10.1016/j.aquaculture.2006.06.044

Miller LM, Kapuscinski AR, Senanan W, Paul S (2004) A biosafety approach to addressing risks posed by aquaculture escapees. WorldFish Center, Penang, pp 56–65

Moccia RD, Bevan DJ (2005) Environmental issues concerning water use and wastewater impacts of land-based aquaculture facilities in Ontario, Final Version edn. Ontario Sustainable Aquaculture Working Group, Ontario

Morgan F, LaFary EW (2009) GIS, ecosystems and urban planning in Auckland, New Zealand: technology, processes and people. In: Gatrell JD, Jensen RR (eds) Geotechnol ed. Springer Science+Business Media B.V., pp 57–77

Morris DJ, Adams A (2006) Transmission of freshwater myxozoans during the asexual propagation of invertebrate hosts. Int J Parasitol 36(3):371–377. https://doi.org/10.1016/j.ijpara.2005.10.009

Muir WM (2004) The threats and benefits of GM fish. EMBO Rep 5(7):654–659. https://doi.org/10.1038/sj.embor.7400197

Murray AG (2015) Does the use of salmon frames as bait for lobster/crab creel fishing significantly increase the risk of disease in farmed salmon in Scotland? Prev Vet Med 120:357–366. https://doi.org/10.1016/j.prevetmed.2015.04.020

Nair CM, Salin KR (2012) Current status and prospects of farming the giant river prawn *Macrobrachium rosenbergii* (De Man) and the monsoon river prawn Macrobrachium malcolmsonii (H.M. Edwards) in India. Aquac Res 43(7):999–1014. https://doi.org/10.1111/j.1365-2109.2011.03074.x

Nair CM, Salin KR, Joseph J, Aneesh B, Geethalaksmi V, New MB (2014) Organic rice–prawn farming yields 20% higher revenues. Agron Sustain Dev 34(3):569–581. https://doi.org/10.1007/s13593-013-0188-z

Navas JM, Telfer TC, Ross LG (2011) Spatial modeling of environmental vulnerability of marine finfish aquaculture using GIS-based neuro-fuzzy techniques. Mar Pollut Bull 62(8):1786–1799. https://doi.org/10.1016/j.marpolbul.2011.05.019

Naylor R, Burke M (2005) Aquaculture and ocean resoources: raising tigers of the sea. Annu Rev Environ Resour 30(1):185–218. https://doi.org/10.1146/annurev.energy.30.081804.121034

Neto RM, Ostrensky A (2015) Nutrient load estimation in the waste of *Nile tilapia Oreochromis niloticus* (L.) reared in cages in tropical climate conditions. Aquac Res 46(6):1309–1322. https://doi.org/10.1111/are.12280

New MB, Nair CM, Kutty MN, Salin KR, Nandeesha MC (2008) Macrobrachium: the culture of freshwater prawns. Macmillan, Delhi

Nobre AM, Musango JK, de Wit MP, Ferreira JG (2009) A dynamic ecological-economic modeling approach for aquaculture management. Ecol Econ 68(12):3007–3017. https://doi.org/10.1016/j.ecolecon.2009.06.019

Nostbakken OJ, Hove HT, Duinker A, Lundebye AK, Berntssen MHG, Hannisdal R, Lunestad BT, Maage A, Madsen L, Torstensen BE, Julshamn K (2015) Contaminant levels in Norwegian farmed Atlantic salmon (*Salmo salar*) in the 13-year period from 1999 to 2011. Environ Int 74:274–280. https://doi.org/10.1016/j.envint.2014.10.008

Nugent C (2009) Review of environmental impact assessment and monitoring in aquaculture in Africa, 527 edn. FAO, Rome, pp 59–151

Nyerges TL, Jankowski P (2010) Regional and urban GIS: a decision support approach. Guilford Press

OECD (2008) Gender and sustainable development: maximising the economic, social and environmental role of women

OECD/FAO (2015) OECD-FAO Agricultural outlook 2015–2024. OECD Publishing, Paris

OIE (2015) Aquatic animal health code (2015), 2015 edn. Organisation Mondiale de la Sante Animale

Olafsen JA (2001) Interactions between fish larvae and bacteria in marine aquaculture. Aquaculture 200(1–2):223–247. https://doi.org/10.1016/S0044-8486(01)00702-5

Olalo C (2001) Production, accessibility and consumption patterns of aquaculture products in the Philippines. In: Production, accessibility, marketing and consumption patterns of freshwater aquaculture products in Asia: a cross-country comparison, vol FAO Fisheries Circular No. 973. Food and Agriculture Organisation of the United Nations, Rome, p 275

Olanrewaju OS, Kader ASA, Wan Nik WB, Peter SW, Oladimeji O (2013) Apply Safety Risk and Reliability Analysis of Marine System. Xlibris Corporation

Overmars KP, Helming J, van Zeijts H, Jansson T, Terluin I (2013) A modelling approach for the assessment of the effects of common agricultural policy measures on farmland biodiversity in the EU27. J Environ Manage 126:132–141. https://doi.org/10.1016/j.jenvman.2013.04.008

Ozbay G, Blank G, Thunjai T (2014) Impacts of aquaculture on habitats and best management practices (BMPs). In: Hernandez-Vergara MP, Perez-Rostro CI (eds). Intech Open, pp 274–274. https://doi.org/10.5772/57471

Paez-Osuna F (2001) The environmental impact of shrimp aquaculture: causes, effects, and mitigating alternatives. Environ Manage 28(1):131–140. https://doi.org/10.1007/s002670010212

Page FH, Losier R, McCurdy P, Greenberg D, Chaffey J, Chang B (2005) Dissolved oxygen and salmon cage culture in the Southwestern New Brunswick Portion of the Bay of Fundy. Hdb Env Chem 5(July):1–28. https://doi.org/10.1007/b136002

Parker R (2011) Aquaculture science. Cengage Learning

Pearsons TN, Fritts AL (1999) Maximum size of Chinook Salmon consumed by Juvenile Coho Salmon. N Am J Fish Manag 19(1):165–170. https://doi.org/10.1577/1548-8675(1999)019<0165:MSOCSC>2.0.CO;2

Pelletier N, Tyedmers P (2010) Life cycle assessment of frozen tilapia fillets from indonesian lake-based and pond-based intensive aquaculture systems. J Ind Ecol 14(3):467–481. https://doi.org/10.1111/j.1530-9290.2010.00244.x

Phanna N (2011) Mass balance for water and carbon (C) in Pangasius Ponds, Mekong Delta, pp 47–47

Pillay TVR (1997) Economic and social dimensions of aquaculture management. Aquacult Econ Manag 1(1–2):3–11. https://doi.org/10.1080/13657309709380199

Prapaiwong N, Boyd CE (2012) Effluent volume and pollutant loads at an inland, low-salinity, shrimp farm in Alabama. Aquacult Eng 48:1–5. https://doi.org/10.1016/j.aquaeng.2011.12.004

Rabalais NN, Turner RE, Sen Gupta BK, Boesch DF, Chapman P, Murrell MC (2007) Hypoxia in the northern Gulf of Mexico: does the science support the plan to reduce, mitigate, and control hypoxia? Estuaries Coasts 30(5):753–772. https://doi.org/10.2307/27654714

Radiarta IN, Saitoh SI, Miyazono A (2008) GIS-based multi-criteria evaluation models for identifying suitable sites for Japanese scallop (*Mizuhopecten yessoensis*) aquaculture in Funka Bay, southwestern Hokkaido, Japan. Aquaculture 284(1–4):127–135. https://doi.org/10.1016/j.aquaculture.2008.07.048

Ragbirsingh Y, De Souza G (2005) Site suitability for aquaculture development on the Caroni River Basin, Trinidad West Indies using GIS, pp 661–673

Rajitha K, Mukherjee CK, Vinu Chandran R, Prakash Mohan MM (2010) Land-cover change dynamics and coastal aquaculture development: a case study in the East Godavari Delta, Andhra Pradesh, India using multi-temporal satellite data. Int J Remote Sens 31(16):4423–4442. https://doi.org/10.1080/01431160903277456

Ranchan V (1984) A general review of the freshwater fish culture in India. In: Srivastava UK, Vathsala S (eds). Concept Publishing Company, pp 195–202

Raptis CE, Pfister S (2016) Global freshwater thermal emissions from steam-electric power plants with once-through cooling systems. Energy 97:46–57. https://doi.org/10.1016/j.energy.2015.12.107

Rawson MV, Chen C, Ji R, Zhu M, Wang D, Wang L, Yarish C, Sullivan BJ, Chopin T, Carmona R (2002) Understanding the interaction of extractive and fed aquaculture using ecosystem modeling. In: Stickney RR, McVey JP (eds). CAB International, pp 263–296

Reantaso MB (2008) Application of risk analysis in aquaculture production: outcomes of the desk study and the FAO/NACA expert workshop on understanding and applying risk analysis in aquaculture production. FAO Aquaculture Newsletter

Rico A, Satapornvanit K, Haque MM, Min J, Nguyen PT, Telfer TC, van den Brink PJ (2012) Use of chemicals and biological products in Asian aquaculture and their potential environmental risks: a critical review. Rev Aquacult 4(2):75–93. https://doi.org/10.1111/j.1753-5131.2012.01062.x

Ross LG, Telfer TC, Falconer L, Soto D, Aguilar-Manjarrez J, Asmah R, Bermúdez J, Beveridge MCM, Byron CJ, Clément A (2013) Carrying capacities and site selection within the ecosystem approach to aquaculture, pp 19–19

Sahu SC, Kumar M, Ravindranath N, H. (2016) Carbon stocks in natural and planted mangrove forests of Mahanadi mangrove wetland, East Coast of India. Current Sci 110(12):2253–2260

Salam MA, Khatun NA, Ali MM (2005) Carp farming potential in Barhatta Upazilla, Bangladesh: a GIS methodological perspective. Aquaculture 245(1–4):75–87. https://doi.org/10.1016/j.aquaculture.2004.10.030

Saleh M (2008) Capture-based aquaculture of mullets in Egypt. In: Lovatelli A, Holthus PF (eds), pp 109–126

Sammut J (1996) Amelioration and management of shrimp ponds in acid sulfate soils: key researchable issues. In: Smith PT (ed). Australian Centre for International Agricultural Research, Songkhla, pp 102–106

Sanil NK, Vijayan KK, Kripa V, Mohamed KS (2010) Occurrence of the protozoan parasite, *Perkinsus olseni* in the wild and farmed Pearl Oyster, *Pinctada fucata* (Gould) from the Southeast coast of India. Aquaculture 299(1–4):8–14. https://doi.org/10.1016/j.aquaculture.2009.12.007

Scott C, Wilde A (2006) Measuring democratic governance: a framework for selecting pro-poor and gender sensitive indicators. UNDP

Secretan PAD, Bueno PB, van Anrooy R, Siar SV, Olofsson A, Bondad-Reantaso MG, Funge-Smith S (2007) Guidelines to meet insurance and other risk management needs in developing aquaculture in Asia, vol 53, FAO Fisher edn. FAO, Rome. https://doi.org/10.1017/CBO9781107415324.004

Sepúlveda F, Marín SL, Carvajal J (2004) Metazoan parasites in wild fish and farmed salmon from aquaculture sites in southern Chile. Aquaculture 235(1–4):89–100. https://doi.org/10.1016/j.aquaculture.2003.09.015

Sethi SA (2010) Risk management for fisheries. Fish Fish 11(4):341–365. https://doi.org/10.1111/j.1467-2979.2010.00363.x

Shang YC (1981) Aquaculture economics: basic concepts and methods of analysis. Westview special studies in agriculture/aquaculture science and policy. Westview Press, Boulder

Shang YC, Tisdell CA (1997) Economic decision making in sustainable aquacultural development. In: Bardach J (ed). Wiley, NJ, pp 127–148

Silva C, Ferreira JG, Bricker SB, DelValls TA, Martín-Díaz ML, Yáñez E (2011) Site selection for shellfish aquaculture by means of GIS and farm-scale models, with an emphasis on data-poor environments. Aquaculture 318(3–4):444–457. https://doi.org/10.1016/j.aquaculture.2011.05.033

Singh VP (1980) Management of fish ponds with acid sulphate soils. Asian Aquacult 5:4–6

Siriwardena SN (2007) Freshwater fish seed resources and supply: Asia regional synthesis. In: G B-RM (ed), pp 59–90

Slater MJ, Mgaya YD, Mill AC, Rushton SP, Stead SM (2013) Effect of social and economic drivers on choosing aquaculture as a coastal livelihood. Ocean Coast Manag 73:22–30. https://doi.org/10.1016/j.ocecoaman.2012.12.002

Small HJ, Pagenkopp KM (2011) Reservoirs and alternate hosts for pathogens of commercially important crustaceans: a review. J Invertebr Pathol 106(1):153–164. https://doi.org/10.1016/j.jip.2010.09.016

Solberg MF, Zhang Z, Glover KA (2015) Are farmed salmon more prone to risk than wild salmon? Susceptibility of juvenile farm, hybrid and wild Atlantic salmon *Salmo salar* L. to an

artificial predator. Appl Anim Behav Sci 162:67–80. https://doi.org/10.1016/j.applanim.2014.11.012
Soon JM, Baines RN (2012) Aquaculture farm food safety and diseases risk assessment (AquaFRAM): development of a spreadsheet tool for salmon farms. Aquacult Eng 49:35–45. https://doi.org/10.1016/j.aquaeng.2012.02.002
Southall T, Telfer T, Hambrey J (2004) Environmental capacity modelling in aquaculture development, vol 1. Tropeca. Nautilus Consultants Ltd, Edinburgh, Scotland, UK., Institute of Aquaculture, University of Stirling, Scotland, UK., Hambrey Consulting, UK
Sowles J (2003) Aquaculture task force discussion paper on bio-physical carrying capacity. Government of Maine Department of Marine Resources, Maine
Spalding M, Kainuma M, Collins L (2010) World atlas of mangroves. Taylor & Francis Group, UK
St-Hilaire S, Kent ML, Iwama GK (1998) Factors affecting the health of farmed and wild fish populations: a perspective from British Columbia. Canadian Stock Assessment Secretariat, Ottawa
Staples D, Funge-Smith S (2009) Ecosystem approach to fisheries and aquaculture: implementing the FAO code of conduct for responsible fisheries
Stenton-Dozey JME, Busby AJ, Jackson LF (1999) Impact of mussel cultire on Macrobenthic community structure in Saldanha Bay, South Africa. Mar Pollut Bull 39:357–366
Stevenson JR, Irz XT, Alcalde R-G, Morrisens P, Petit J (2007) An empirical typology of brackish-water pond aquaculture systems in the Philippines: a tool to aid comparative study in the sector. Aquacult Econ Manag 11(2):171–193. https://doi.org/10.1080/13657300701370358
Stigebrandt A (2011) Carrying capacity: general principles of model construction. Aquac Res 42 (Suppl. 1):41–50. https://doi.org/10.1111/j.1365-2109.2010.02674.x
Stigebrandt A, Aure J, Ervik A, Hansen PK (2004) Regulating the local environmental impact of intensive marine fish farming: III. A model for estimation of the holding capacity in the modelling-ongrowing fish farm-monitoring system. Aquaculture 234(1–4):239–261. https://doi.org/10.1016/j.aquaculture.2003.11.029
Strain PM, Hargrave BT (2005) Salmon aquaculture, nutrient fluxes and ecosystem processes in Southwestern New Brunswick. Environ Eff Mar Finfish Aquacult 5(July):29–57. https://doi.org/10.1007/b136003
Subasinghe R, Ahmad I, Kassam L, Krishnan S, Nyandat B, Padiyar A, Phillips M, Reantaso M, Miao W, Yamamoto K (2012) Protecting small-scale farmers: a reality within a globalized economy? In: Subasinghe RP, Arthur JR, Bartley DM et al (eds) Farming the waters for people and food, 2012, pp 705–717
Subasinghe R, Phillips M (2007) Aquaculture certification: a challenge for the small farmer? FAO Aquacult Newslett 38(September):34–36
Swaminathan MS (2012) Aquaculture and sustainable nutrition security in a warming planet, keynote address. In: Subasinghe RP, Arthur JR, Bartley DM et al (eds) Farming the waters for people and food. FAO/NACA, pp 3–19
Taberima S, Nugroho YD, Murdiyarso D (2014) The distribution of carbon stock in selected mangrove ecosystems of wetland Papua: Bintuni, Teminabuan and Timika Eastern Indonesia. In: Paper presented at the international conference on chemical, environmental and biological sciences (CEBS-2014), Kuala Lumpur, Sept 17–18
Tabthipwon P (2008) Aquaculture development toward the sustainable and current situation of aquaculture. Food Fertil Technol Center, Taipei, pp 45–55
Tacon AGJ, Metian M (2008) Global overview on the use of fish meal and fish oil in industrially compounded aquafeeds: trends and future prospects. Aquaculture 285(1–4):146–158. https://doi.org/10.1016/j.aquaculture.2008.08.015
Taranger GL, Boxaspen KK, Madhun AS, Terje S (eds) (2011) Risk assessment—environmental impacts of Norwegian aquaculture. Institute of Marine Research
Tekinay A, Guroy D, Cezik N (2009) The environmental effect of a land-based trout farm on Yuvarlakçay, Turkey. Ekoloji 73(19):65–70. https://doi.org/10.5053/ekoloji.2009.738

Teng S-K (2008) Risk analysis of the soil salinization due to low-salinity shrimp farming in central plain of Thailand (trans: FI). Reports and Studies—Joint Group of Experts on the Scientific Aspects of Marine Environmental Protection (UN/GESAMP). Food and Agriculture Organisation of the United Nations, Rome

Tidwell J, Coyle S (2008) Impact of substrate physical characteristics on grow out of freshwater prawn, *Macrobrachium rosenbergii*, in ponds and pond microcosm tanks. J World Aquacult Soc 39(3):406–413. https://doi.org/10.1111/j.1749-7345.2008.00162.x

Tlusty M (2002) The benefits and risks of aquacultural production for the aquarium trade. Aquaculture 205:203–219

Trottier B (1987) Women in aquaculture production in West Africa. In: Nash CE, Engle CR, Crosetti D (eds) Women in Aquaculture, Rome

Tsangaris C, Kormas K, Strogyloudi E, Hatzianestis I, Neofitou C, Andral B, Galgani F (2010) Multiple biomarkers of pollution effects in caged mussels on the Greek coastline. Comparat Biochem Physiol C Toxicol Pharmacol 151(3):369–378. https://doi.org/10.1016/j.cbpc.2009.12.009

Tucker CS, Hargreaves JA, Boyd CE (2008) Aquaculture and the environment in the United States. In: Tucker CS, Hargreaves JA (eds). Blackwell, pp 3–54

Uki N (2006) Stock enhancement of the Japanese scallop *Patinopecten yessoensis* in Hokkaido. Fish Res 80(1):62–66. https://doi.org/10.1016/j.fishres.2006.03.013

UNWomen (2014) World survey on the role of women in development 2014: gender equality and sustainable development, vol I. https://doi.org/10.4324/9781315686455

USAID (2013) Sustainable fisheries and responsible aquaculture: a guide for USAID staff and partners. U.S. Agency for International Development, Washington, DC

Valiela I (2006) Global coastal change. Wiley-Blackwell, Malden

Van Meijl H, Van Rheenen T, Tabeau A, Eickhout B (2006) The impact of different policy environments on agricultural land use in Europe. Agr Ecosyst Environ 114(1):21–38. https://doi.org/10.1016/j.agee.2005.11.006

Vandergeest P, Unno A (2012) A new extraterritoriality? Aquaculture certification, sovereignty, and empire. Political Geogr 31(6):358–367. https://doi.org/10.1016/j.polgeo.2012.05.005

von Lompe M (2003) The world agricultural trade simulation system WATSIM an overview. Institut für Agrarpolitik, University of Bonn, Bonn

Wada KT (1998) The present status of genetic conservation of cultured aquatic species in Japan. In: Harvey B, Ross C, Greer D, Carolsfeld J (eds) Action before extinction: an international conference on conservation of fish genetic diversity. IDRC, Vancouver, pp 225–230

Waite R, Beveridge M, Brummett RE, Castine S, Chaiyawannakarn N, Kaushik S, Mungkung R, Nawapakpilai S, Phillips M (2014) Improving productivity and environmental performance of aquaculture, pp 1–60. https://doi.org/10.5657/FAS.2014.0001

Walker TR, Grant J, AaM Weise, McKindsey CW, Callier MD, Richard M (2014) Influence of suspended mussel lines on sediment erosion and resuspension Lagune de la Grande Entrée, Îles-de-la-Madeleine, Québec, Canada. Aquaculture 433:450–457. https://doi.org/10.1016/j.aquaculture.2014.07.006

Wang Q, Zhuang Z, Deng J, Ye Y (2006) Stock enhancement and translocation of the shrimp *Penaeus chinensis* in China. Fish Res 80(1):67–79. https://doi.org/10.1016/j.fishres.2006.03.015

Waples RS, Hindar K, Hard JJ (2012) Genetic risks associated with marine aquaculture. NOAA Technical Memorandum NMFS-NWFSC-119. U.S. Department of Commerce, National Oceanic and Atmospheric Administration, National Marine Fisheries Service

Weinert M, Mathis M, Kröncke I, Neumann H, Pohlmann T, Reiss H (2016) Modelling climate change effects on benthos: distributional shifts in the North Sea from 2001 to 2099. Estuar Coast Shelf Sci, vol 175. https://doi.org/10.1016/j.ecss.2016.03.024

Weir LK, Grant JW (2005) Effects of aquaculture on wild fish populations: a synthesis of data. Environ Rev 13(4):145–168. https://doi.org/10.1139/a05-012

Weir LK, Hutchings JA, Fleming IA, Einum S (2005) Spawning behaviour and success of mature male Atlantic salmon (*Salmo salar*) parr of farmed and wild origin. Can J Fish Aquat Sci 62 (5):1153–1160. https://doi.org/10.1139/f05-032

Wells G (1996) Hazard identification and risk assessment. Institution of Chemical Engineers

White P, Diego-McGlone MLS (2008) Ecosystem-based approach to aquaculture management. Sci Diliman 20(2):1–10

White PG (2013) Environmental consequences of poor feed quality and feed management. In: Hasan MR, New MB (eds) FAO, Fisher edn. FAO, Rome, pp 553–564

WHO (2003) Patient safety: rapid assessment methods for estimating hazards—report of the WHO Working Group meeting Geneva, 17–19 Dec 2002, Geneva

Williams L (2011) Impact of climate change on fisheries and aquaculture in the developing world and opportunities for adaptation. International Fund for Agricultural Development, Rome

Wilson AG, Birkin M (1987) Dynamic models of agricultural location in a spatial interaction framework. Geogr Anal 19(1):31–56

WorldBank (1998) Shrimp farming and the environment—can Shrimp farming be undertaken sustainability? In: A discussion paper designed to assist in the development of sustainable shrimp aquaculture. World Bank, USA

WorldBank (2015) The worldwide governance indicators. http://info.worldbank.org/governance/wgi/index.aspx#home. Accessed 18 July 2016

WorldFish (2011) Gender and aquaculture: sharing the benefits equitably. World Fish Center, Penang

WorldFish C (2007) The threat to fisheries and aquaculture from climate change

WRC-Report (2010) A manual for rural freshwater aquaculture. WRC Report No. TT 463/P/10. Rural Fisheries Programme, Rhodes University. Water Research Commission, Pretoria, South Africa

WWAP (2016) The United Nations World Water Development report 2016: water and jobs (United Nations World Water Assessment Programme), UNESCO, Paris, France

Yoe C (2012) Principles of risk analysis: decision making under uncertainty. CRC Press, Florida

Yoo KH, Boyd CE (2012) Hydrology and water supply for pond aquaculture. Springer, US

Young N, Matthews R (2011) The aquaculture controversy in Canada: activism, policy, and contested science. UBC Press

Zhang YB, Li Y, Sun XL (2011) Antibiotic resistance of bacteria isolated from shrimp hatcheries and cultural ponds on Donghai Island, China. Mar Pollut Bull 62(11):2299–2307. https://doi.org/10.1016/j.marpolbul.2011.08.048

Zhao YP, Li YC, Dong GH, Gui FK, Teng B (2007) Numerical simulation of the effects of structure size ratio and mesh type on three-dimensional deformation of the fishing-net gravity cage in current. Aquacult Eng 36(3):285–301. https://doi.org/10.1016/j.aquaeng.2007.01.003

Chapter 2
Sustainable Aquaculture: Socio-Economic and Environmental Assessment

Bishal Bhari and C. Visvanathan

Abstract One of the goals of the sustainable development is to minimize or eliminate the environmental externalities and target social and economic development. Socio-Economic and Environmental Assessment (SEEA) deal with assessing the socio-economic and environmental issues that can potentially be a threat to the existing condition. SEEA also deals with developing a proper alternative or management techniques. As the world capture type of fishing is stagnant or declining, the growth of the aquaculture is inevitable as it fills the gap between declining natural production and increasing market demand. Aquaculture is the only viable way of raising the production of seafood and freshwater fish. Thus, the sustainable development of aquaculture industries has been the necessity. This chapter highlights the different socio-economic and environmental issues that aquaculture leads to and also presents the impact areas, mitigation and monitoring plans that can be adopted to ensure sustainability of the aquaculture.

Keywords Sustainable aquaculture · Environmental assessment
SEEA · Environmental impact

2.1 Introduction

Aquaculture also known as aquafarming, is the farming (breeding, rearing and harvesting) of both aquatic plants and animals in various water environments like ponds, rivers, lakes and the ocean under controlled condition. The conditions are designed to increase the production of the organisms beyond the natural capacity

B. Bhari · C. Visvanathan (✉)
Environmental Engineering and Management Program,
Asian Institute of Technology, Bangkok, Thailand
e-mail: visu@ait.asia

B. Bhari
e-mail: bishal.bhari@gmail.com

F. I. Hai et al. (eds.), *Sustainable Aquaculture*, Applied Environmental Science and Engineering for a Sustainable Future,
https://doi.org/10.1007/978-3-319-73257-2_2

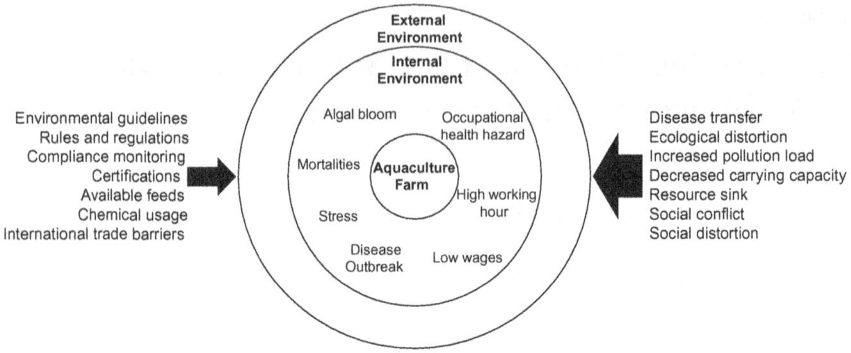

Fig. 2.1 Internal and external environment of the aquaculture farm

and involves cultivation of both marine and freshwater species and utilizes natural resources and interacts with the environment.

Aquaculture differs from the conventional approach of capture type of fishing and refers to more planned and technical approach of farming which is a more labor intensive process. Aquaculture helps to sustain many farmers and is one of the major sources of income to many households. It also indirectly affects the social and economic aspects of many stakeholders who are indirectly involved in it. Thus, a considerable socio-economic impact can be associated with it, in additional to the environmental aspect of aquaculture. As aquaculture industry utilizes resources to cultivate the stock beyond the natural carrying capacity, ecological and environmental impacts are the major concern in the aquaculture industry. As presented in Fig. 2.1 external environment which majorly includes market demand, governmental regulations and institutional capacity of the countries plays a significant role to control the internal environment of the aquaculture. However as presented in Fig. 2.1 these forces are smaller than the raising social and environmental issues. Uncontrolled external environment can worsen the internal environment leading to many social and environmental issues.

With the aim of achieving sustainable aquaculture production while exerting minimum environmental degradation, prior assessment of socio-economic and the environmental component is needed. Sustainable aquaculture implies socially and economically sound aquaculture industry where the environmental damages are minimized or avoided. Socio-Economic and Environmental Assessment (SEEA) is one of the methods to harmonize social, economic and environmental conditions for sustainable growth of the aquaculture industry.

The primary objective of the SEEA is to identify the activities that hamper the lives of people. SEEA performs detail study and analysis and helps to predict direct, indirect and cumulative impacts of the project. Another main objective of SEEA is to mitigate these impacts either by avoiding it, remedying it or by compensating the effects of the impacts. SEEA can act as an important mechanism to ensure the sustainability of the aquaculture. However, the success of the socially, environmentally and economically sound aquaculture practice depends on the attitude of the three key players: Proponent, Stakeholders, and Decision Makers.

Proponents include the entrepreneur's/companies/government departments, etc., who plans to carry out the project and are responsible for complying with the imposed rules, regulation, standards, etc. In additional to making a profit from the , the proponent is also responsible for harmonizing the social aspect of the project, which is important for long-term planning. Stakeholders are the one who either benefits or affected by the project and includes institutes, governmental agencies, businesses, labors, associated business, etc. who are to be benefitted or affected by the project. The proponent has a responsibility to harmonize the stakeholders to avoid any chaotic conflicts. The last key players are the decision makers who have the legislative power of licensing, regulating standards, etc. and pushes the proponent to adopt practices that are more socially and environmentally acceptable. These three key players play a major role in the sustainable aquaculture and it is important for any project to harmonize their concerns and interest. SEEA can also act as a tool to harmonize these key players.

Thus, SEEA aims to harmonize these three key players by providing them with the following information:

i. Information regarding the current socio-economic and environmental scenario within the virtual project influencing boundary, within which the project impacts can be predicted to be felt with high magnitude.
ii. Description of the key socio-economic and environmental parameters that will be potentially impacted due to the established project.
iii. Impact identification, prediction, and evaluation due to the implications of the project.
iv. Highlights of the major environmental impact and plans to mitigate the effects.
v. Monitoring plans to ensure the compliance of the outcomes of the SEEA study.

SEEA tends to focus on the avoidance of adverse impacts and optimization of the beneficial impacts. The beneficial impacts of the project generally include rise in living standard due to increased employment opportunity and economic activity; improved business opportunity; improved infrastructure as the project matures, while the adverse impact might include loss of endemic species due to the introduction of exotic species, loss of farmland, loss of traditional business, etc. SEEA also targets to study the interaction of various impacts, which could be synergistic and irreversible in nature. Moreover, it predicts indirect impacts. All the identified impacts are evaluated and only significant impacts with high magnitude are mitigated. It provides an important platform for the decision maker to rationalize their decision based on the findings of the SEEA.

2.2 Types of Aquaculture

Aquaculture can be categorized based on intensification level, species cultivated and technology used. The socio-economic and environmental issues related to it also differ according to the types of aquaculture. This chapter highlights the aquaculture based on the intensification level. Based on the intensification level of the aquaculture, it can be divided into extensive, semi-intensive and intensive aquaculture. The productivity and the type of food requirements vary according to the level.

2.2.1 Extensive Aquaculture

Extensive aquaculture utilizes natural productivity of the environment for the growth. Under the extensive aquaculture, no additional food is added for the growth and there is very little control over the stocks. It can be done in freshwater, brackish and marine environment using several techniques like multiple mesh, trapping nets, pond culture, etc. Since the growth conditions like temperature, pH, nutrients, etc. cannot be altered, an extensive form of aquaculture strongly relies on the surrounding conditions. This form of aquaculture also has detrimental impacts and proper management is essential. If not managed properly, it can lead to the damage in the surrounding natural habitat. The organic waste from the cultured area can potentially deplete dissolved oxygen level and reduce the benthic habitat population. In addition, it can also introduce (in the form of escapes) foreign species or less tolerant genetically modified species in the natural environment, which can reduce the adaptive capacity of the indigenous species as they interbreed with these less tolerant cultured species.

2.2.2 Semi-intensive Aquaculture

Semi-intensive techniques utilize different culture techniques like raceways, sea-cages and require to supplement the stock with additional food. However, semi-intensive system is partially dependent upon the natural productivity. Thus, it requires less space than the extensive system to have the same yield. Its environmental risk is similar to the extensive and intensive aquaculture.

2.2.3 Intensive Aquaculture

Intensive aquaculture is a highly dense farming and involves the total addition of the food. It is also a technology driven process which focuses on maximizing the yield by maintaining palatable growth condition for the target species. The stock is fully dependent on the artificial food provided and involves many activities that could lead to environmental issues.

As the environment variables (pH, temperature, oxygen level, feed, etc.,) are completely controlled and managed by the skilled workforce, higher yield can be obtained which is one of the advantages of the intensive aquaculture. However, it has a higher environmental impact and its magnitude and significance vary according to the technology used. One of the general issues related to the intensive system is effluent management. The effluent of the intensive system are rich in nutrients (Pullin 1989) (both organic and inorganic) and if not properly managed can lead to eutrophication in the natural environment causing a threat to the indigenous species.

Table 2.1 presents the impact associated with different intensification level and technology used. Moreover, it must also be noted that some impacts are location specific.

2.3 Socio-Economic Impacts of Aquaculture

As aquaculture business deals with the usage of environmental resources and human resources for the extraction and production of the consumable products, it will inevitably cause distortion in the social and economic conditions of the project area. This distortion often termed as 'impacts' can be both beneficial (positive) as well as adverse (negative) in nature as presented in Fig. 2.2. However, the ultimate goal of any project is to maximize the positive impact and minimize or eliminate the negative impact of proper technological and operational measures. Some of the beneficial and adverse socio-economic impacts are presented below in Sects. 3.1 and 3.2, however, it must also be noted that the intensity of these impact varies according to the species, location of the farm, farm yield and technology used.

2.3.1 Beneficial Impacts

Aquaculture has many socio-economic benefits. Some of the socio-economic benefits are as follow.

2.3.1.1 Food Security

Food security is the current global problem to be addressed as it is estimated that the world will need 70–100% more food by 2050 (The World Bank 2007; Baulcombe et al. 2009). Sustainable Development Goal (SDG) for 2030 has also targeted goals to achieve food security, improve nutrition and promote sustainable agriculture.

Table 2.1 Different levels of aquaculture and socio-economic and environmental impact associated with different technology adopted and level adopted

	Type of aquaculture*	Damage to reefs	Mangrove destruction	Saltwater intrusion	Eutrophication problem	Wastewater hazards	Social conflict	Social disruption	Public health risk	Creation of resource sink	Loss of traditional jobs	Increase in jobs	Export earnings
Extensive	Coastal bivalve culture (mussels, oysters)	✓					✓	✓	✓			✓	✓
	Coastal fishpond (shrimps, tilapias)	✓	✓				✓	✓	✓	✓	✓	✓	✓
	Cage culture in rich benthos (carps, catfish, tilapias, etc.)				✓		✓	✓				✓	✓
Semi-intensive	Fresh and brackish water pond (shrimps & prawns, tilapias, catfish, etc.)	✓	✓	✓	✓	✓	✓	✓	✓	✓	✓	✓	✓
	Integrated agricultural aquaculture (rice-fish, poultry-fish, vegetable-fish, etc.)					✓			✓			✓	✓
	Cage culture in rich benthos (carps, catfish, tilapias, etc.)				✓		✓	✓				✓	✓

(continued)

Table 2.1 (continued)

	Type of aquaculture*	Damage to reefs	Mangrove destruction	Saltwater intrusion	Eutrophication problem	Wastewater hazards	Social conflict	Social disruption	Public health risk	Creation of resource sink	Loss of traditional jobs	Increase in jobs	Export earnings
Intensive	Freshwater, brackish water & marine pond (shrimps, finfish especially carnivorous, etc.)	✓	✓	✓	✓	✓	✓	✓	✓	✓	✓	✓	✓
	Freshwater, brackish water & marine cage culture (finfish especially carnivorous, and some omnivorous)	✓	✓	✓	✓	✓	✓	✓	✓		✓		✓
	Re-circulating system					✓							✓
	others—raceways, silos, tanks, etc.				✓	✓	✓	✓	✓				✓

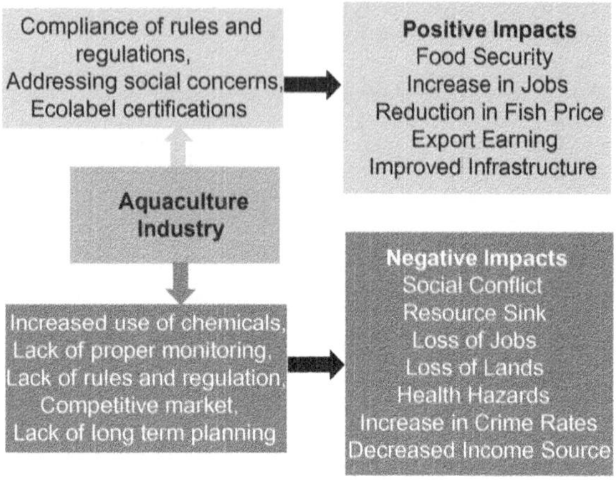

Fig. 2.2 Causes of socio-economic impacts

However, the key barrier to increasing the food is the availability of productive land due to rapid urbanization, increased competition of land, loss of productive land due to natural hazards, increased competition from biofuels. Thus, the prominent solution under these circumstances is to increase the agricultural productivity. Aquaculture, which has a higher yield than the conventional fishing is indeed one the method which can help to attain the SDG 2030 goals.

When analyzing the food security from the aspect of nutrients available, the health benefits from the finfish and shellfish are well known, as it is rich in vitamins and protein. In 2010, 16.7% of the global population's intake of the animal protein was covered by the consumption of fish. Fish protein has been the vital source (around 50%) of animal protein intake to many poorer island and coastal states (FAO 2014a, b). Moreover fish are cheaper than the other source of animal protein, which makes it affordable to low-income groups as well. Consumption of aquatic food is found to be higher in the developing and least developed countries and so aquaculture plays an important role to ensure both the quantity and quality of food.

2.3.1.2 Increase in Jobs

Aquaculture increases the job opportunity at several levels. Jobs are created in the whole supply chain from the production to the supply. Low-income group and rural communities are the ones who benefits significantly from the employment created. The ability of aquaculture to create jobs in the rural areas is one of the reasons for governments to promote aquaculture. The potential for job creation is not limited to the fishing industry itself but other associated industries (like net industry, boat building, food processing, etc.) are also benefited from the aquaculture. With all the jobs created, cumulating the job holders and their dependent, fisheries and aquaculture support estimated livelihood of around 10–12% of the world's population (FAO 2012). These indirect benefits can also be experienced by increased trade and inflow of the traders into the area, as more jobs are created to manage the basic needs of these traders.

2.3.1.3 Reduction in Fish Price

Fish price will continue to decrease in the future and will be more affordable to the low-income groups. Aquaculture is a promising method of growing stocks at cheaper operating cost due to the possibility of increasing its yield beyond the natural productivity. As aquaculture rises in the future, the production cost of these fish will get cheaper affecting the market cost. Considering the time frame from 1990–2010, the overall decline in the fish price was observed due to the sharp decline in price in some of the species which was able to radically change due to aquaculture development (The World Bank 2013). Thus in future, as technological advancement are achieved with technologies such as aquaponics, aquaculture production cost will lower down making it more accessible to all.

2.3.1.4 Export Earning

Unlike to the traditional catch type of fishing, aquaculture provides more opportunities for the farmers to increase their production capacity following the increased demand for seafood worldwide. For Asian countries like Thailand, Bangladesh, Indonesia, Vietnam, etc. export earnings from the seafood industries is significant. As the production capacity of capture type fishing has remained stagnant, the only method to increase the production of finfish and shellfish has been the expansion of the aquaculture industry. These industries are generating a GDP both locally and in the form of exports. Shrimp industry is the third largest exporting industry in Bangladesh and it plays a crucial role in the GDP of the Bangladesh generating export earnings of 544 million USD in 2013 (Kabir 2013).

2.3.1.5 Improved Infrastructure in Rural Areas

Aquaculture can have the indirect benefit of improved roads, governmental facilities, harbor, etc., which increases the productivity of aquaculture. The rural communities are benefitted from the improvement in infrastructure.

2.3.2 Negative Impacts

Aquaculture has several negative impacts which need to be considered for the smooth operation of the business. As aquaculture has many negative environmental impacts, social issues ripple through its effects.

2.3.2.1 Conflict Over Resource Usage

Resource usage has always been an issue for the aquaculture industry. One of such conflicts that is normally seen is between the shrimp farmers and other farmers (crops, freshwater fish) who lose their yield due to the environmental impact created by the shrimp culture. The shrimp farm salinizes the freshwater bodies and crop lands which cause conflicts over the usage of the resources. In the sub-Saharan Africa where the water is scarce, conflicts have also arisen due to the conflict over the use of water between tobacco farmers and fish farmers (Subasinghe 2006). Also as the cage and pen culture of aquaculture is also dependent upon the natural food, conflicts are seen over the artisanal fishers and the aquaculture farmers.

Social Issues of Shrimp Farming in Khulna, Bangladesh

Khulna is the leading producer of the Bangladesh's vast shrimp industry. Shrimp farming has certainly employed a lot of people in Khulna but it has also increased the vulnerability group and reduced the coping capacity of the farmers. Some of the social issues observed in Khulna are:

i. **Loss of productivity of the land**: The shrimp farming has affected the fertility of the nearby lands due to the leaching of sediments. The lands are now barren and traditional rice farming is not possible. Cattle raising is also impossible as the lands are barren. There are hardly any environmental monitoring and big farmers hardly cares for the environmental impact to the community.
ii. **Illegal land acquisition**: Most of the shrimp farming in the area is being done by the immigrants. These immigrants/big farmers often with the help of the local regulatory bodies, illegally control or occupies the land of the locals. Shrimp farming has raised the corruption level in the area.
iii. **The increase of vulnerability group**: Although shrimp farming has provided jobs to many, a lot of farmers associated with it are paid very less for their effort. These vulnerable groups are associated in the catching of juvenile shrimp from the local rivers. Moreover, the market of the shrimp farming is so intense in the area that they can hardly engage in other areas for income. Malnourishment is commonly observed in these vulnerable groups.
iv. **Loss of jobs**: Fishman who traditionally caught fish in the river are severely affected as they can hardly find any fish in the rivers now. The juvenile shrimp are caught using a very fine net. These nets also trap juvenile fishes which are then discarded in the land. Loss of juvenile fish has affected the fish population to a great extent.

v. **Increased use of pesticides**: Driven by the market need, the use of the chemical is uncontrolled in the region. Farmers use chemicals, most of which have already been banned in many countries due to its health hazards. The effluent from the shrimp farm has risked the health of the local people.

vi. **Increased rate of crime**: As the workers in the shrimp farm are mostly immigrant, they have the least responsibility toward the community. The crime rate in the Khulna has also increased due to the Shrimp farming.

vii. **Conflicts**: There is a long battle between the local habitants who have lost their lands and occupations to the powerful shrimp industry.

Source: Link TV. (2005, Jan 5) & Environmental Justice Foundation. (2014, Aug 14).

2.3.2.2 Creation of a Resource Sink

The opportunity cost of aquaculture development must also be evaluated as a significant amount of capital and labor is required. The failure in the market can adversely impact the rural areas where the aquaculture is more concentrated and have the least adaptive capacity. Thus, a careful evaluation is needed under the existing economic and resource potential to evaluate aquaculture in terms of long-term profitability. Lack of planning and management can lead to a resource sink, which implies low resource and labor productivity. One of the examples is the aquaculture development in the Sub-Saharan Africa where nearly 100 million USD was invested, however, little benefit was generated from it (Neiland et al. 1991).

2.3.2.3 Loss of Traditional Occupation's

As aquaculture creates new job employment opportunities, traditional occupations are also lost in the process. It leads to loss of traditional skills that were sustainably utilized for income generation. Switching jobs to more income generating activity are economically sound but can be vulnerable to the 'Boom and Bust Cycle.[1] Any possible market failure of aquaculture will not only result in the loss of jobs in future but will also result in loss of capability to revert back to the traditional jobs. Thus, aquaculture can have an impact on the traditional values of the societies as well.

[1]Boom and bust cycle: It is a process of economic growth and contraction, which occurs frequently and is the key characteristic of capitalist economies. Boom phase of the growth creates numerous job opportunities while the bust phase of the cycle collapses these jobs.

2.3.2.4 Health Hazards

The aquaculture industry is associated with many occupational hazards which are found more prominently in developing nations due to lack of policies. Further to highlight 87% of the aquaculture production is done in developing nations (Waite et al. 2014), which imply only a small or negligible portion of the aquaculture can be regarded as complying with proper occupational safety measures. As aquaculture uses several chemicals (pesticides, inorganic fertilizers, antibiotics, etc.) for the growth of the stock, aquaculture practitioners are more prone to the potential detrimental effects of it. Labor are more vulnerable to the skin diseases, respiratory diseases (asthma, bronchitis, etc.) and allergies. Further, long-term and chronic diseases are being attributed to the aquaculture (Erondu and Anyanwu 2005). The wastewater generated from aquaculture if not properly treated can also potentially cause a threat to the local communities.

2.4 Environmental Impacts of Aquaculture

Previously aquaculture was considered too small an industry to have any significant impact on the environment. However, the remarkable growth of the aquaculture industry in many countries (China, Vietnam, Thailand, Indonesia, etc.,) over the past decades has also increased the adverse impact of it on the environment. Aquaculture focuses on growing stocks beyond the environmental carrying capacity by the use of inputs like fertilizers, antibiotics, pesticides, etc. which negatively impacts the ecology. In such systems resources are pumped in, used up, and pumped out in a linear fashion, rather than being recycled. This leads to accumulation of wastes in the recipient ecosystems, often causing severe and irreversible environmental problems. Aquaculture technology/practice requires high inputs of protein and phosphorus diets, and a high rate of water exchange. A large portion of nutrients becomes waste, which is then directly discharged to the surrounding waters causing rapid deterioration of water quality.

Some of the environmental impacts caused by the aquaculture are discussed below. However, the nature, magnitude, and significance of these impacts varies according to the species cultivated, intensity of the farm, carrying capacity, the geography of the farm, etc.

Socio-Economic and Environmental Impact of Shrimp Farming
Most of the shrimp production (55%) is through the aquaculture (WWF 2016). As shrimp farming is profitable, intensive aquaculture methods have been adopted to increase the yield of the shrimp. Menasveta and Fast (1998) estimated the production level of the intensive shrimp farming to be greater than 6000 kg/ha/yr which was found to more than the semi-intensive (600–1800 kg/ha/yr) and extensive (100–300 kg/ha/yr). However, the use of

intensive shrimp farming techniques has double the environmental impact when compared with the less intensive system (Cao et al. 2011). Some of the environmental impacts of the intensive shrimp farming are as follow:

i. **Loss of lands**: Marine shrimp aquaculture leads to the loss of lands. It has caused the loss of thousands of hectares of mangrove and wetlands. Moreover, it causes soil acidification as the waste of the shrimp is dumped to the land.
ii. **Destruction of other juvenile species**: During the harvesting of the shrimps, often juvenile shellfish, shrimps, finfish, macrozooplankton animals are caught which disrupts the ecosystem. It disturbs the entropy of the eco-system causing biodiversity loss and reduction of the food for other species in the food chain.
iii. **Impacts of excessive feeding**: Shrimp farming often excessively use the nutrients (fertilizers) to naturally grow the food for the shrimp or uses supplemental feeding. Utilized nutrients, feed, and excreta in the shrimp farm increases the nutrient loading, reduces oxygen in the pond water supplies and increases the sedimentation. This wastewater discharge from such pond can cause eutrophication and death of animal and plants in the receiving water bodies.
iv. **Impacts due to the chemical dosing**: Various chemicals are used during the shrimp farming to control the pathogens causing diseases. These chemicals contaminate the surrounding environment, as well as negatively affect human health. Excessively used antibiotics can also make the disease more resistance to the antibiotic causing more problem in its treatment in the future.
v. **Ground water depletion**: Shrimp farming uses a lot of fresh water to maintain appropriate salinity level for the shrimps. The aquifer used for this purpose becomes vulnerable to drying out causing the risk of salt-water intrusion in the ground water source.
vi. **Abusive land seizure**: Shrimp farms are often associated with human right issues like the seizure of land without any compensation. Land encroachment by powerful companies has jeopardized the traditional farming practices to the risk of extinction and has left many farmers landless. Shrimp farming is often done in coastal areas where no formal land rights exist.
vii. **Labor right violation**: Shrimp farms often pay very low wages to the laborers to maximize their profit and labor rights are always violated. In developing nations, the issue of human trafficking is commonly seen in shrimp farming due to weak governmental policy.

2.4.1 Loss of Mangroves Areas

Mangrove forest destruction is one issue at the forefront of environmental concerns in tropical areas. Lack of ownership supported by policy gaps leads to these lands being exploited for the aquaculture by small farmers, which majorly includes the shrimp farmers. Mangrove ecosystem is a reservoir, refuge, feeding ground and nursery for many useful plants and animals. Several tropical countries have lost extensive mangrove areas due to clearing and conversion to fish and shrimp ponds (Barg 1992). In Thailand, 16–32% of the total loss of mangrove between 1979 and 1993 was attributed to shrimp farming alone (Dierberg and Kiattisimkul 1996). The mangrove areas are important for the sediment and coastline stabilization, trapping of water, providing habitats and food for animals hence destruction or alteration of the mangroves leads to the adverse impacts to the benthic communities, microbial flora, phyto- and zooplanktons and other wild fish stock, and animals (Rosenthal 1992). Moreover, the reclaimed mangrove area is acidic in nature. Jayasinghe (1995) reported that oxidation of pyrite (FeS_2) occurs during pond-bottom drying which results in the release of sulfuric acid into the pond water and adjacent water bodies causing acidification and generation of highly toxic soluble aluminum phosphate.

2.4.2 Intensive Water Uses and Pollution

Aquaculture is water-intensive sector and uses a lot of water which is then polluted by the usage of chemicals. Waite et al. (2014) estimates that in 2010 the usage of freshwater in the aquaculture industry was 2% of the global agricultural water consumption.

2.4.3 Impacts of the Chemical Waste

The use of chemicals in aquaculture is obvious due to the high market demand for it. Chemicals are important for aquaculture industries to ensure the high yield of the stock. Some of the commonly used chemicals in the aquaculture are shown in (Table 2.2).

These chemicals increase the chemical waste causing various impacts in the ecosystem (Fig. 2.3). Commonly used chemicals in aquaculture are formalin, malachite green, potassium permanganate, copper sulfate, medicated feed, and local herbs. Aquaculture also causes water pollution as discharges consist of excess nutrients, fish waste, antibiotic drugs, pesticides, hormones and inorganic fertilizers. These pollutants affect the entropy of the natural aquatic habitat, leads to eutrophication in the nearby water bodies and cause diseases in the natural species. It increases mortality in the endemic species and also causes sub-lethal effects.

Table 2.2 Different type of chemicals used and its impacts

Types	Example	Environmental impact
Fertilizers	Chicken manure, animal manure, ammonium phosphate, urea, solophos	Eutrophication and damage to benthic population
Soil and water treatment	Alum, EDTA, lime, zeolite, gypsum	Sediment contamination
Disinfectant	Sodium or calcium hypochlorite and chloramine, benzalkonium chloride (BKC), formalin, iodophores, ozone	Localized biological effects
Pesticides and herbicides	Saponin, rotenone, ammonia, gusathion, Sevin, organophosphates, organotins, carbaryl, ivermectin	Affects the local ecosystem where the wastewater is discharged; Death of non-targeted species: occupational hazard
Antibacterial agents	Nitrofurans, phenicols, erythromycin, chloramphenicol, oxolinic acid, sulphonamides, tetracyclines, quinolones	Increased resistance in the pathogen; Sediments contamination; Transfer to the endemic species and benthic environment
Other therapeutants	Formalin, acriflavine, malachite green, methylene blue, potassium, copper compound, permanganate, Trifluralin	Long-term exposure to it is carcinogenic; Affects health of workers and consumers
Feed additives	Immunostimulants, preservatives and anti-oxidants, feeding attractants, vitamins, carotenoids, ethoxyquin	Not known
Anesthetics	Benzocaine, quinaldine, metomidate, carbondioxide	Used in limited amount hence least environmental impact
Hormones	Corticosteriods, anabolic steroids, growth hormones, serotonin	Consumer health risk

2.4.4 Saltwater Intrusion

Aquaculture farming can potentially lead to the saltwater intrusion in the nearby freshwater sources. The impact is generally caused by the pond type of aquaculture practice, which commonly occurs in the mangrove zones. These areas are affected by surface and subsurface salt-water intrusions generated by the aquaculture ponds. This may lead to changes in the salinity of the freshwater supplies used for irrigation and potable water sources. (Dierberg and Kiattisimkul 1996). Intensive Shrimp farming has been strongly related to the declining health of farmers due to the salt water intrusion in the drinking water source in Bangladesh (Joanna 2016)

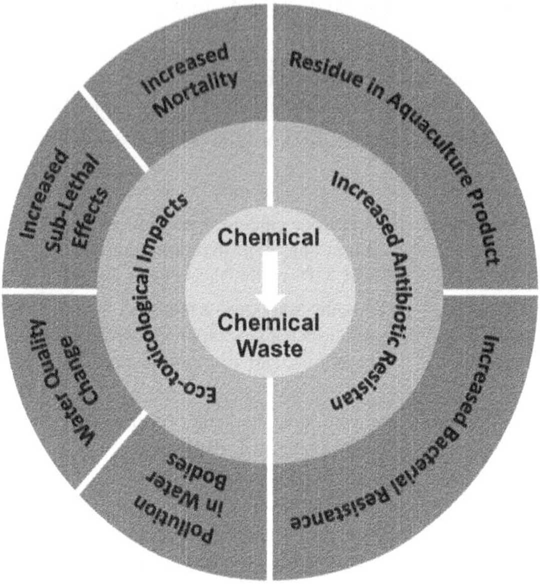

Fig. 2.3 Impacts of chemical waste from aquaculture

2.4.5 *Effluent and Sediment Management*

Sediment management in the pond culture is an environmental issue as improper management can lead to deterioration of ecosystem. The bottom of the pond is usually constructed with fine subsoil, which is impervious in nature. As the water is added for the cultivation, sediment formation is inevitable due to sedimentation of uneaten food, excess nutrients, excreta from stock, dead phytoplankton and zooplanktons, dead stocks, inorganics added etc. The accumulation of the sediments in the bottom of the ponds causes trapping of feeds and creation of anaerobic zones which results in the death of benthic organisms, increased pollution load in the discharged effluents, etc. Thus, the sediments are removed periodically from the pond and the frequency of cleaning varies with the type of the species cultivated. Accumulation of the sediments in the shrimp farming is substantial and was reported to be 157–290 tons/ha in Thailand (Boyd 1992) and Senarath and Visvanathan (2001) reported 5–10 cm of sediments disposal for Sri Lanka.

Disposal of accumulated sediments leads to increased nutrient loads to the discharged water bodies as these pond wastes are often drained in the process of sediment cleaning. The sedimentation unit, which functions to collect sediments might not be present as it requires land and money to operate. These effluents affect the local ecosystem. The treatment and disposal of the sediment is a costly process thus avoided by farmers as environmental laws and regulations of pond culture is more often least monitored in developing countries. Recirculation system (Fig. 2.4) that can treat the effluent with biological and physical treatment process and reuse the treated effluent back to the system has also gained popularity over past decade.

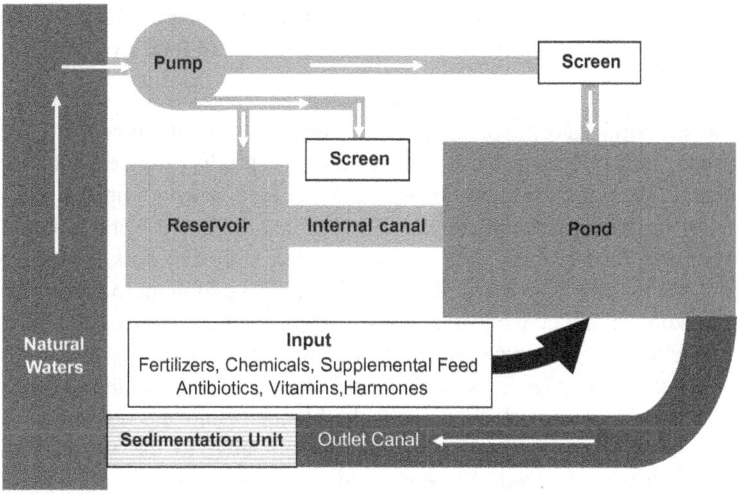

Fig. 2.4 Water management approach using recirculation of water

2.4.6 Consumer Health

The chemical fertilizers, lime, flocculants, algaecide, disinfectants, and chemotherapeutics are widely used in aquaculture and are persistent in nature. They are considered to be hazardous from the perspective of food safety as some of these compounds are biomagnified (Erondu and Anyanwu 2005). These compounds might have a detrimental effect on the consumer's health.

2.4.7 Introduction of Non-endemic Species Causing Ecological Imbalance

Non-endemic/exotic fish from the farm can escape from the aquaculture facilities and cause a threat to the endemic species. In Norway from 2001 to 2009, 3.93 million Atlantic salmon, 0.98 million rainbow trout and 1.05 million Atlantic cod was estimated to have escaped from the farm (Jensen et al. 2010). These juvenile, as well as adult fishes, are lost from the aquaculture through holes in the nets and operational errors. These fish can breed causing genetic impact in the adaptive capacity of the endemic/wild species (Thorstad et al. 2008). The offspring from such breeding has been found to be less adaptive to the environmental changes. Interbreeding between the farm and wild stock may lead to the reduction of the population of fish or lead to the extinction of the vulnerable groups (Naylor et al. 2005). It can also outcompete the endemic species. Thus, a huge ecological imbalance can also be associated with the escaped fish from the farm as it leads to competitive interactions for food and affects the levels of food availability.

Environmental advantages of Finfish over other meat sources

Finfishes are capable of converting more of the product they eat into edible products. Thus, the efficiency of the fishes is high when compared to other animals like beef, pork and chicken as illustrated in Figs. 2.5 and 2.6. Finfish can convert the feed with 30% efficiency while the beef, pork, and chicken with 5, 13 and 25% efficiency. Beef requires 31.7 kg of grain to produce 1 kg of the edible product while the finfish on average requires only 2.3 kg to convert into 1 kg of edible product. Moreover, the edible portion of the finfish is higher than the other livestock which makes it easier from the perspective of waste management. Fish being cold blooded animal spends very less energy to maintain its body temperature compared to the warm-blooded livestock, hence, the feed can be utilized more efficiently.

Livestock also causes higher environmental emission than finfish with the exception of poultry. As illustrated in the Fig. 2.7 beef has the highest nitrogen and phosphorus emission followed by pork. Finfish and chicken have the lowest nitrogen (360 and 300 kg/ton protein produced respectively) and phosphorous emission (48 and 40 kg/ton protein produced respectively).

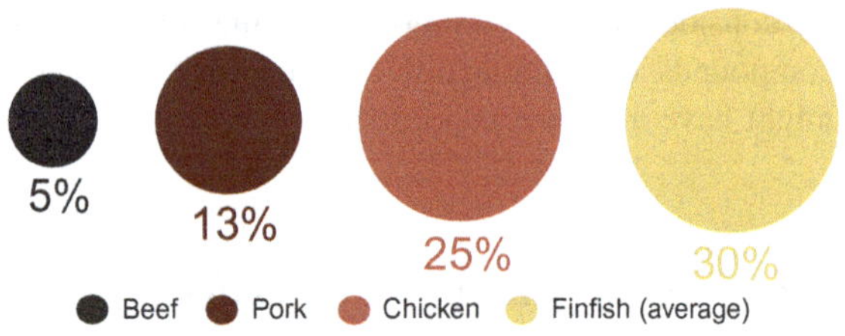

Fig. 2.5 Protein efficiency (%) of various meat sources. *Data source* Hall et al. (2011)

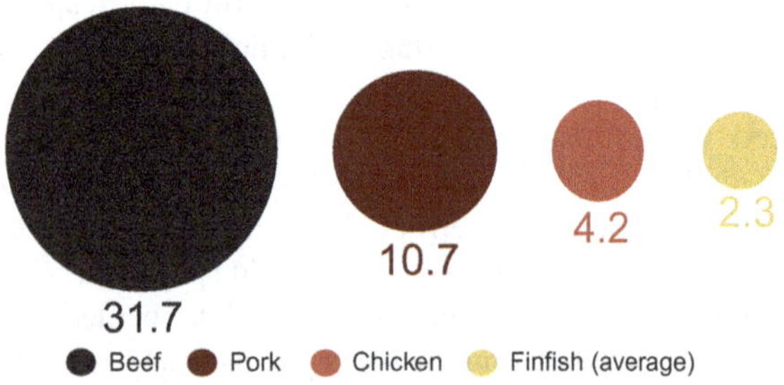

Fig. 2.6 Food conversion in kg feed/kg edible weight of various meat sources. *Data source* Hall et al. (2011)

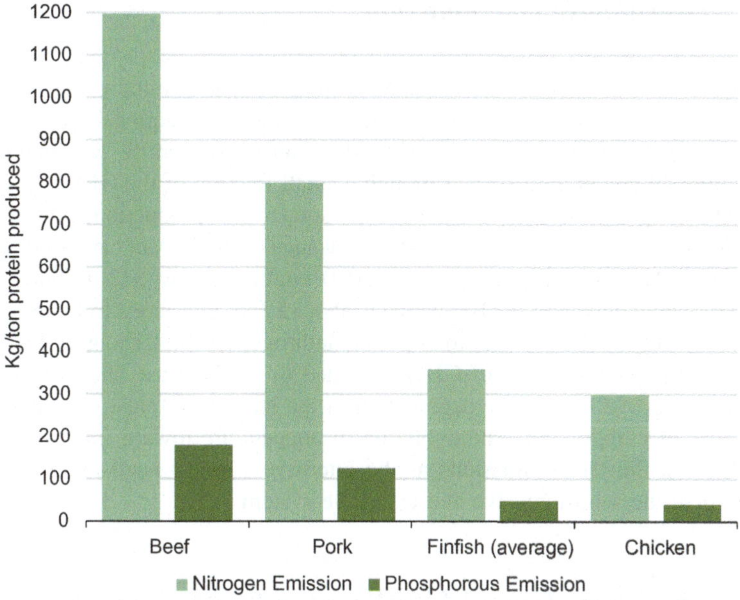

Fig. 2.7 Nitrogen and Phosphorous emission of various meat source. *Data source* Hall et al. (2011)

2.4.8 Spread of Diseases from the Aquaculture

Aquaculture is also one of the potential threats to the transfer of diseases to the surrounding environment. Uncontrolled aquaculture, intentional or unintentional management errors, lack of knowledge about the disease, etc., might be the factor contributing to the outbreak of disease. The additional absence of buffer zone around the open aquaculture system like sea-cage attracts much wild fish due to the availability of the food which can lead to the transfer of diseases like sea lice to the native species. Escaped fish from the farm cages can also act as a vector for diseases and parasites. Moreover, the risk of transmission of the disease is high for the intensive type of aquaculture due to high stock density (FAO 2014a) and since the world aquaculture is trending toward the intensive aquaculture system, driven by the market demand, more disease can be predicted to be transferred under normal circumstances. As the aquaculture products are traded from one country to another, the disease can also be transferred from one country to the other like *Haplosporidium nelsoni* in the Pacific oysters was unintentionally transferred from Japan to eastern oysters in the United States (Burreson et al. 2000) and the Sabellid worm was transferred from South African Abalone to the Californian Abalone (Kuris and Culver 1999).

2.4.9 Greenhouse Gas Emission

Greenhouse Gas Emission Activity such as energy use to maintain water level and quality, production of feed, transportation, processing of the aquaculture, packaging of the products, disposal of the waste, etc., cause greenhouse gas emission in aquaculture (Waite et al. 2014). Although a small fraction of the GHG emission is attributed to aquaculture, but with the raising aquaculture production and increasing concerns about the climate change, the significance of the impact can be considered to be high. Aquaculture production in 2010 emitted nearly 332 million tons of carbon dioxide equivalent (CO_2e) which is about 5% of emissions from agricultural production and less than 1% of total global anthropogenic emissions (FAO 2014a, b). Another potential source of GHG is related to the land use change associated with the mangrove forest. The degradation of the mangrove forest ultimately leads to the loss of carbon sink. The evolution of aquaculture toward intensive system will also add the GHG emissions as the intensive systems need more energy to operate than the semi-intensive and extensive system.

2.4.10 Fishmeal Trap: Added Pressure to the Fisheries

Sustainable aquaculture demands sustainable feeds but the raising concern for the aquaculture industry is the culture of carnivorous species like Salmon, which further add pressure to the wild fisheries for fishmeal[2] and fish oil. As the aquaculture industry expand the fishmeal and fish oil will be scarcer because as discussed earlier, capture type of fishing has already reached its saturation point and can no longer expand and a significant portion of the wild fish captured are the ones (small bony fish) utilized for the fishmeal. The sustainability of such farming is also questionable as about 6 kg of wild fish are required to produce 1 kg of the farm fish (Schipp 2008). Thus either a sustainable feed (alternative to the current fishmeal and fish oil) or the aquaculture of the herbivorous breed is required for sustainable aquaculture. The opportunity cost of these captured fish could be high for the wild fish productivity.

2.5 Assessment of Impacts

Impact assessment is one of the key processes of the SEEA. It requires the involvement of experts and stakeholders. The hired experts/consultant/practitioner also needs to be unbiased in impact identification. Different methods can be used to

---

[2]Fishmeal, which is derived from wild capture is the processed meal for the aquaculture carnivorous fish. It is majorly processed from fresh wild captured small, bony/oily fish and a small fraction is processed from the other fish trimmings (or fish waste). These kind of captured fish and by-products are not suitable for direct human consumption.

identify the impacts. Methods used depend on the experience of the consultant hired and also depend on the size, location and nature of the project. However, the method used must be simple and easy to interpret as a different level of decision makers will be later involved in the decision-making process. Some of the commonly used methods are the matrix, checklist, network, mathematical modeling, stakeholder consultation, expert judgment, etc. Some of the specific methods involved in the impact identification of the aquaculture project are as follow.

2.5.1 *Stakeholder Analysis*

Stakeholder analysis emphasizes on the individual interest of the stakeholders. Stakeholders that might be involved in an aquaculture project are governmental agencies, business associations, non-governmental agencies, community bodies, community leaders, religious bodies and local residents. This analysis helps to understand and anticipate the role and impacts of stakeholder with the introduction of the project. Figure 2.8 is a form of stakeholder analysis, which includes different stakeholders who has better understanding of the local environment. This kind of analysis not only helps to explore the impacts and its causes, but it also helps understand stakeholder interest to some extent. Since, aquaculture involves activities and actions that need to use the natural resources of the community, key stakeholders and their role (positive or negative) need to be understood. Stakeholder analysis identifies both the beneficiaries and affected groups and focuses on the active group (who have an economic interest in the project) of stakeholders as they can affect the project. Stakeholder analysis identifies their interests, examines the conflicts and explores trade-offs (Cordell et al. 2009).

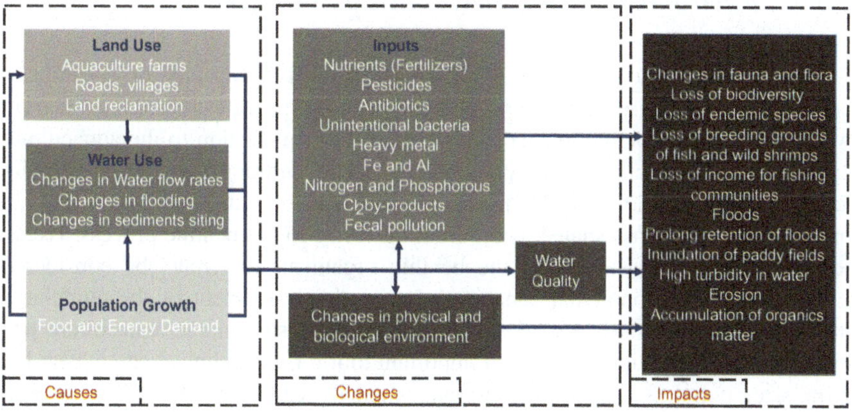

Fig. 2.8 Pollution aspects of the mangroves and the salt marshes generated from stakeholder analysis in Puttalam Lagoon and Dutch Bay, Sri Lanka due to shrimp farming. Reproduced from Senarath and Visvanathan (2001)

2.5.2 Rapid Rural Appraisal and Participatory Rural Appraisal

Rapid rural appraisal (RRA), and participatory rural appraisal (PRA) is a tool to promote sustainable development and is widely used in sustainable aquaculture and fisheries. This technique facilitates the interaction between stakeholders, researchers, and planners to exchange information and opinions. With the use of maps, matrices, details of past events, etc., brainstorming exercises are carried out to draw project impact and appropriate solutions to it. RRA and PRA technique ensures the incorporation of impact identified or predicted by the public. In addition, it also acts as a tool to generate mitigation measures from the public and provide them the opportunity to be involved in the decision-making. This planning process also helps to gain public acceptance and additionally RRA and PRA is an effective tool to utilize the local knowledge in the decision-making process.

2.5.3 Remote Sensing and Geographic Information System

Remote Sensing (RS) and Geographic Information System (GIS) have been widely used in the environmental and socio-economic analysis of the baseline information in sustainable aquaculture planning. However, there are plenty of limitations associated with the use of RS and GIS as it is a more costly process. In this scenario, PRA and RRA can be a more effective mechanism for planning aquaculture projects.

2.5.4 Environmental Capacity and Limit to Change

Environmental capacity also referred to as absorptive capacity or assimilative capacity is the ability of the environment to accommodate a particular activity without any unacceptable impact (GESAMP 1996). In relation to the aquaculture, environmental capacity can play a crucial role in defining the rate of nutrition addition and organic flux. Nutrition addition causes eutrophication while the organic flux can be associated as the limiting factor to the benthic process. Excess feed and organic waste affects the benthic organisms and must be considered. Evaluation of environmental capacity helps in the assessment of the cumulative impacts. This analysis is also useful to calculate the sustainable aquaculture production rate. In addition to being an important tool for the technical parameters like farm size, population size, and carrying capacity can also be applied to more regional issues like an ecosystem and watershed management (Byron and Costa-Pierce 2013).

Estimation of environmental capacity is expensive and to be cost-effective preliminary scoping about the impacts relevant to the type of aquaculture and technology must be performed. For example, shellfish breeding causes reduction of phytoplankton while the finfish cultivation will cause nitrogen, phosphorous and organic matter pollution. Environmental capacity can be calculated using various models. It has been used to evaluate: impacts caused by phytoplankton's by bivalve cultivation, the impact of nitrogen inputs from salmon cultivation, impact of organic matter input to seabed's, impact of organic matter input to benthic population, etc.

2.6 Identifying Mitigation Measures

One of the main purposes of the SEEA is to propose mitigation measures based on the social and environmental condition of the project area. Hence, it is not necessary that the mitigation measures appropriate for an area be appropriate for another area. The identified impacts can vary in nature as some impacts are beneficial (e.g.: the creation of jobs to local people) while some are adverse (e.g.: loss of biodiversity due to eutrophication). The mitigation measures focus on either enhancing the beneficial impacts or mitigating the adverse impact with the principle of avoiding first, then reduce, then propose remedy measures and if nothing is possible to mitigate by compensation.

The identified mitigation measures should be an integral part of the project approval and must be implemented during different phases of the project to mitigate the project impacts. Usually, the mitigation measures are incorporated in the contract/terms of condition documents so that it is implemented during the planning, construction and operational stages of the aquaculture. Some of the mitigation measures that can be taken at different stages of the aquaculture are presented in Table 2.3.

2.7 Monitoring

Monitoring is an important step involved in the socio-economic and environment assessment. As the sustainable or environmentally friendly practices are adopted it becomes necessary to monitor the adopted measures and the effectiveness of it. Monitoring is done with the following aim:

- To ensure that the mitigation measures adopted are incorporated in the project design and in the tender document
- To keep the record of the changes, that follows after the execution of the projects
- To ensure the achievement of the targeted standards

Table 2.3 Presents the impacts that commonly occurs in different stages of aquaculture and the potential mitigation measures that could be taken

Project activity	Impacts	Mitigation measures
Site selection		
Conflicts with existing site users	Competition for the use of resources	Adoption of relevant land uses planning
		Consultation and mutual agreement with the beneficiaries
Change in livelihood of the local inhabitants	Rise in social conflicts	Participation of local people in aquaculture projects
Ecologically sensitive site in the project area demarcation	The potential loss of biodiversity	Consideration of sensitive zones during the site selection with integration of aquaculture into integrated coastal zone management (ICZM)[a]
		Physical demarcation of ecological sensitive zone and inclusion of it in the management plan
Natural hazards like typhoons, flooding, hurricanes	Destruction or damage to the aquaculture's physical facilities and loss of harvest	Consideration of catastrophic events during site selection
		Designing of climate-resilient structures
Effluent generation from aquaculture	Deterioration of water quality causing reduction/loss of production	Consideration of carrying capacity as a key parameter during evaluation of appropriate site (Alternative analysis can be done to select the appropriate site)
		Adoption of ICZM to keep the water pollution within the carrying capacity
Disease in the fish	Loss of harvest, loss of production and possible infection to the nearby indigenous wild fish	Expert consultation
		Nearby farm survey for the detail information regarding types, frequency and occurrence of the disease to develop preventive measures for the risk avoidance
		Planning of risk management strategies to reduce risk before the operation of the project

(continued)

Table 2.3 (continued)

Project activity	Impacts	Mitigation measures
Design of the farm		
Project design of the farm	Lack of experience and poor understanding of the project components can result in negative environmental impacts	Proper design of the farm with proper consultation (public and expert)
		Designing with the principles of sustainability
Construction		
Change in socio-economic condition	Raise in social conflict	Public involvement in all the stages (planning, design, operational) of the project
		Priority to local employment. Enhancing the opportunity of locals by capacity building and training
Use of natural resources in the project area	Hampers traditional occupation	Locate the site away from the traditional users
		Create and monitor buffer areas between farm and other users
Construction and operation of physical facilities in the project area	Deterioration of aesthetic beauty in project area	Siting the farm away from the local inhabitant
		Adoption of designs and technology like low profile cages which minimize the uses of unsightly structures
		Considering local architecture while constructing physical facilities
Construction of aquaculture farm	Various environmental impacts due to poor construction practice	Built it with standard engineering and construction practice
	Disturbance to the wildlife and benthos ecosystem during construction	Maintenance of buffer zone and minimizing the construction disruption to the construction area only
Farm operation and management		Adoption of 'Best Management Practice' and ecolabel schemes
Solid waste disposal	Impacts on benthos wildlife due to decreased oxygen level	Collection and safe disposal of the non-organic solid waste materials
Wastewater/effluent discharge	Deterioration in water quality level of the streams where effluent is discharged causing impact to the population of other species	Adopting best management practices available
		Efficient feeding practices (optimizing the quantity of fish food)
		Locating farm in the area with adequate tidal flow

(continued)

Table 2.3 (continued)

Project activity	Impacts	Mitigation measures
Use of chemicals	Possibility of negative effects on worker's health	No use of chemicals or avoid the use of chemicals
		Use of safety measures by workers
	Decrease in quality of the product due to deteriorated water quality	Avoid or minimize the use of chemicals
		Adoption of preventive management system
Rearing of exotic or farmed stocks	Escape of farmed or exotic species can have negative impact on the ecosystem as well in the gene of wild stocks	Development of hatcheries
		Designing the farm to avoid any escapes
		Designing the farm to be resilient to natural damage (e.g. strom)
		Introduction of exotic species following the Code of Practice of ICES/FAO (Turner 1988)
Outbreak of disease	Impact to the endemic species due to the dispersion of disease	Preventive management system
		Regular monitoring of the water and harvest
		Sanitary disposal of the dead or infected harvest
Occurrence of natural events like storm	Loss of harvest	Preventive approach against the storm
		Developing strategy to deal with the occurrence of unlikely events
		Routine monitoring and maintenance of nets, mooring, etc.
		Climate resilient design, technology, and practice
Interference of predators and wildlife	Decline in productivity of the aquaculture	Consideration of predators and wildlife during site selection
		The introduction of relevant management plans to cope with it. Eg. double net

[a]The European Commission defines ICZM as "a dynamic, multidisciplinary and iterative process to promote sustainable management of coastal zones. It covers the full cycle of information collection, planning (in its broadest sense), decision making, management and monitoring of implementation. ICZM uses the informed participation and cooperation of all stakeholders to assess the societal goals in a given coastal area, and to take actions towards meeting these objectives. ICZM seeks, over the long-term, to balance environmental, economic, social, cultural and recreational objectives, all within the limits set by natural dynamics. 'Integrated' in ICZM refers to the integration of objectives and also to the integration of the many instruments needed to meet these objectives. It means integration of all relevant policy areas, sectors, and levels of administration. It means integration of the terrestrial and marine components of the target territory, in both time and space"

- To measure the accuracy of the predicted impact
- To monitor the effectiveness of the mitigation measures adopted and to provide scope of adopting better adaptive measure through the feedback mechanism
- To provide data for environmental audit
- To maintain the threshold set by the project which is often guided by governmental standards and policies
- To identify, measure and mitigate unanticipated impacts.

In addition to targeting minimum environmental impact monitoring also provides scope to increase public acceptance. Through regular and systematic monitoring activities, the project will have minimum impact on the social and environment component of the project area which will reduce the chances of social conflict.

Some of the methodologies used in monitoring are shown in Table 2.4.

Selection of monitoring parameters needs to be chosen considering various factors like legal standards, nature of the impacts identified initially, the technology used for the culture, species grown, etc. Monitoring parameters should also consider quantifying the positive impact in addition to the negative impacts.

2.8 Environmental Certification to Sustainable Aquaculture

Over the past decades, the use of market-based management approaches like codes of conduct, best management practices, eco-labelling and certification which targets both the aquaculture and capture type of fishing has grown. These voluntary approaches target both the socio-economic and environmental aspects of the fishery industry. As the regulatory approaches have a high implementation, monitoring, and enforcement cost (USAID 2013) voluntary approaches can be considered a cost-effective means to achieve sustainable targets.

Codes of conduct refer to the guideline that incorporates the socio-economic and environmental aspects and is designed to minimize negative impacts, ensure safety, increase benefits and optimize production. Adoption of these best management practices are voluntary in nature, however, efforts are given at national level to advocate the benefits of it.

Certification and Ecolabelling are another widely used voluntary method, which targets to disseminate information for the consumers to make the appropriate decision. The certifications and ecolabel required a set of criteria to be fulfilled and these criteria focus on making the product environmentally and socially sound. These ecolabels on the product help consumers to make purchasing of the environmentally friendly products and allows the consumers to create demand for sustainable goods. In developed regions such as North America and Europe where green consumerism has flourished, greater demand for the certified aquaculture products is observed at the supermarkets and restaurant chains (Waite et al. 2014).

Table 2.4 Monitoring methods

SN	Methodology	Monitoring components
1	Walkthrough survey	General overview of the changes compared to the baseline scenario
2	Questionnaire survey, key informant interview, secondary data collection, etc.	Social conflicts, economic status, environmental problems, etc.
3	Video survey	Approximate sediment thickness; sediment color; sediment consistency; surface consolidation; gas bubbles; presence of feed and feces; macro-fauna/flora; presence of detritus and fouling organisms
4	Sediment sampling	Solids deposited in the core due to the aquaculture
5	Water quality sampling	Water quality parameters (often according to the governmental standards) Eg: Redox, pH, DO, TVS, TDS, TOC, Zn, Cu, etc.
6	Sampling of various components	Biophysical characteristics, microfauna abundance
7	Modeling (Models like DEPOMOD can be used)	Area of maximum impact from culturing operation
8	Echo sounder monitoring	Bathymetric profile
9	Visual inspection	Disease, vectors, fungus and others that can cause disease leading to loss of productivity of the harvest

Thus, aquaculture also gains an advantage by increasing the marketability of the product by adding value to the quality of the product. Some of certifications and ecolabel for the aquaculture are:

I. Aquaculture Stewardship Council (ASC): ASC ecolabel certifies the farmed seafood that has taken measures to reduce the environmental damage. In addition to demonstrating environmental responsibility, the aquaculture also needs to demonstrate social responsibility toward the workers to be certified by ASC.

II. Marine Stewardship Council (MSC): MSC certification was initiated by WWF and Unilever in 1997 and is an independent body that certifies the sustainability of the industry. The industry has to undergo the MSC auditing process and comply with all its standards to obtain the certification to use its ecolabel. The use of MSC ecolabel certifies that the aquaculture industry linked to the product has adopted sustainable and responsible practices.

III. The EU Eco-label: was launched in 1992 by the European Commission with the motive of developing a Europe-wide trustworthy labeling scheme that consumer could believe to have minimum environmental stress. As of September 2015, it had 44,771 products and 2031 services licensed under it (European Commission 2016). The licenses give companies the right to use the EU Ecolabel logo on their product group. Aquaculture products with EU eco-label in the European market have more demand than the other.

IV. Best Aquaculture Practices (BAP): BAP ecolabel is commonly used for the shrimp farm and hatcheries and seafood processing plants. The use of BAP ecolabel reflects the standards that are specifically directed toward the protection of biodiversity and workers right.

The use of ecolabelling can reflect the sustainability of the industry, the market for sustainable goods is also an important component for its success. As the majority of the aquaculture production and consumption occurs in developing countries (Eg. China, Thailand, Vietnam, Indonesia, etc.,) the demand for the sustainable aquaculture product is low. Currently, the private ecolabelling schemes only certify 5% of the global production (Bush et al. 2013).

2.9 Concluding Remarks

With the increasing demand, aquaculture has provided new means to increase the aquatic products, which have reached its saturation limit in nature. Over the past decade, the aquaculture industry has kept expanding and has over-exploited at socio-economic and environmental cost. Aquaculture production is now focused on increasing the yield at the lowest possible cost to be competitive in the market. This has driven the aquaculture industry to a more intensive system with more impacts related with it. This trend tends to neglect the environmental and socio-economic aspect, which forms the pillar of sustainable development.

The competitive market, market demand for low price products and overly ambitious aim of the producers to maximize the profit are the constraints to the sustainability of this industry. Apart from the benefits such as increased jobs, food security, increased trade, etc. aquaculture leads to several adverse socio-economic impacts like loss of traditional occupation, social conflicts, food safety, etc. It also has environmental cost as it pollutes the nearby environment, hampers wild fishes, spreads diseases, and causes genetic variation in the ecosystem. These impacts vary geographically and according to species cultivated and technology used. It further depends on the intensity of the farm. Thus, it is necessary to identify these impacts and take necessary mitigation measure with strong management practice.

SEEA can act as an important mechanism to achieve the sustainability of aquacultures. Moreover, as the intensive aquacultures are predicted to grow in the future, SEEA will play a more vital role. Sustainable aquaculture might need a little more effort and spending but it could be a viable option to ensure long-term profitability of the industry.

References

Barg UC (1992) Guidelines for the promotion of environmental management of coastal aquaculture development. FAO fisheries technical paper no. 328. FAO, Rome

Baulcombe D, Crute I, Davies B, Dunwell J, Gale M, Jones J, Toulmin C (2009) Reaping the benefits: science and the sustainable intensification of global agriculture. The Royal Society, London

Burreson EM, Stokes NA, Friedman CS (2000) Increased virulence in an introduced pathogen: Haplosporidium nelsoni (MSX) in the eastern oyster Crassostrea virginica. J Aquat Anim Health 12(1):1–8. 10.1577/1548-8667(2000)0122.0.co;2

Bush SR, Belton B, Hall D, Vandergeest P, Murray FJ, Ponte S, Oosterveer P, Islam MS, Mol APJ, Hatanaka M, Kruijssen F, Ha TTT, Little DC, Kusumawati R (2013) Certify sustainable aquaculture? Science 341(6150):1067–1068. https://doi.org/10.1126/science.1237314

Boyd CE (1992) Shrimp pond bottom soil and sediment management. In: Wyban J (ed) Proceedings of the special session on shrimp farming. Orlando, Florida, May 25–26 1992, pp 166–180

Byron CJ, Costa-Pierce B (2013) Carrying capacity tools for use in the implementation of an ecosystems approach to aquaculture. In: Ross LG, Telfer TC, Falconer L, Soto D, Aguilar-Manjarrez J (eds) Site selection and carrying capacity for inland and coastal aquaculture, pp 87–101. FAO/Institute of Aquaculture, University of Stirling, Expert Workshop, 6–8 Dec 2010. Stirling, the United Kingdom of Great Britain and Northern Ireland. FAO fisheries and aquaculture proceedings no. 21. Rome, FAO. 282 pp

Cao L, Diana JS, Keoleian GA, Lai Q (2011) Life cycle assessment of Chinese shrimp farming systems targeted for export and domestic sales. Environ Sci Technol 45(15):6531–6538. https://doi.org/10.1021/es104058z

Cordell D, Drangert JO, White S (2009) The story of phosphorus: global food security and food for thought. Glob Environ Change 19(2):292–305. https://doi.org/10.1016/j.gloenvcha.2008.10.009

Dierberg FE, Kiattisimkul W (1996) Issues, impacts, and implications of shrimp aquaculture in Thailand. Environ Manage 20(5):649–666. https://doi.org/10.1007/BF01204137

Environmental Justice Foundation (2014) Impossibly cheap: abuse and injustice in Bangladesh's shrimp industry [video file]. Retrieved from https://www.youtube.com/watch?v=ylFJDvHYnt8 . Accessed on 19 June 2016

Erondu ES, Anyanwu PE (2005) Potential hazards and risks associated with the aquaculture industry. Afr J Biotech 4(13):1622–1627

European Comission (2016) Facts and figures. Retrieved from http://ec.europa.eu/environment/ecolabel/facts-and-figures.html. Accessed on 14 June 2016

FAO (2012) The State of the world fisheries and aquaculture 2012. FAO, Rome

Food and Agriculture Organization (FAO) (2014a). The state of world fisheries and aquaculture. Food and Agriculture Organization of the United Nations, Rome

Food and Agriculture Organization FAO (2014b) FAO Regional Conference for Asia and the Pacific: sustainable intensification of aquaculture for food and nutritional security in the Asia-Pacific region. Information note, 32nd session, Ulaanbaatar, Mongolia, 10–14 Mar 2014

GESAMP (IMO/FAO/UNESCO-IOC/WMO/WHO/IAEA/UN/UNEP Joint Group of Experts on the Scientific Aspects of Marine Environmental Protection) (1996) The contributions of science to integrated coastal management. In: GESAMP Reports and Studies, No. 61. FAO, Rome, 66 pp. http://www.fao.org/3/a-w1639e.pdf. Accessed on 12 June 2016

Hall SJ, Delaporte A, Phillips MJ, Beveridge M, O'Keefe M (2011) Blue frontiers: managing the environmental costs of aquaculture. The WorldFish Center, Malaysia

Jayasinghe JMPK (1995) Report on a regional study and workshop on the environmental assessment and management of aquaculture development. FAO Network of Aquaculture Centres in Asia-Pacific, Bangkok, Thailand, April 1995

Jensen Ø, Dempster T, Thorstad EB, Uglem I, Fredheim A (2010) Escapes of fishes from Norwegian sea-cage aquaculture: causes, consequences and prevention. Aquac Environ Interact 1(1):71–83. https://doi.org/10.3354/aei00008

Joanna Lovatt (2016) The Bangladesh shrimp farmers facing life on the edge. Retrieved from http://www.theguardian.com/global-development-professionals-network/2016/feb/17/the-bangladesh-shrimp-farmers-facing-life-on-the-edge. Accessed on 12 June 2016

Kabir SH (2013) Sea food export from Bangladesh and current status of traceability. Retrieved from https://www.unescap.org/sites/default/files/6-%20%20Sea%20Food%20Export%20from%20Bangladesh-Kabir.pdf. Accessed 19 June 2016

Kuris AM, Culver CS (1999) An introduced sabellid polychaete pest infesting cultured abalones and its potential spread to other California gastropods. Invertebr Biol 118(4):391–403. https://doi.org/10.2307/3227008

Link TV (2005) Murky waters: shrimp farming in Bangladesh [video file]. Retrieved from https://www.youtube.com/watch?v=JlX7nkxAqWY. Accessed on 19 June 2016

Menasveta P, Fast A (1998) The evolution of shrimp culture and its impact on mangroves. INFOFISH Int 1(98):24–29

Naylor R, Hindar K, Fleming IA, Goldburg R, Williams S, Volpe J, Whoriskey F, Eagle J, Kelso D, Mangel M (2005) Fugitive salmon: assessing the risks of escaped fish from net-pen aquaculture. BioScience 55(5):427–437. https://doi.org/10.1641/0006-3568(2005)055[0427:FSATRO]2.0.CO;2

Neiland AE, Shaw SA, Bailly D (1991) The social and economic impact of aquaculture: a European review. Aquac Environ 16:469–482

Pullin RS (1989) Third-world aquaculture and the environment. NAGA 12(1):10–13. Retrieved from http://pubs.iclarm.net/Naga/na_3238.pdf. Accessed 19 June 2016

Rosenthal H (1992) Environmental impacts: negative and positive aspects in workshop, fish farm effluents and their control in EC countries. Congress Center Hamburg, 23–25, Nov 1992

Schipp G (2008) Is the use of fishmeal and fish oil in aquaculture diets sustainable? Technote no: 124. Northern Territory Government, Australia. https://dpir.nt.gov.au/__data/assets/pdf_file/0003/233562/tn124.pdf. Accessed 19 June 2016

Senarath U, Visvanathan C (2001) Environmental issues in brackish water shrimp aquaculture in Sri Lanka. Environ Manage 27(3):335–348. https://doi.org/10.1007/s002670010153

Subasinghe R (2006) State of world aquaculture 2006. FAO Fisheries Technical Paper, FAO, Rome

The World Bank (2007) World development report 2008: agriculture for development (no. E14 231). World Bank Group, Washington, DC

The World Bank (2013) Fish to 2030: prospects for fisheries and aquaculture. World Bank Report Number 83177-GLB. World Bank Group, Washington, DC

Thorstad EB, Fleming IA, McGinnity P, Soto D, Wennevik V, Whoriskey F (2008) Incidence and impacts of escaped farmed Atlantic salmon Salmo salar in nature. NINA special report 36

Turner GE (1988) Codes of practice and manual of procedures for consideration of introductions and transfers of marine and freshwater organisms. EIFAC/CECPI occasional paper no. 23. FAO, Rome

United States Agency for International Development (USAID) (2013) Sustainable fisheries and responsible aquaculture: a guide for USAID and partners. USAID, Washington, DC

Waite RI, Beveridge MA, Brummett RA, Castine SA, Chaiyawannakarn NU, Kaushik SA, Mungkung RA, Nawapakpilai SU, Phillips MI (2014) Improving productivity and environmental performance of aquaculture. Working paper, installment 5 of creating a sustainable food future. World Resources Institute, Washington, DC

World Wildlife Fund (WWF) (2016) Overview. Retrieved from http://www.worldwildlife.org/industries/farmed-shrimp. Accessed on 12 June 2016

Chapter 3
Sustainable Fishing Methods in Asia Pacific Region

Sudath Terrence Dammannagoda

Abstract After the Second World War, marine fishery industry was developed to a commercial industrial fishery by war torn countries to boost their economies. As a result, compared to traditional pole and line fishery, much larger fishing gears such as trawls and purse seines, and much larger and more powerful fishing vessels were built, and deployed to traditional fisheries grounds in the southern hemisphere. Although this led to an increase in the global marine catch in several folds, many fisheries around the world, however, have collapsed and depleted due to the over effort and overexploitation. Further, many marine resources have been destroyed as non-targeted by catch by ill-designed industrial scale fishing gears such as bottom trawls and purse seines. World-wide annual marine bycatch is around 27 million tonnes, and for bottom trawls 66–93% of the catch consists of bycatch while this is 64–79% for purse seine. Moreover, benthic marine environment around the world has been affected drastically by the bottom trawl fleet. Today, reduced industrial commercial fleet, and tough fishing regulations in developed countries have made a considerable progress towards reducing the fishing effort and hence the reduced bycatch and discards. Although modifications of industrial bottom trawls and purse seines have made a progress in reducing the bycatch to a certain extent, these fishing gears are fundamentally unsustainable. Customer interest on 'sustainably caught' fish and hence marine stewardship is increasing in developed countries. To this end, pole and line fishery should be propagated around the world as one of the best sustainable fishing method. Also, Ecosystem based fisheries management and small scale regional fisheries management should be the future approach in fisheries management as local knowledge on the local marine ecosystem can be used together with the participation of local fisher folks.

Keywords Sustainable fishing · Industrial fishing · Gear selectivity
By catch · Quota policy

S. T. Dammannagoda (✉)
School of Biomedical Sciences, Griffith University, Gold Coast Campus,
Parklands Dr, QLD 4215 Southport, Australia
e-mail: t.dammannagodaacharige@griffith.edu.au

F. I. Hai et al. (eds.), *Sustainable Aquaculture*, Applied Environmental
Science and Engineering for a Sustainable Future,
https://doi.org/10.1007/978-3-319-73257-2_3

3.1 Introduction

For thousands of years, marine fisheries have been the main livelihood for individuals residing in coastal countries around the world. By the year 2012, as an example, there were an estimated 39 million fishers of capture fisheries around the world, with 84% of them living in Asia, followed by Africa, and Latin America and the Caribbean (FAO 2014). Taken together, these figures suggest that around 163 million people (5 members per family on average) rely on fishing as their primary source of income in coastal Asian countries.

In the coastal population around the Indian Ocean especially, marine fish has typically been the main animal protein source. For example, explorer Ibn Batuta wrote in his travel records that even in the early 14th century, tuna fishing was a lucrative industry in the South Asian coastal countries. Thus, it is evident that these fishing communities have been practising sustainable fishing for thousands of years while attaining their animal protein requirement and main source of livelihood. The world's marine fishery production in 2011 was 93.7 million tonnes, which shows a general declining trend since 1996. The largest marine fish production which accounts for 70% of total global marine fish production is from the Pacific, which is over 84 million tonnes. The highest capture fishery was recorded from Northwest Pacific with 21.4 million tonnes (26% of global marine catch), followed by the Southwest Pacific with 12.3 million tonnes (15%), the Western Central Pacific with 11.5 million tonnes (14%), and the Northeast Atlantic with 8 million tonnes (9%) (FAO 2014). Capture fishery production per fisher is highest in Europe (24.2 tonnes/year), followed by North America (19.7 tonnes/year), Oceania (10.4 tonnes/year), Latin America and the Caribbean (6.2 tonnes/year), Asia (1.6 tonnes/year), and Africa (1.5 tonnes/year). The European Union (member countries) is the largest single market for imported fish and fishery products, and in 2012 represented 36% of world fish exports worth an estimated US$47 billion (FAO 2014). Comparatively, United States of America and Japan are the largest single importers. Also of interest is that per capita fish consumption is rising in developing countries (from 5.2 kg in 1961 to 1.8 in 2010), but is still the highest is in developed countries (FAO 2014).

The total marine fish production in the Eastern and Western Indian Oceans shows a continuous growth which in 2012 was approximately 7.4 and 4.5 million tonnes, respectively. Majority of the marine fish production in the Indian Ocean is for domestic consumption for the over 200 million of coastal population in this region. Industrial marine fish catching is prominent in the Western Indian Ocean, mainly by Spanish and French fleet, using gears such as purse seines and long lines. This production is mainly exported to European countries.

In the past, fishers always employed sustainable fishing gears and practices with great knowledge and experience about the marine environment, marine fauna and their ecology, biology, behaviour and interactions with marine communities. Thus, these fishermen's fishing gears and practices were based on vast knowledge and experience, leading to very simple but efficient, highly selective, smartly designed fishing gears and methods. These fishing gears and methods were categorised as artisanal fishing. Artisanal fishing in the Asia-Pacific region was sufficient in providing the main animal protein source for millions of coastal population for thousands of years without compromising the 'health' of the marine ecosystem.

Prominent environment friendly artisanal fishing practices in the Asia-Pacific region include pole and line fishing or live-bait fishing, trolling or handlines, small scale multi-filament gill nets, various types of trapping, harpooning and angling.

3.2 Development of Fisheries and Fishing Gear from Artisanal to Industrial and Commercial Fishery in the World

After the Second World War, World Food and Agricultural Organisation (FAO) focussed on the development of world industrial marine fisheries as part of rebuilding collapsed economies of war torn European countries. Between 1945 and 1950, the first FAO technical committee identified major problems affecting marine fisheries in the Northern hemisphere including resource depletion, discarding practices, and lack of data for high seas management (Hall 1999). At the same time, this committee identified that resources in the Southern hemisphere were underutilized, and hence encouraged the expansion of long-range fisheries. While the world marine fish catch started rapidly increasing due to these long-range expansions, by early 1950s a number of important fishery resources in the northern hemisphere including the Hokkaido sardine, the North sea and Atlanto-Scandian Herring, and the Californian pilchard were collapsed, mainly due to over-exploitation (Hall 1999).

Between the 1960s and 1970s, developing countries and financial institutions were encouraged to expand world fisheries, while long distance fleets from many nations expanded their operations by means of subsidy schemes and technological developments such as large scale nets with synthetic fibres, and powerful engines. As a result, an industrial fisheries sector emerged in a select few developing Asian countries including shrimp fishery in India, Pakistan, and Kuwait. While these industrial fisheries raised foreign exchange, they also put increasing pressure on traditional fishing grounds in developing countries. For example, Peruvian anchoveta fishery collapsed from 12 to 2 million tonnes in the early 1970s.

Alarmingly, between the 1970s and 1990s the rate of growth in the world's industrial fishing fleet was twice that of the global catch with both the total tonnage and the number of vessels doubling in that period (Safina 1995). Between the 1970s and 1980s many coastal states extended their jurisdiction to 200 nautical miles in anticipation of international legislation. By the early 1980s, final expansion of distant water fleets occurred into the Indian Ocean, the South Pacific, and the Southwest Atlantic targeting high value species such as tuna, shrimp and cephalopods. For example, Taiwan, China, Korea, Japan, and Russia expanded their industrial long line fishery in the Indian and Western Pacific Oceans targeting tuna. While the overall production was increased, concerns were raised around overexploitation and the potentially damaging effects of demersal industrial fishery. In parallel, artisanal fisheries were developed and modernised increasing the fishing effort on traditional inshore fishing grounds.

3.3 Fishing Practices in the Asia-Pacific Region

Fishing practices in the Indo-Pacific and Atlantic regions can be classified as artisanal, commercial, and recreational. The main fishing practices are artisanal and commercial, as recreational fishing has not yet become widespread in the Asia and South Pacific regions. Artisanal fishing practices are characterised by small fishing vessels and gears that are operating in relatively low depth inshore areas. These artisanal fishing gears include mainly small scale gillnets, pole and line fishing or live-bait fishing, beach seines, cast nets, trolling, traps, and angling. In the past, the main type of fishing craft used for fishing within in-shore areas were wind sail powered wooden canoes. Today, the artisanal fishing sector uses crafts that range from motorised canoes, outboard motor fibre-glass boats to 20 m long inboard motor boats.

3.4 Main Types of Fishing Gear Around the World

A comprehensive description of principal types of commercial fishing gears and their operation can be found in Sainsbury (1996). Fishing gears can be classified using a number of methods. Main methods are based on

(a) Nature of operation (towed, encircling, static, and other)
(b) Place or depth of operation
(c) Artisanal or industrial.

Towed or dragged gear are mainly trawl nets that can be operated in the mid water column or on the sea bottom. Sometimes paired trawl nets or a single large trawl net are used depending on the fishery and the power of the fishing vessel. Moreover, dredging on the sea bed is also a recognized towing fishing method.

Encircling gears are used to encircle a dense school of fish on or near the surface of the water column with a large wall of net. Purse seines are the main type of commercial encircling gears and also are the largest commercial fishing gears around the world. Beach seines are the traditional types of small scale encircling gears.

Static gears are the main traditional fishing gears that set out in a particular location. While gill nets and long lines are the main types of static gear, pots and traps are also included in this type. In saying this however, there are a variety of other fishing methods which are used around the world. Some main types are pole and line bait fishery, trolling, harpooning, lines or angling, lift nets, and diving (including collecting and spear fishing).

Many fishing gears are designed or modified to suit their environment of operation (e.g. inshore, offshore, surface of the water column, mid layer of the water column, or bottom of the ocean). Gillnets and trawl nets are operated in different stratum of the water column and accordingly, there are drift gill nets, mid water column gill nets, and bottom gill nets. Trawl nets also operate in the middle of the water column or on the bottom of the ocean.

After the Second World War, many European countries started designing very large, industrial scale fishing fleets targeting offshore and deep sea marine resources mainly in the southern parts of the world. This led to an emergence of very large vessels with very large capacities, fish processing units and cold/freezing rooms which are powered by highly powerful inboard engines and equipped with modern navigational and fish finding equipment. In some cases, there is a large mother vessel with fish processes, and cold storage facilities, while a number of sister vessels fish and bring the catch to the mother vessel. To carry out large scale fishing, very large drift gill nets that can stretch 10–25 km, very large purse seines which are in size of several square kilometres, or industrial long lines that consist of several thousands of hooks were produced and put in operation.

3.5 Selectivity of Each Fishing Gear and Method, Associated Bycatch, and Impact on the Marine Environment

In general, selectivity of traditional artisanal fishing gears is very high meaning there is almost no bycatch or non-target catch. The impact on the marine environment and on other living communities is also at a minimum from the artisanal fishing gears and methods. Traditional artisanal fishing gears have evolved for many thousands of years through the addition of fishermen's knowledge and experience about the biology, behaviour, and ecology of the target species, and also the vast knowledge about the marine ecosystem as a whole on the design, operation, operational place and time. Artisanal fishing has been carried out not only as a source of livelihood for coastal communities, but also to fulfil their animal protein source. As there was no requirement of overproduction or large scale commercial production, artisanal fishing gears are designed to catch a small portion of the target population with a high selectivity on target species and adult animals.

3.5.1 Pole and Line Fishery

Pole and line tuna fishery, for example, is one of the best artisanal fishing methods based on sustainability and high selectivity (Fig. 3.1). Pole and line fishery targets feeding schools of adult tuna thus alleviating juveniles and other non-target species. Only adult tuna feeding schools come close to the water surface, and they can be easily located by flocks of marine birds or sea gulls as these birds attempt to feed on them. After locating feeding tuna schools, fishers can easily attract them by throwing small live fish or live bait to the water. This action is mimicked by the splashing of surface water towards the tuna school using bamboo splices. Attracted tuna are then caught one at a time by a crowd of fishermen using a pole and line that is equipped with a less curved hook. Through this method, adult tuna can be caught quite quickly and with minimum damage, an approach which preserves both the tuna's energy and meat quality. Skilled fishermen can catch hundreds of tuna in less than half an hour using this method. As the whole adult school tuna population is not swept by this fishing method (i.e. juveniles are not caught) the continuity of the tuna population is assured by pole and line fishing method. There is also almost no bycatch although non-target tuna species that are associated with the target school can occasionally be caught. Reason for this is that tuna feeding schools can be associated with other fish species including other tuna and fish species. On the whole, pole and line fishing methods aim to minimize harm to the marine environment, and other marine communities. It is clear therefore, that traditional fishermen were equipped with knowledge, expertise on life history, biology, ecology, and behaviour of each fish species/communities, and used this wisely in designing

Fig. 3.1 Small scale pole and line fishing in Sri Lanka. Photograph by Sudath T. Dammannagoda

fishing gears and methods while giving priority to the sustainability of the marine resources and marine environment.

Another artisanal method is traditional traps and pots which exploit the behaviours of particular fish species such as eels and other bottom dwelling fish, lobsters, and crabs. As fish can be collected in their live state, untargeted species and juveniles can be released back to their environment without harm. The same minimization of harm approach applies in the use of traditional traps and pots. This is at clear odds with beach seines which are a main artisanal fishing gear that targets inshore fish species, or fish species that migrate to inshore areas for spawning. As the beach seine is dragged along the bottom of inshore areas, it can disturb and destruct the bottom environment and fauna, spawning grounds, and nursery grounds. However, most of the beach seines around the world are operated seasonally and therefore this damage is very unlikely to be chronic.

3.5.2 Bottom Trawls and Dredges

In contrast, most of the Industrial commercial fishing gears invented in the past century lack gear selectivity, are expensive to build and operate, and are destructive to the overall marine community and the marine environment. Best examples to this end are the bottom trawl nets and industrial purse seines.

Bottom trawling has been practiced in all oceans and major seas around the world, in tropical and temperate regions, deep and shallow areas, including Europe, Australia, New Zealand, Indonesia, India, Thailand, and North & South America. Bottom trawling has been subjected to many complaints and high criticism since at least the 14th century due to both its direct and indirect destructive effects on the marine environment and marine resources. One main consequence of bottom trawling and dredging is that it scrapes, and in turn directly destroys, the sea bed, thus damaging the spawning and nursery grounds of the numerous species that inhabit these sea beds. Another significant problem associated with bottom trawling is the extremely high bycatch (non-target species) and associated discards as there is almost no gear selectivity in this method.

Dredges are rake like devices that use a large bag to collect demersal creatures. Dredges mainly target benthic molluscs such as scallops and oysters but occasionally, are used to catch crustaceans, demersal finfish, and echinoderms. Design of the dredge can be different based on the nature of the substratum (e.g. soft muddy bottom or rocky bottom) and the target fauna. The basic design consists of a steel frame opening that can span up to 2 m with the base of the frame equipped with a blade with teeth. Large, offshore dredges used to catch sea scallops can have a mouth opening width up to 4.5 m, and can weigh from 500 to 1000 kg (NRC 2002). These dredges use hydraulic pumps that inject water into the sediments to disturb the sea bed and dislodge scallops for capture in the net.

In an effort to increase the efficiency of towed gear, towing vessels have also become larger and more powerful which has helped the evolution of dredges to bottom trawls. First, bottom trawls were beam trawls that consisted of a larger and lighter steel frame connected to a larger and funnelled shape net. Heavy tickler chains are attached to the steel frame, and these chains disturb the sea bed to stimulate an escape response in bottom dwelling fish. As the gear is towed at a high speed, escape of fish is unlikely and the catch efficiency is high. Current large beam trawls have mouth openings of 15–20 m, and are towed from both sides of high-powered engine trawlers. Beam trawlers are mainly used to catch flat fish and other ground fish species in Northern Europe, shrimp and other demersal species in United States.

In an attempt to increase the horizontal opening of the trawl mouth, otter trawlers were designed which involved the same principles but without the heavy rigid trawl mouth. The ground gear can be as long as 200 m, thus enormously increasing the swept area. The bottom trawl net is funnel shaped, with a blind end referred to as the cod end. Mesh size varies depending on the fish species caught with small mesh sizes for shrimp and small fish, and large mesh size to catch large fish. In the United States, all coastal states use bottom trawls to catch demersal fish. Large bottom trawl nets are dragged by a pair of vessels.

Consequences of trawling and dredging can be divided into main two sections, direct and indirect impacts. Direct impacts include,

- Population mortality as part of the catch, bycatch, discards, destruction of benthic and demersal species and making them vulnerable to scavengers and other predators.
- Increased food availability by discards, bycatch, and dead benthic organisms.
- Loss of habitat due to destruction of sea bed.

Indirect effects are secondary to the long term effects of direct effects including reduction in total biomass, effects on prey, predator, and competitors' dynamics and equilibrium, and broadly, effect on seafloor community structure. Indirect effects include the impact on non-living systems that include changes in the flow of material and energy through the ecosystem and shifts in the balance among the processes of primary production, primary consumption, and secondary production.

The impacts of bottom trawling have been documented in various parts of the world including Europe (Lindegarth et al. 2000; Kaiser and Spencer 1994, 1999, 2002), Australia (Gibbs et al. 1980; Butcher et al. 1981; Hutchings 1990), New Zealand (Saxton 1980; Bradstock and Gordon 1983), Indonesia and Thailand (Chong et al. 1987), India (Meenakumari and Pravin 2008; Kumar and Deepthi 2006; Jagadis et al. 2004) and North America (Dolah et al. 1987; Goude and Loverich 1987). However, the impacts of bottom trawling on the physical, chemical & biological environment of the marine ecosystem and the diversity and quantity of by-catch and discards remain poorly documented for the tropical waters. As the marine biodiversity and productivity is higher in tropical waters compared to temperate waters, there is comparatively a much higher species impact from bottom trawling than has been observed in temperate fishing areas (Aish et al. 2003).

As already mentioned, the ecological impact of bottom trawling has been studied in various parts of the world. Many of these studies have shown that bottom trawling is among the most destructive of fishing methods is it causes direct and indirect disturbances and damages to sea bed and their living communities (Jones 1992; Kaiser 1998, Kaiser et al. 1999, 2002; Zacharia et al. 2006; Kurup et al. 2004; Raman 2006). Likewise, a large volume of studies have shown that bottom-trawling affects the biomass and production of benthic invertebrate communities (Jennings and Kaiser 1998; Hall 1999), in addition to changing energy flow through benthic ecosystems, leading to large scale shifts in benthic community structure and regime shifts (Hiddink et al. 2006). Alongside this ecological impact on marine fauna, bottom trawls also cause a large quantum of bycatch and discards including sea grasses, benthic fauna, corals, larval stages, juveniles and sub-adults of whole marine fauna.

Physical impact

Different types of bottom trawl nets can stir up sediments disarraying benthic fauna, removing hard sediments such as coral rubbles and gastropod shells, breaking live corals, and changing the texture and structural heterogeneity of bottom substratum—an important factor contributing to the diversity of benthic fauna (Krost et al. 1990). Studies at the Gulf of Maine in the Northwest Atlantic Ocean have shown that bottom trawls and dredges altered the physical structure and complexity of benthic habitats (Auster et al. 1996). Physical structure was altered by direct removal of benthic fauna associated structures including sponges, hydrozoans, bryozoans, amphipod tubes, holothurians, shell aggregates, and sedimentary structures (e.g., sand waves, depressions). Reductions in habitat complexity may lead to increased predation on juveniles of harvested species and ultimately recruitment to the harvestable stock (Auster et al. 1996).

McConnaughey et al. (2005) carried out an examination of chronic trawling effects on soft-bottom benthos of the Eastern Bering Sea by sampling and analysing previously unfished (UF), and heavily fished (HF) areas. Using statistical methods they showed that species diversity of sedentary macro fauna (anemones, soft corals, sponges, bryozoans, ascidians), overall diversity, and niche breadth was high in UF areas.

Hiddink et al. (2006) developed linked 'state' and 'pressure' indicators that show the impact of bottom-trawling on benthic communities in the Southern North Sea (between England and Netherland). Based on bottom-trawling intensity in 2003, they showed that 53.5% of the Southern sea was trawled too frequently for biomass to reach 90% of its pristine benthic biomass. This is because the time taken for recovery to 90% of pristine biomass was estimated to be anywhere between 2.5 to 6 years (Hiddink et al. 2006).

According to NRC (2002), in the eastern Atlantic, the highest intensity of effort, based on rough estimates of the number of times a reporting area is swept, occurs in the fishing grounds of the Gulf of Mexico and New England regions. In contrast, bottom trawling in the mid-Atlantic, Pacific, and North Pacific regions is relatively light, with only one tow per year in some of these reporting areas. However, throughout the 1990s and into 2001 there were significant reductions in the intensity and spatial extent of bottom trawling in the USA through effort reductions, area closures, and gear restrictions instituted by fisheries managers in response to problems with declining fish stocks, bycatch, or interactions with endangered species.

Bio-geo chemical impact

Studies have shown that bottom trawling can affect carbon mineralisation and biogeochemical fluxes thereby destabilising sediment community structure and functions (Duplisea et al. 2001). Bottom trawling can cause an increase in turbidity reducing the level of photosynthesis in sea grass and algae. Further, it decreases dissolved oxygen content in the water column due to oxygen reduction by resuspended bacteria in sediments (Reimann and Hoffman 1991). Turbid sediment clouds can further change physio-chemical parameters of the water column (Main and Sanger 1990).

Bivalves and many other benthic organisms are filter feeders of which respiration is directly affected by resuspension of sediments from the dredging activity of bottom trawls (Caddy 1973). While there are no detailed studies on biogeochemical impact of bottom trawling in tropical waters, Thomas et al. (2004) reported an increase in water temperatures and nitrates, and decrease in dissolved oxygen, organic matter and organic carbon as immediate effects of the stirring action of bottom trawls along the Kerala (Southwest Indian) coast.

Benthic invertebrates are the major food source for many commercially exploited fish species, and have a significant role in supporting ecosystem processes such as bentho-pelagic coupling and nutrient cycling (Choi et al. 2004; Lohrer et al. 2004; Widdicombe et al. 2004).

Deng et al. (2005) showed that trawl vessel monitoring system (VMS) data can be used to examine trawl track, trawling intensity, and fish stock depletion due to bottom trawling. They showed that the catch per unit effort and cumulative catch in areas with highly aggravated trawl effort is not proportional to the overall target species biomass. Further, this high number of trawl efforts in productive areas has been shown to have a marked impact on its benthic fauna.

The impacts of frequent trawling are cumulative and vary among taxa (Poiner et al. 1998; Tanner 2003). Some benthic taxa such as sponges can be severely affected by a single bottom trawl effort, while other benthic fauna including algae, bryozoans, and pennatulacea may withstand several repeated trawling efforts, with the cumulative impact increasing with successive trawls (Poiner et al. 1998).

Wilson (1979) showed that bottom trawling breaks corals (i.e. Lophelia sp.), and unfortunately, corals die when they are not in contact with the substrate. Repeated trawling over the same coral patch can therefore eliminate coral species. The effects of systematic destruction by bottom trawlers on bryozoan beds in Tasman Bay, New Zealand which provides habitat for juvenile snapper (*Pargus auratus*) and tarakihi (*Nemadactylus macropterus*) have been reported by Saxton (1980) and Bradstock and Gordon (1983) whereby they noted a decline in juvenile fish with the removal of bryozoan beds.

Bottom trawling can also have an indirect impact on the fishing mortality on benthic fauna. For e.g. McLoughlin et al. (1991) reviewed the indirect fishing mortality rates on scallop beds by dredges in the Bass Strait in Australia. Nine months after the dredging, almost all scallop stock was destroyed due to a bacterial infection that resulted from decomposing scallops damaged by dredging (McLoughlin et al. 1991).

Sainsbury (1988) has reported changes in fish catch composition as a secondary impact of bottom trawling. Due to a significant reduction in sponges together with alcyonarians and gorgonians on the Australian Northwest shelf bottom trawling, *Lethrinus* and *Lutjanus* fish species which are frequent in habitats rich with large epibenthos have significantly declined. Consequently, this has caused an increase in the biomass of fish species *Nemipterus* and *Saurida* which are inhabited in the open sand areas. Thus, an overall decrease in species diversity due to bottom trawling has been reported in several studies. Also, continuous bottom trawling has changed the benthic fauna species composition and abundance of particular benthic invertebrates. For example, Reise (1982) reported an unusual increase in polychaetes in the Wadden Sea (in the North Sea) over a period of 112 years, as a long term impact of bottom trawling.

Changes to the sea bed by bottom trawling might affect the fisheries as most of the trophic levels are associated and interact with benthic environment and benthic fauna (Bradstock and Gordon 1983; Sainsbury 1988). Studies have shown that fish and benthic organisms were more prevalent before intensive bottom trawling began (Frid and Clark 2000; Greenstreet and Hall 1996).

In summary, a meta-analysis of 57 published studies by Collie et al. (2000) reveal that the common impacts of trawling and dredging on sea floor communities are, reduced habitat complexity, discernible changes in benthic communities, reduced productivity of benthic habitats, and a heightened vulnerability to fishing gear disturbance among fauna that live in low natural disturbance regimes.

3.5.3 Industrial Purse Seines with Artificial FADs

In the open ocean, many species including tunas associate with objects drifting on the surface, such as logs or branches. This is highly advantageous to purse seine fishing as floating objects aggregate sparsely distributed schools, are more easily spotted than tuna swimming freely beneath the surface, stabilise schools and reduce the speed at which they travel, making them comparatively easy to catch. Consequently, fishing around floating objects is associated with a higher successful haul, compared to targeting free swimming schools. In the 1970s, countries like The Maldives used artificial FADs made of natural substances like bamboo rafts and coconut leaves. Since the 1980s, industrial fishers have started using electronic buoys, and today these electronic buoys are equipped with echo sounders that transmit data on an hourly or daily basis about the fish biomass underneath.

The latest FADs reduce the fish searching time drastically, and allow for the catching of fish even in the dawn. Although FADs are evidently useful fishing tools, their use has been associated with several potential negative ecosystem impacts, including the catching of juvenile tunas, the bycatch of vulnerable non-target species (Bromhead et al. 2003), (Amande et al. 2010), and consequences of modification of the pelagic habitat on tuna biology. Currently there is little control on these FADs, and over capacity of FADs may lead to overexploitation of tuna and associated stocks (Fonteneau et al. 2013). Potential ecological impacts of FADs are reviewed in Dagorn et al. (2012).

FADs have facilitated extremely high catches of tuna in every ocean, including the Indian Ocean, thus causing a reduction in spawning stock biomass by overfishing and a loss in potential yield by catching smaller fish and reducing the number of large breeding individuals in the stock. Skipjack tuna (SJT) (*Katsuwonus pelamis*), yellowfin tuna (*Thunnus albacares*), and bigeye tuna (*Thunnus obesus*) are the main tuna species inhabiting in FADs. Skipjack tuna makes up the largest catch (5–82%) using this fishing practice across all four oceans (Dagorn et al. 2012). Although SJT is a highly fecund species with a high growth rate, given the current rate of over capitalisation on modern FADs, there is a concern of over exploitation of SJT stocks. Also, as juveniles and small fish of yellow fin tuna and bigeye tuna are abundant around FADs, purse seine catches around FADs can cause loss of potential yield through a reduction in the number of large spawning fish in the stock.

A much larger ecological impact associated with FADs is the by-catch of non-target species. A whole community of different fish species gather around FADs that consist from small fish to predatory large fish such as sharks (Bromhead et al. 2003; Romanov 2002). Sharks, rays, billfishes are caught by purse seining around FADs in a significant amount. As these species show slow growth rate, late maturity, and few progeny, their vulnerability to population decline is high. In addition, turtles also frequently entangle in hanging purse sein nets around FAD rafts (ghost fishing). Dagorn and colleagues (2012) show that the bycatch of purse seines around FADs are three to seven times higher than the bycatch in fishing free swimming schools. Another ecological impact is that by employing a large number of FADs and attracting tunas into these artificial habitats it will effect natural habitats, dispersal and migratory patterns of tunas in addition to exposing them to predators (ecological trap hypothesis) (Hallier and Gaertner 2008). This change in natural ecosystem and the impact from proliferated buoys is yet to be studied in depth.

3.6 Destructive Fishing Gears and Practices—Case Studies

3.6.1 Small Bottom Trawls Over Capitalisation, Overfishing, and Poaching—Southern India

Palk Bay and Gulf of Mannar, situated between Sri Lanka and Southeast India, have gained huge attention in regional and even in global media in recent times due to large scale poaching and high intensity illegal fishing in Sri Lankan waters by Southern Indian (Tamil Nadu State) large bottom trawling fishing fleet.

In early 1970s mechanised bottom trawlers were introduced to Tamil Nadu (TN) state through DANIDA (Danish International Development Agency) and SIDA (Swedish International Development Agency) funded Bay of Bengal Programme (BOBP 1987). During this time, bottom trawl fleet in Southern India (Tamil Nadu state) were drastically increased due to very generous incentives by the Tamil Nadu state government which enabled trawlers to be economically viable. The number of mechanized vessels increased by more than eight times along the SE Indian coast during 1961–1998, which is higher than the increase along the NW and SW Indian coasts (Vivekanandan et al. 2005). In the 1970s a marked revolution occurred in fishing in the Palk Bay with the introduction of mechanised bottom trawlers (so called blue revolution) and the emergence of prawn fishery targeting international market (so called pink gold rush) (Kurien 1978). This led to an increase of over 400% in the total fish catch in Tamil Nadu state from the Palk Bay (Kumaraguru et al. 2008). Currently, the bottom trawl fishing fleet in Tamil Nadu state itself represents the second largest trawl fleet (16.4%) of total bottom trawl fishing fleet in India (Marine Fisheries Census-India 2010).

Fishery resources were drastically depleted on the Indian side of the Palk Bay and Gulf of Mannar (GoM) due to intensive bottom trawling of this over capitalised fishing fleet (Kumaraguru et al. 2008; Venkataraman et al. 2002; Hettiarachchi 2007). Taking the fishing restriction on Sri Lankan fishermen in their territory as an advantage, Southern Indian bottom trawling fishing fleet began poaching into Sri Lankan waters looking for high market value shrimps, green tiger prawns (*Penaeus semisulcatus*) and sea cucumbers on the Sri Lankan side of the Palk Bay and adjacent Sri Lankan waters. By the year 2005, there were an estimated 5300 bottom trawls (CMFRI 2005) in the Tamil Nadu state itself, with over 60% of them regularly fishing in Palk Bay and adjacent waters on the Sri Lankan side (Vivekanandan 2004). According to conserved estimations, the annual loss to Sri Lanka from Indian poaching is approximately USD 47 million (Amarasinghe 2011) to USD 79 million (Kumara 2014) and for this reason poaching has become a large-scale socio-political problem between Sri Lanka and India. In addition to these large economic losses faced by Sri Lanka, bottom trawlers also cause detrimental damage to sea grass beds, coral reefs, and associated marine fauna of Palk Bay and GoM by scraping the ocean floor and catching all the marine creatures indiscriminately. Further, TN trawlers use trammel nets which are detrimental to marine fauna as these nets consist of three layers with different mesh sizes in each which consequentially trap the whole fauna without any selection. Recent developments which aim to increase the catch quickly involve the use of paired trawlers with larger trawl nets to increase the sweeping area and power. A conservative estimate is that over 5000 bottom trawls have been fishing on a single day in Palk bay, a very small area of approximately 17,000 km^2 (Sivalingam 2005), and adjacent waters (Vivekanandan 2004). Such a high intensive, continuous bottom trawling effort should have caused significant sediment imbalance and detrimental chronic impact on the eco system of Palk Bay and adjacent waters. According to some estimates, barely one third of the fishing pressure of existing fishing fleet can be sustained in the Palk Bay region (Pramod 2010).

Currently, bottom trawling in the Palk Bay mainly targets shrimp fishing which has a high market value. Shrimp trawling is one of the most indiscriminate kinds of fishing because the small mesh used to retain the shrimp allows few other animals to escape (Kelleher 2005). In the Palk Bay, small cod end mesh size of trawl fishing leads to the capture of small body size species, juvenile prawns, and sub adults hence resulting in reduction of average size of prawns.

The second largest problem from the bottom trawl fleet in this region is the quantum of bycatch and discards. According to Pramod (2010), discards at sea by mechanised trawlers in Tamil Nadu state per year ranges from 179,274 to 246,665 mt (average 212,969 mt). In the Palk Bay region, due to the sweeping of the sea bed by bottom trawl nets, 83% (Manickam et al. 1987) to 94% (Kumara 2014) of the catch makes by bycatch. This bycatch mainly consists of sea grasses and sea weeds, juvenile stages of all kinds of prawns and fish in addition to eggs and larval forms of all kinds of fish.

In recent years, fish catch composition in the Palk Bay and along Tamil Nadu coast has changed drastically. In recent years, 20% of fish catch consisted of low trophic level oil sardines (*Sardinella longiceps*), while the second and third largest commodities were silver bellies (*Leiognathus* spp.), and other sardines (*Sardinella* spp.) respectively (CMFRI 2014). It is suggested that overexploitation of the top predators have caused rapid increase of low trophic level species like sardines.

According to Davies et al. (2009), the Indian trawl bycatch represents 56.3% of the Indian estimated total marine catch while one third of the non-shrimp trawl catch (which is 600,000 mt) was discarded. Menon and Pillai (1996) assessed the bottom trawl bycatch in Indian waters and found that approximately 30% of the bycatch is discarded, and approximately 12% of the by-catch is a heterogeneous species mix belonging to the bottom fauna. Further, bottom trawl commonly catches large quantities of juveniles and sub adults of a variety of demersal fishes, for example sciaenids, catfishes, flatfishes, flatheads, silver bellies, perches, whitefish, promfets, of which most are discarded (Menon and Pillai 1996). This has often led to recruitment overfishing and conservation problems while nemipterids, saurids, and flatheads have faced overexploitation.

Central Marine Fisheries Research Institute (CMFRI), India recommends a reduction in the fishing pressure of mechanised trawlers by 40% in order to achieve a healthy fishery. Moreover, sharks, skates, and rays stock status in TN are stated as depleted, declined, and less abundant, respectively, and may be due to high levels of juvenile bycatch by trawlers (CMFRI 2014).

3.6.2 Gillnets

Gillnets have been used in the Asia and Pacific marine fisheries for centuries. In the past, gill net fishery was an artisanal small scale fishery that was limited to inshore areas. Fishermen used multi-filament gillnets made with degradable material. With the development of technology of gillnet fishery however, mono-filament invisible gill nets were produced which are not so easily degradable. Gillnets in general catch not only fish that fit to mesh size, but also small juveniles—especially when the mesh size is small. Mono-filament invisible gill nets add more bycatch to this end. Trawls targeting tuna fish in the Asia and Pacific use gillnets extensively, hence adding a considerable amount of bycatch including juveniles. However, as quota and target fishery regulations are not strictly implemented in this region, fishermen do not discard non-target fish catch. In addition, almost all species are edible in the Asia-Pacific region, and hence non-target catch and bycatch has a market.

With the industrialisation of gill net fishery, very long gill nets especially drifting gillnets have been manufactured in developed countries. In 1992, the United Nations adopted a global moratorium on large-scale drift net fishing on the high seas. Despite this moratorium, in the Mediterranean, drift net fishing uses 10–12 km long drift nets for swordfish and albacore, and takes in a lot of cetaceans as a bycatch. Currently, in the European Union, drift nets that are longer than 25 km are banned.

3.6.3 Explosives and Diving

Dynamiting has been employed by artisanal fishers in the Asian and African regions illegally, especially to catch fishes associated with coral reefs and mangroves. Dynamites that are manufactured for use in mines or quarries are modified by fishers to make sure that these 'fish bombs' explode in the middle of the water column. Dynamiting indiscriminately kill every living being in the vicinity of the blast in the water column. Not only are a few square meters of corals destroyed in every explosion, but also the shock waves kills most fishes, and bottom dwelling invertebrates within a radius of 50 m or more (Moore and Jennings 2008). The problem with this approach is that most of the fish implicated in the explosion are never collected by fishermen as they sink to the bottom. Dynamiting has been intensively carried out in some parts of the Southeast Asian coral reef systems converting them into rubbles, and reducing the diversity of fauna to a large extent.

Diving and collecting fish resources is a very selective fishing method, and is a healthy fishing method when it is carried out in a sustainable way. However, today diving is carried out at very large scales to collect marine ornamental reef fishes, associated ornamental invertebrates, and ornamented soft corals. This overexploitation results in a loss of marine biodiversity, imbalance in functioning of marine ecosystem, and risk of extinction of commercially highly demanding species.

Some countries use modern poisons such as sodium cyanide, bleach, and DDT to extract fishes from tropical coral reef systems or mangroves. According to estimates of World Wild life Fund, over 6000 divers use 150,000 kg of poison (mainly cyanide) that affects 33 million coral heads annually (Moore and Jennings 2008). For example, in the Phillippines these poisons are used extensively to illegally catch humphead and Napolean wrasses which are highly demanding delicacies in Southeast Asia.

3.6.4 Ring Nets With Light as a FAD

Some parts of South Asia use a combination of diving, and a modified ring net to illegally surround ecologically rich coral reef systems. These nets ('Laila' and 'Surukku') are functionally similar to a small purse seine, and have varying mesh sizes depending on the type of fish caught or location that is employed. Commonly, these nets are used to surround egg spawning large coral reef fish assemblages. On occasion, 20–25 fishermen and scuba divers are put in operation using several small trawler boats and in some instances even dynamites are used in combination with this destructive fishing practice. This is a sweeping of spawning adult population of reef fishes. Sometimes a variation of these nets with small mesh sizes are used to catch very small prey fish populations in shallow sea. Another variation of these nets are used to surround tuna schools in the pelagic sea, and sometimes in the night with a combination of powerful lights to attract all types of large and small fish.

3.6.5 Beach Seine

Beach seines have been used around the world for thousands of years as an artisanal fishery practice. Beach seines are very long surrounding nets that are operated in inshore areas at a particular season of the year. Beach seines mainly target the adult fish that migrate towards inshore areas for spawning. As beach seines scrape the bottom, and catch any fish indiscriminately, it is not considered to be an environment friendly fishing method. However, as this method is used only at a particular season of the year, and is operated in small scale, there are enough resources and time left to recover and replenish the inshore marine community.

3.6.6 Industrial Bottom Trawls and Dredges

Prominent Industrial bottom trawl fleets have been operating in Northern Australia, Thailand, North Pacific, especially for shrimp industry. Dredges are used to catch blue crabs in the mid-Atlantic region during the winter, and with two dredges used in a tandem (National Research Council 2002). The 10 cm long teeth of the front blade dig the crabs out of the bottom and the same gear is used in Chesapeake Bay to catch Whelk in summer and mussels in fall.

There are bottom trawl fisheries for demersal fisheries on all U.S. coasts. In the Northeast, 15–50 m vessels use small mesh nets to catch northern shrimp, silver hake, butterfish, and squid. Large mesh trawls catch cod, haddock, flounders, and other large species. These trawls typically are rigged with long ground wires that create sand clouds on the sea bed, herding the fish into the trawl mouth. In the Southeast and Gulf coast areas, small mesh trawls catch shrimp. Southern shrimp trawl vessels tow two to four trawls from large booms extended from each side of the vessel. On the West coast, stern trawlers catch shrimp, flatfish, and rockfish species. Factory trawlers from 50–100 m long, catch, process, and freeze their products on board.

After the passing of the Magnuson-Stevens Act in 1976, the USA promoted the expansion and efficiency of the domestic fishing fleet. This included changes in trawl and dredge gear that increased the capacity of fisheries to cover large areas and to reach deeper and rougher habitats. Larger, more powerful boats pulling gear with wider sweeps increased the amount of area potentially affected. New gear, however, has been designed specifically to reduce habitat damage, for example, the raised footrope trawl used in the Gulf of Maine whiting fishery (NRC 2002).

3.6.7 *Industrial Purse Seines With Artificial FADs for Tuna in the Western Indian Ocean*

Mozambique Channel is rich with natural FADs due to the large number of rivers opening to the sea. As the fishing productivity was high around these natural fish aggregates, industrial purse seiners such as the French and Spanish fleet rapidly increased the number of artificial FADs through 1990s–2000s. To match the high fishing capacity around increasing artificial FADs, much bigger purse seiners such as 'super-seiners' (>2000 gross tonnage) and even 'super super-seiners' (>3500 gross tonnage) were built (Davies et al. 2014). These very large seiners become more reliant on proliferated FADs to be cost efficient and remain profitable (Campling 2012). Figure 3.2 shows that Spain, The Maldives, and France accounts for the largest tuna catches respectively, in the Western Indian Ocean for last three decades among other coastal countries in this region. It should be noted, however, that this tuna catch of Spain and France is by industrial purse seines, while The Maldives' main fishing method is pole and line, and trolling which are considered as sustainable fishing methods.

3.6.8 *Over Effort, Overexploitation of Fleet, Bycatch, and Discards in Industrial Fishing*

Definitions of by catch and discards vary considerably from region to region depending on the cultural and marine food habit differences, types of fishing practices and associated regulations and laws. Amongst these regional differences,

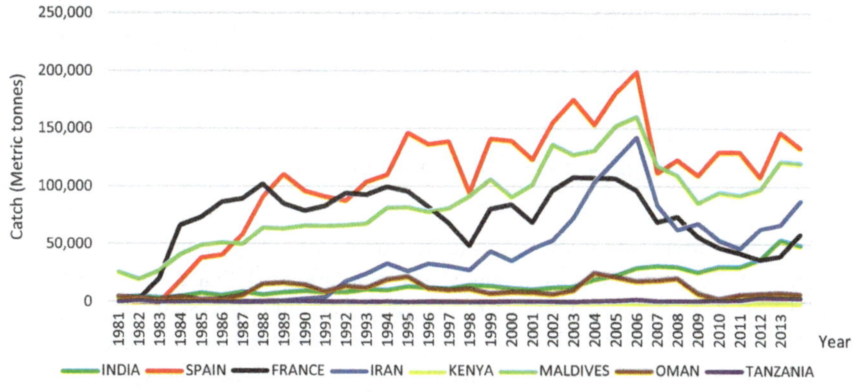

Fig. 3.2 Tuna catch (total of skipjack tuna, yellowfin tuna, and bigeye tuna) in the Western Indian Ocean (FAO fishing area 51) by major fishing countries. *Data source* Indian Ocean Tuna Commission data set, 2016

McCaughran (1992) has provided a comprehensive description on the definition of terms for by catch and discards.

It is known that the discards in tropical Asia are much less compared to other temperate regions in the world. The reason behind this is that most of the untargeted species caught in marine fishery in Asia are edible for the people in this region. Comparably, in the Northwest Pacific, discards account for just one third of the total (global) discards which is 9.1 million tonnes mainly from cod and shrimp fishery. This is followed by 3.7 million tonnes discarded in the Northeast Atlantic Ocean. The third largest discards are from the West Central Pacific, largely from shrimp fisheries (Hall 1999). Interestingly, Southeast Pacific ranks fourth, not because of the fisheries (anchoveta and pilchards which has 1–3% discards) but simply due to the enormous size of the total catch. Overall, by catch and discards are relatively low in tropical regions. In the tropics, the main by catch is from shrimp trawl fishery. Alverson (1994) study provides a detailed global view of bycatch and discards of various types of fisheries around the world.

While the discards in the tropics are dominated by small body size adult fish, in the temperate regions, discard is dominated by sub-legal and legal sizes of commercially importent larger bodied fishes which are part of the future commercial catch (Hall 1999). The shrimp fisheries have the highest discards on both, a species and gear type basis. Crab fisheries have the second highest discards.

Serious public ethical concerns are raised around the by catch of sharks, marine mammals, turtles and marine birds. Sharks are caught in high numbers in coastal and high seas long lines, purse seine around logs and FADs, and drift net fisheries for tuna and bill fish. Among shark by catch in high seas fleet, blue shark, oceanic white tip, silky shark, short fin macko, and thresher shark are dominant. Shark by catch is detrimental as sharks show slow growth, late age at maturity, and low fecundity. Marine mammal by catch including dolphins, and dugongs are higher for large gill nets that are many kilometres long.

Marine discards have caused some indirect ecological problems, such as increase of some marine bird populations, and benthic scavengers. A good example is Audion's gull (*Larus audoinii*) an endemic sea bird in the Mediterranean region. In the early 1970s its population was very small and threatened (Cramp et al. 1985). Because of discarding by local trawler fleets however this bird colony become the world's largest bird colony (Ruiz et al. 1996). On the other hand, if this discarding practice is stopped, it will be detrimental to these bird populations as they rely very much on these discards.

3.6.9 Quota System Policy for One Species

As described previously, many countries in Europe, and American continent follow a one target species quota system policy. The non-target fish that are unintentionally caught have to be discarded although it is a commercially important edible species. Further, small trawler boats are not required to keep records of such discards. This policy creates a large wastage of marine resources and can cause depletion of marine resources.

3.6.10 Transferring of Flagship and Associated IUU and Fisheries Management Problems

Transferability of flagship to foreign countries and individual transferable quotas can create management problems for fisheries and illegal, unreported and unregulated (IUU) fishery problems due to difficulties in surveillance and implementing regulations. For example, recently Sri Lanka lost its fish exports quota to the European Union due to an IUU issue associated with transferred flagship from Sri Lanka to a foreign country. Similarly, there are uncontrolled foreign fleets in the western Indian Ocean that employed under transferred flagships. There are some cases where transferred flagship vessels tend to catch over the quota and then abandon some of the fishing nets purposely in the sea to pretend a lower catch. These abandoned fishing gear cause ghost fishing and many other devastating impacts on marine fauna.

3.6.11 Cultural and Regional Issues with Definition of Discards, and Eating Habits

Definitions and understanding of bycatch and discards vary to a considerable degree from region to region in the world. Reasons behind these are cultural and food habit differences, differences in fisheries policies and regulations, and differences in economies and fishery markets. In the tropical region or developing countries, there are no strict fisheries quota regulations, and generally fishers can catch any fish with no need to discard 'untargeted' fish. On the other hand, almost all types of fish caught in the tropical region are edible to the people living in this region. Further, people within tropical and developing countries eat even smaller fish such as anchoveta and shrimps. There is a market therefore for undersized incidental catch, and even the non-edible bycatch that comes from bottom trawling in this region has a market for animal food in the poultry industry. On the other hand, in the temperate region and developed countries, discards are high due to single species quota policy, market preferences, and food habits. Because of these reasons, the quantum of discards are minimum in both the tropical region and developing countries compared to the high level of discards in temperate region fisheries and developed countries.

3.7 Towards Sustainability

3.7.1 Sustainable Fishing Gears and Practices

The main characteristics of a sustainable fishery should be that the fishing gear and the method to catch fish is highly selective, there is no bycatch or minimum level bycatch, there is no sweeping of populations, no catching of juveniles or undersized fish, no over effort and over exploitation, and no or minimal impact on the marine environment and other fauna. Traditional artisanal fisheries generally fulfil all or most of these characteristics as they have evolved over thousands of years with the vast knowledge and experience on the marine environment, ecology, biology, behaviour and life history of most of the marine animals. In addition, traditional artisanal fishers fish for their livelihood, not for large profit making as we see in the current industrial marine fishery. As described previously, pole and line or live bait fishery can be considered as the number one sustainable fishing method as it fulfils all the above factors of a sustainable fishery. Trolling also can be considered as a sustainable fishing method. Trolling aims to mimic the actions of moving bait as the lines move very fast. This mimic attracts adult fish that come to surface searching for food. Thus trolling is highly selective, and mainly target adult tuna and bill fishes. Fish caught by trolling or pole and line is taken on board quickly, and has no any damage compared to fish caught by a trawl net or purse seine. Therefore, the meat quality and post-harvest quality of these fish is very high. Traps and pots are also a sustainable fishing method with a high selectivity. The main advantage of traps and pots are that any undersized and non-target other species which are caught can be released back to the sea alive, without any harm to animal. Pole and line fishery, trolling, and traps/pots do not sweep populations and leave a portion of the adult population for spawning and continuity of the fish stock. Further, and most importantly, these three fishing methods shows the highest catch quality as physical damage, frightening, and loss of energy is minimal for the animal in these fishing methods. Comparatively, purse seines and trawl nets surround or shoal a large fish school into a large net. Crowded and trapped fish school are more likely to get frightened, exhausted and damaged because of the fish net and confinement, and in turn lose much of their energy while trying to escape. The amount of Adenosine Tri phosphate (ATP) or amount of energy stored in caught fish is directly related to its 'freshness' and post-harvest quality (Sikorski 1990). Thus, fish caught from pole and line fishery, trolling, and traps/pots are in a high post-harvest quality, as they are caught in a more humane way.

3.7.2 Prohibition of Some Fishing Gears and Fishing Practices

Severely destructive fishing methods such as dynamiting, poisoning, and trammel nets, which are already banned, should be thoroughly implemented and enforced. It is clear that a lack of resources, poverty, corruption, and intervening of politics in these countries are the reasons behind the lack of enforcement of fishery laws and regulations in the Asia, Africa and Pacific region. Secondly, regulations of mesh sizes of gill nets used in pelagic sea targeting tuna should be implemented and enforced as gill nets with small mesh sizes increase the bycatch significantly.

In industrial fisheries, bottom trawl nets, purse seines and purse seining around heavily proliferated FADs are severely destructive fishing methods due to very large bycatches, under-sized fish and huge quantities of discards, and damage to the marine eco system. The best solution therefore would be a ban of these destructive fishing methods. The best long term sustainable fishing practice would be pole and line fishery targeting free swimming schools around FADs that are well controlled in numbers. Bottom trawling, especially in the tropical regions, result in much higher damage to the benthic environment and benthic fauna as the marine biodiversity is several fold higher in tropical regions compared to temperate regions. To this end, setting up of artificial reefs is a firm and long term solution to tackle the bottom trawling problem, and also to recover and enhance already depleted marine populations in the tropical region. There should be a wider, continuous awareness and education program for fishermen together with the fishery regulation enforcement. Also, over capitalised fishing fleet should be reduced through a government funded buyback program.

Today the over exploitation of marine fish by the super industrial sector in developed countries, as evidenced by their catch, bycatch and discards, are several fold high compared to the marine fish catch, bycatch, and discards of artisanal fisheries in developing countries. It is well known that the main purpose of marine fisheries in developing countries is to secure the livelihood of millions of coastal population and support the feeding of hundreds of millions of people in coastal countries in Asia, Africa, and the Pacific. On the other hand, the main purpose of industrial fisheries carried out in international waters by developed nations is mere profit making. For example, in European Community landings, 25% of fish catch is from outside home waters (Moore and Jennings 2008). As a result, developed nations have access to a variety of marine food commodities and high consumption. Thus, per capita consumption of captured marine food in developed nations is several fold higher compared to that of developing nations in coastal countries.

3.7.3 Reduction of Fishing Effort

Fishing at sustainable levels of assessed stocks has decreased from 90% in 1974 to 71.2% in 2011. Thus, 28.8% of fish stocks were estimated as overfished and fully fished stocks accounted for 61.3% in 2011 (FAO 2014).

Effort reduction is the cornerstone of managing the effects of fishing, including, but not limited to, effects on habitat. Today, the main problem in the world marine fishery, whether it is industrial or artisanal, is the over capitalised fishing fleet. Over capitalised fishing fleet cause over effort, over exploitation, and thereby very high bycatch and discards due to badly designed fishing gears with destructive fishing methods. In addition to very high bycatch, the impact on the physical marine environment and marine fauna by these destructive fishing gears and methods is enormous. Many industrial fisheries around the world were pushed to limit their fishing fleet and fishing voyages due to reduced catch per unit effort and hence unprofitable business. For example, Japan and Taiwan have reduced their long line fishery due to reduced catch per unit effort.

Management of effort in dredge fisheries in USA is generally achieved with time and area closures and with restrictions on the size (blade width and weight) of dredges, the number of dredges, and the size and horse-power of the towing vessels (NRC 2002).

3.7.4 Reduction of Overcapitalized Fishing Fleet

A very good example of overcapitalized fishing fleet due to generous government incentives, lack of regulations and regional politics is the bottom trawl fleet in Tamil Nadu State, Southern India. As described previously, only one third of over 5000 of these trawl boats have sufficient sea area for sustainable fishing (Pramod 2010). Due to scarcity of marine fish in Tamil Nadu waters due to over exploitation, and due to extended closed seasons and closed areas by the Indian government, Tamil Nadu bottom trawlers are compelled to poach into neighbouring countries to remain profitable. Currently, the Indian government is implementing a 45 day per year fishing ban. This policy, however, has limited impact as fishermen move to other states to avoid the fish ban season, or alternatively, they overfish just before or after the banned season. Scientists and policy makers in the region urge India to implement a trawl buyback program.

3.7.5 Extended Closed Seasons, Extended Closed Areas, and Marine Protected Areas/Marine Reserves

Closed areas are necessary to protect a range of vulnerable, representative habitats. Closures are particularly useful for protecting biogenic habitats (corals, bryozoans, hydroids, sponges, seagrass beds) that are disturbed by even minimal fishing efforts.

The optimal combination of these management approaches will depend on the characteristics of the ecosystem and the fishery-habitat type, resident species, frequency and distribution of fishing effort, gear type and usage, and the socioeconomics of the fishery. Each characteristic should be considered during development of management plans for mitigating the impacts of fishing.

Extended closed seasons have been implemented around the world to synchronise with breeding, and spawning times of marine animals. Together with extended closed seasons, extended closed areas are also implemented that are major breeding, spawning or nursery grounds, major migratory paths, and sometimes areas with bio diversity hot spots. A further step to closed areas is marine reserves and marine protected areas that are significantly rich in marine biodiversity. Extended closed seasons and marine reserves have been proven, to a certain extent, to help significantly in mitigating the effects of biodiversity loss caused by fisheries (Halpern 2003).

3.7.6 Change of Policy of Giving Quota for a Single Species

Today, a significant quantum of discards is caused by a weakness in fishery policies within countries that give quota for a single species. Because of this policy, fishermen have to discard non-targeted species even though they are edible and/or marketable. The reality is that no one can catch a single fish species, and fish are caught as multiple species. If the fish caught are undersized, they should be released alive, if possible. Otherwise, if the fish caught are edible or marketable, those fish should be retained, and hence the fishery quota should take into consideration this 'non-target' catch. Also, it should be compulsory to keep records of all targeted, untargeted and discarded fish. In this way, a large loss of marine fishery resources can be prevented.

3.7.7 Modification of Fishing Gear and Practices

The design of some fishing gears can be modified or improved to minimise bycatch and the impact to the marine environment to a certain extent. For example, some trawl nets are equipped with by-catch reduction devices that involve a metal mesh with large holes which fits into the cod end of shrimp trawls to let finfish escape. With this modification, by-catch was reduced by 50% in US shrimp trawls, but loss of shrimp at the same time through this made the modification unpopular (Moore and Jennings 2008). Otter trawls can be modified by varying the size and the shape of the mesh panels used in different parts of the net and the circumference and length of the cod end. Mid-water trawls, the largest trawl fishery in USA, were able to reduce juvenile wall eye Pollock by 75% by using a single layer of mesh instead of two (Moore and Jennings 2008). These escaping fish, however, suffer damages due to trapping and close confinement, and die easily or are prone to predators, according to some studies (Chopin and Arimoto 1995). In a similar way, Turtle excluding devices and cetacean excluding devices are commonly included within USA and Australian trawls now. Purse seiners implement a program to rescue dolphins trapped in large purse seines. In long line fishery, a large number of sea gulls become casualties by attempting to feed on the bait on the hooks. This happens during day time operations at which sea gulls are active. By operating the fishery in the night time, the impact to sea gulls can be stopped. Industrial gill nets that are larger than 10 km should be banned and strictly enforced as such a long net can kill many sharks, cetaceans and sea birds. These fishing devices can be fitted with acoustic devices that can generate different types of sound frequencies to avoid large animals such as cetaceans. This type of research should be encouraged to find whether such devices can be used routinely in commercial fisheries.

3.7.8 Lessons from Traditional Cultures

It is high time to go back to traditional sustainable fishing methods such as pole and line fishery, trolling, and traps to sustain marine resources for future generations. On the other hand, fishermen's knowledge and experience should be used to study gear impacts and to develop new gear technology. Their active engagement in research will help ensure that mitigation strategies are practical, enforceable, and acceptable to the fishing community (NRC 2002).

Tuna fish are perhaps the most economically important marine fisheries around the world, but face a major threat to sustainability. Pole and line fishery should replace at least a portion of purse seining of tunas, especially around FADs, as these industrial purse seining are not ecologically sustainable. Tuna fish caught from pole and line fishery is getting popular among consumers in developed countries, who demand sea food certification schemes. Pole and line fishery has been practised in the Indian Ocean, developing coastal countries especially, Maldive Islands, Sri Lanka, and some Eastern African countries for thousands of years. Tropical Islands of the Pacific Ocean have also being practicing this sustainable fishing method for centuries. It is therefore important to recognise this fishing method as the most sustainable fishing method for tuna fishery, by developed nations, with the help of world leading organisations such as FAO. There should be technology transfer programs between countries and joint venture programs preferably facilitated through United Nations and FAO. Current major constraints to the wide spread of this sustainable fishing method is the lack of recognition among developed industrial fishing nations, and lack of facilitation and encouragement by worldwide organisations such as FAO to adopt and practise this fishing method in industrial tuna fisheries. Some challenges to this end are finding sufficient amount of bait fish, and training fishermen to acquire these fishing skills.

3.7.9 Ecosystem Based Fisheries Management and Small Scale Regional Fisheries Management

Large marine ecosystems (LMEs) are regions of the world's oceans that includes coastal areas, river basins and estuaries from the land area to the seaward boundaries of continental shelves and major ocean current systems. The US National Oceanic and Atmospheric Administration (NOAA) has identified 64 such LMEs around the world for conservation purposes enabling ecosystem-based fisheries management as these LMEs produce 95% of the world's annual marine fishery biomass yields. Many research are carrying out seeking feasibility for Large Marine Ecosystems (LME) as regional units for the implementation of management actions leading to sustained and predictable yields of living resources (Hall 1999; Gable 2004). Small scale regional fisheries management is a much better approach in fisheries management as local knowledge on the local marine ecosystem can be used together with the participation of local fisher folks.

3.7.10 Role of Conservation Organisations and Governments

Environmental certification of fishing practices and fish products

The latest satellite technology can be used to monitor fishing vessels for IUU activities that cross international boundaries, fishing in closed areas or marine reserves. This satellite vessel tracking technology has been already implemented in North America, Europe, Southern Ocean, and a number of developing countries.

Fish products can be traced back to their source of origin or fish stock using genetic technology. Accurate species identification, population assignment and supply chain traceability using this technology will provide a great support in enforcement of fishery regulations (Ogden 2008).

Sustainable sea food guides

Today, consumer awareness on sustainably caught fish is increasing. European community countries are very strong on this aspect, and as a result, tuna caught in purse seines around FADs are becoming unpopular to buy. These countries are demanding sustainably sourced fish and therefore sea food certification schemes such as Marine Stewardship Council (MSC) are popular (Jaffry et al. 2016). Further, some studies have shown that MSC-certified seafood is three to five times less likely to be subject to harmful fishing than uncertified seafood (Gutiérrez et al. 2012).

The US Court of International Trade recently imposed a ban on importing prawns from 52 countries that do not have a program to facilitate escaping of turtles from shrimp trawl nets or a turtle conservation program (Moore and Jennings 2008).

References

Aish A, Trent S, Williams J (2003) Squandering the Seas: How shrimp trawling is threatening ecological integrity and food security around the world. Environmental Justice Foundation, London, UK

Alverson DL (1994) A global assessment of fisheries bycatch and discards (no. 339). Food & Agriculture Organisation

Amandè MJ, Ariz J, Chassot E, de Molina AD, Gaertner D, Murua H, Chavance P (2010) Bycatch of the European purse seine tuna fishery in the Atlantic Ocean for the 2003–2007 period. Aquat Living Resour 23(4):353–362

Amarasinghe O (2011) Fisheries conflict in Palk Bay: is there a way out? From a Sri Lankan view point. Working report. University of Ruhuna

Auster PJ, Malatesta RJ, Langton RW, Watting L, Valentine PC, Donaldson CLS, Babb WG (1996) The impacts of mobile fishing gear on seafloor habitats in the Gulf of Maine (Northwest Atlantic): implications for conservation of fish populations. Rev in Fish Sci 4(2):185–202

BOBP (1987) High-opening bottom trawling in Tamil Nadu, Gujarat and Orissa: a summary of effort and impact. BOBP/REP/37. Madras, India

Bradstock M, Gordon DP (1983) Coral-like bryozoan growths in Tasman Bay, and their protection to conserve commercial fish stocks. NZ J Mar Freshwat Res 17(2):159–163

Bromhead D, Foster J, Attard R, Findlay J, Kalish J (2003) A review of the impacts of fish aggregating devices (FADs) on tuna fisheries. Final report to the fisheries resources research fund. Bureau of Rural Sciences, Canberra, ACT, Australia

Caddy JF (1973) Under water observations on tracks of dredges and trawls: some effects of dredging on a scallop ground. J Fish Res Bd Can 30:173–180

Campling L (2012) The tuna 'commodity frontier': business strategies and environment in the industrial tuna fisheries of the Western Indian Ocean. J Agrarian Change 12(2–3):252–278

Choi JS, Frank KT, Leggett WC, Drinkwater K (2004) Transition to an alternate state in a continental shelf ecosystem. Can J Fish Aquat Sci 61(4):505–510

Chong KC, Dwiponggo A, Ilyas S, Martosubroto P (1987) Some experiences and highlights of the Indonesian trawl ban: bioeconomics and socioeconomics. In: Indo Pacific fisheries commission, Papers presented at the symposium on exploitation and management of marine fishery resources in Southeast Asia held in conjunction with the twenty-second session of the Indo-Pacific fishery commission, Darwin, Australia, pp 16–26

Chopin FS, Arimoto T (1995) The condition of fish escaping from fishing gears—a review. Fish Res 21(3):315–327

CMFRI (2005) Central marine fisheries research institute. Annual report, India, p 2005

CMFRI (2014) Central marine fisheries research institute. Annual report, India, p 2014

Collie JS, Escanero GA, Valentine PC (2000) Photographic evaluation of the impacts of bottom fishing on benthic epifauna. ICES J Mar Sci J du Conseil 57(4):987–1001

Cramp S, Brooks DJ, Dunn E, Gillmor R, Hall-Craggs J, Hollom PAD, Simmons KEL (1985) The birds of Western Palearctic. vol 4. Oxford University Press

Dagorn L, Holland KN, Restrepo V, Moreno G (2012) Is it good or bad to fish with FADs? What are the real impacts of the use of drifting FADs on pelagic marine ecosystems? Fish Fish 14 (3):391–415

Davies RWD, Cripps SJ, Nickson A, Porter G (2009) Defining and estimating global marine fisheries bycatch. Mar Policy 33(4):661–672

Davies TK, Mees CC, Milner-Gulland EJ (2014) The past, present and future use of drifting fish aggregating devices (FADs) in the Indian Ocean. Mar Policy 45:163–170

Deng R, Dichmont C, Milton D, Haywood M, Vance D, Hall N, Die D (2005) Can vessel monitoring system data also be used to study trawling intensity and population depletion? The example of Australia's northern prawn fishery. Can J Fish Aquat Sci 62(3):611–622

Duplisea DE, Jennings S, Malcolm SJ, Parker R, Sivyer DB (2001) Modelling potential impacts of bottom trawl fisheries on soft sediment biogeochemistry in the North Sea. Geochem Trans 2 (14):112–117

Poiner I, Glaister J, Pitcher R, Burridge C, Wassenberg T, Gribble N, Elli, N (1998) Final report on effects of trawling in the far northern section of the great barrier reef: 1991–1996. CSIRO Marine Research, Cleveland, Australia

FAO (2014) The state of world fisheries and aquaculture. Food and Agricultural Organisation of the United Nations, Rome

Fonteneau A, Chassot E, Bodin N (2013) Global spatio-temporal patterns in tropical tuna purse seine fisheries on drifting fish aggregating devices (FADs): taking a historical perspective to inform current challenges. Aquat Living Resour 26(1):37–48

Frid CLJ, Clark RA (2000) Long-term changes in North Sea benthos: discerning the role of fisheries. Effects of fishing on non-target species and habitats, pp 198–216

Gable F (2004) A large marine ecosystem approach to fisheries management and sustainability: linkages and concepts towards best practices (p 84). US Department of Commerce, National Oceanic and Atmospheric Administration, National Marine Fisheries Service, Northeast Region, Northeast Fisheries Science Center

Gibbs PJ, Collins AJ, Collett LC (1980) Effect of otter prawn trawling on the macrobenthos of a sandy substratum in a New South Wales estuary. Mar Freshw Res 31(4):509–516

Goudey C, Loverich G (1987) Reducing the bottom impact of Alaskan groundfish trawls. In OCEANS'87, pp. 632–637

Greenstreet SP, Hall SJ (1996) Fishing and the ground-fish assemblage structure in the north-western North Sea: an analysis of long-term and spatial trends. J Anim Ecol, 577–598

Gutiérrez NL, Valencia SR, Branch TA, Agnew DJ, Baum JK, Bianchi PL et al (2012) Eco-label conveys reliable information on fish stock health to seafood consumers. PLoS ONE 7(8): e43765. https://doi.org/10.1371/journal.pone.0043765

Hall SJ (1999) The effects of fishing on marine ecosystems and communities. Blackewel Science Ltd

Hallier JP, Gaertner D (2008) Drifting fish aggregation devices could act as an ecological trap for tropical tuna species. Mar Ecol Prog Ser 353:255–264

Halpern B S (2003) The impact of marine reserves: do reserves work and does reserve size matter? Ecol Appl S117–S137

Hettiarachchi (2007) Fisheries in the Palk Bay region: the Indian factor

Hiddink JG, Jennings S, Kaiser MJ (2006) Indicators of the ecological impact of bottom-trawl disturbance on seabed communities. Ecosystems 9(7):1190–1199
Hutchings P (1990) Review of the effects of trawling on macrobenthic epifaunal communities. Mar Freshw Res 41(1):111–120
Jaffry S, Glenn H, Ghulam Y, Willis T, Delanbanque C (2016) Are expectations being met? Consumer preferences and rewards for sustainably certified fisheries. Mar policy 73:77–91
Jagadis I, Menon NG, Shanmugavel A (2004) Observations on the effect of bottom trawling on dislocation of non-edible biota in the Palk Bay and Gulf of Mannar, South East Coast of India. Large marine ecosystem: exploration and exploitation for sustainable development and conservation of fish stocks, pp 607–624
Jennings S, Kaiser MJ (1998) The effects of fishing on marine ecosystems. Adv Mar Biol 34:201–352
Jones JB (1992) Environmental impact of trawling on the seabed: a review. NZ J Mar Freshwat Res 26(1):59–67
Kaiser MJ (1998) Significance of bottom-fishing disturbance. Conserv Biol 12(6):1230–1235
Kaiser MJ, Spencer BE (1994) Fish scavenging behaviour in recently trawled areas. Mar Ecol Prog Ser Oldendorf 112(1):41–49
Kaiser MJ, Cheney K, Spence FE, Edwards DB, Radford K (1999) Fishing effects in northeast Atlantic shelf seas: patterns in fishing effort, diversity and community structure VII. The effects of trawling disturbance on the fauna associated with the tubeheads of serpulid worms. Fish Res 40(2):195–205
Kaiser MJ, Collie JS, Hall SJ, Jennings S, Poiner IR (2002) Modification of marine habitats by trawling activities: prognosis and solutions. Fish Fish 3(2):114–136
Kelleher K (2005) Discards in the world's marine fisheries: an update (no. 470). Food & Agriculture Organisation
Krost P, Bernhard M, Werner F, Hukriede W (1990) Otter trawl tracks in Kiel Bay (Western Baltic) mapped by side-scan sonar. Meeresforschung 32:344–353
Kumar AB, Deepthi GR (2006) Trawling and by-catch: implications on marine ecosystem. Curr Sci 90(8):922–931
Kumara PBTP (2014) Ecosystem health in the Palk Bay. In: International conference on conflict resolution and sustainable fisheries governance in the Palk Bay of South Asia, 21–22 Jan 2014, University of Colombo, Sri Lanka
Kumaraguru AK, Joseph VE, Rajee M, Balasubramanian T (2008) Palk Bay information and bibliography
Kurien J (1978) Towards an understanding of the fish economy of Kerala State. Working paper 68. Centre for Development Studies, Trivandrum
Kurup BM, Thomas JV, Anand V (2004) Status of epifaunal component in the bottom trawl discards along Kerala coast (South India). Fish Technol (India)
Lindegarth M, Valentinsson D, Hansson M, Ulmestrand M (2000) Effects of trawling disturbances on temporal and spatial structure of benthic soft-sediment assemblages in Gullmarsfjorden, Sweden. ICES J Mar Sci J du Conseil 57(5):1369–1376
Lohrer AM, Thrush SF, Gibbs MM (2004) Bioturbators enhance ecosystem function through complex biogeochemical interactions. Nature 431(7012):1092–1095
Main J, Sanger GI (1990) An assessment of the scale of damage to and survival rates of young gadoid fish escaping from the cod end of a demersal trawl. Scottish Fisheries Res Rep 46, p 35
Manickam PES, Arputharaj MR, Rao PV (1987) Exploitation of Juveniles of green tiger prawn *Penaeus semisulcatus* along Palk Bay and its impact on the prawn fishery of the region. National symposium on research and development in marine fisheries. CMFRI Bull 44(1):137–145
Marine Fisheries Census (2010) Part II.4 - Report of Tamil Nadu State
McCaughran DA (1992) Standardized nomenclature and methods of defining bycatch levels and implications. In Proceedings of the national industry bycatch workshop, pp 200–201
McConnaughey RA, Syrjala SE, Dew CB (2005) Effects of chronic bottom effects of chronic bottom trawling on the size structure wling on the size structure of soft-bottom benthic invertebrates. In: American fisheries society symposium vol 41, pp 425–437
McLoughlin RJ, Young PC, Martin RB, Parslow J (1991) The Australian scallop dredge: estimates of catching efficiency and associated indirect fishing mortality. Fish Res 11(1):1–24

Meenakumari UB, Pravin P (2008) Impact of bottom trawling on benthic communities. Fish Technol 45(1):1–22
Menon NG, Pillai CG (1996) Marine biodiversity: conservation and management-introduction 1–3
Moore G, Jennings S (eds) (2008) Commerical fishing: the wider ecological impacts. Wiley
National Research Council (2002) Effects of trawling and dredging on sea floor habitat. National Academy Press, Washington D.C
Ogden R (2008) Fisheries forensics: the use of DNA tools for improving compliance, traceability and enforcement in the fishing industry
Pramod G (2010) Illegal, unreported and unregulated marine fish catches in the Indian exclusive economic zone. Field report, policy and ecosystem restoration in fisheries. Fisheries Centre, University of British Columbia, BC, Vancouver, Canada, p 30
Raman AV (2006) Impact of bottom trawling on benthic communities. Summary of the projects conducted in Kerala, Karnataka and Andhra Pradesh (Back-ground paper). Report prepared by ocean science and technology cell on marine benthos, Ministry of Ocean Development, India
Reise K (1982) Long-term changes in the macrobenthic invertebrate fauna of the Wadden Sea: are polychaetes about to take over? Neth J Sea Res 16:29–36
Riemann B, Hoffmann E (1991) Ecological consequences of dredging and bottom trawling in the Limfjord. Oldendorf 69(1):171–178 (Marine ecology progress series, Denmark)
Romanov EV (2002) Bycatch in the tuna purse-seine fisheries of the western Indian Ocean. Fish Bull 100(1):90–105
Ruiz X, Oro D, Martinez-Vilalta A, Jover L (1996) Feeding ecology of Audouin's Gulls (Larus audouinii) in the Ebro delta. Colonial Waterbirds 68–74
Safina (1995). The world's imperilled fish. Scientific American. 273. 46–53
Sainsbury KJ (1988) The ecological basis of multispecies fisheries and a management of a demersal fishery in tropical Australia. In: Gulland JA (ed) Fish population dynamics, 2nd edn. Wiley, London, pp 349–382
Sainsbury JC (1996) Commercial fishing methods: an introduction to vessels and gears, 3rd edn. The University Press, Cambridge, UK
Saxton F (1980) Coral loss could deplete fish stocks. Catch 7(8):12–13
Sikorski ZE (1990) Seafood: resources, nutritional composition, and preservation. CRC Press
Sivalingam S (2005) General features and fisheries potential of palk bay. Palk strait and its environs
Tanner JE (2003) The influence of prawn trawling on sessile benthic assemblages in Gulf St. Vincent, South Australia. Can J Fish Aquat Sci 60(5):517–526
Thomas JV, Premlal P, Sreedevi C, Kurup BM (2004) Immediate effect of bottom trawling on the physicochemical parameters in the inshore waters (Cochin-Munambum) of Kerala. Indian J Fish 51(3):277–286
Van Dolah RF, Wendt PH, Nicholson N (1987) Effects of a research trawl on a hard-bottom assemblage of sponges and corals. Fish Res 5(1):39–54
Venkataraman K, Srinivasn M, Satyanarayana CH, Prabakar D (2002) Faunal diversity of Gulf of Mannar biosphere reserve. Conserv Area Ser 15:1–77. (Published: Dierector, Zoological Survey of India, Kolkata)
Vivekanandan V (2004) Fishing for a favour, netting a lesson. South Indian Federation of Fishermen Societies, Trivandrum, p P68
Vivekanandan E, Srinath M, Kuriakose S (2005) Fishing the marine food web along the Indian coast. Fish Res 72(2):241–252
Widdicombe S, Austen MC, Kendall MA, Olsgard F, Schaanning MT, Dashfield SL, Needham HR (2004) Importance of bioturbators for biodiversity maintenance: indirect effects of fishing disturbance. Mar Ecol Prog Ser 275:1–10
Wilson J (1979) Patch development of the deep-water coral Lophelia Pertusa (L.) on Rockall Bank. J Mar Biol Assoc UK 59(1):165–177
Zacharia PU, Krishnakumar PK, Durgekar NR, Anoop AK, Muthiah C (2006) Assessment of bycatch and discards associated with bottom trawling along Karnataka coast, India

Chapter 4
Sustainable Aquafeed

Krishna R. Salin, V. V. Arun, C. Mohanakumaran Nair and James H. Tidwell

Abstract The global aquafeed production is expected to grow by 33% to 101.3 million tonnes by 2025 from the current (2015) estimate of 73 million tonnes, closely aligning with the targeted world aquaculture production of 101.8 million tonnes. Aquafeed industry mostly depends on the fish meal and fish oil from capture fisheries to supplement the essential nutrients for optimum growth performance in aquaculture. There has been an increasing trend to incorporate ingredients such as protein meals of plant and animal origin in aquafeeds as a consequence of the limited availability, fluctuating price and the growing concerns on the sustainability of fish meal and fish oil. The algal meal has been successfully incorporated in shrimp diets resulting in growth comparable to fishmeal suggesting potential replacement of fish meal even in shrimp larval feeds. The replacement of fish oil by 40–100% using various plant-based sources such as the marine microalgae, *Schizochytrium* in the diets of salmon, channel catfish, grouper and tilapia among others have also been reported. These results suggest the potential for the formulation of an aquafeed that is completely devoid of fishmeal and fish oil. However, one of the major concerns about the concept of 'vegetarian fish' is related to its taste and nutritional quality, particularly in the content of polyunsaturated fatty acids (PUFA). To sustain the desirable health benefits from fish intake in humans, reduced nutritional quality of farmed fish would demand higher dietary inclusion compared to the currently recommended levels. Genetically modified (GM) yeast, camelina, and metabolically engineered diatoms have been suggested to potentially replace fish oil in aquafeeds for improving the PUFA content in

K. R. Salin (✉)
Aquaculture and Aquatic Resources Management (AARM),
Asian Institute of Technology, Khlong Luang, Pathumthani 12120, Thailand
e-mail: salinkr@ait.ac.th

V. V. Arun
ICAR-Central Institute of Fisheries Education, Mumbai 400061, India

C. Mohanakumaran Nair
Sakalyam, Vikram Sarabai Road, Kochi 682304, India

J. H. Tidwell
Division of Aquaculture, Kentucky State University, Frankfort, KY 40601, USA

F. I. Hai et al. (eds.), *Sustainable Aquaculture*, Applied Environmental Science and Engineering for a Sustainable Future,
https://doi.org/10.1007/978-3-319-73257-2_4

vegetarian fish. However, the ethical, environmental and economic costs of the use of GM organisms as an ingredient in aquafeed need to be evaluated for their recognition as a sustainable alternative in aquafeed.

Keywords Aquaculture feed · Vegetarian fish · Plant-based ingredients
Fish meal replacement · Fish oil

4.1 Introduction

World population is expected to reach 9.7 billion by 2050. Out of the current population of 7.3 billion, 780 million people are estimated to be undernourished in 2015 (SOFIA 2016). A major responsibility is vested with agriculture sector to feed such a huge population and ensure food and nutritional security in a sustainable way, which is a big challenge. Some of the major constraints for sustainable development of agriculture sector are climate change, global warming, scarcity of land and water, outbreaks, lowering oil price, regional conflicts, and instabilities and slow growth rate in global economy. Several strategies have been discussed for ending poverty and hunger, an ambitious target to be achieved by 2030 according to the Sustainable Development Goals set at the United Nations Sustainable Development Summit, 2015 (SOFIA 2016; UN 2015).

Proteins of animal origin in the diet help to alleviate malnutrition as they contain essential nutrients like vitamins, ω-3 fatty acids and minerals. In the next decade a 'nutrition transition' is predicted in developing countries from a calorie rich cereal diet to protein rich meat diet including beef, poultry, pig, sheep and fish, mostly driven by the increased rate of growth in per capita income there. This paradigm shift in consumption pattern would lead to an increased demand for meat products. Population growth and strengthening developed economies will also lead to a higher demand for meat products (OECD/FAO 2016).

4.2 Trends in Global Animal Meat Production

World beef production is expected to register a better average annual growth of 1.38% in the next decade (2015–2025) compared to the previous decade's growth rate of 0.82% (2005–2015). Similar upward trend is expected in the sheep meat production as well (1.44–1.98%). However, average growth rate in all other meat production sectors including pig (1.74–1.07%), poultry (3.19–1.51%) and fish (2.24–1.4%) will slow in the next decade. In general, overall meat production is projected to grow by 16% between 2013–2015 base period and 2025 (OECD\FAO 2016) (Fig. 4.1). However, it is expected that the production of sheep meat

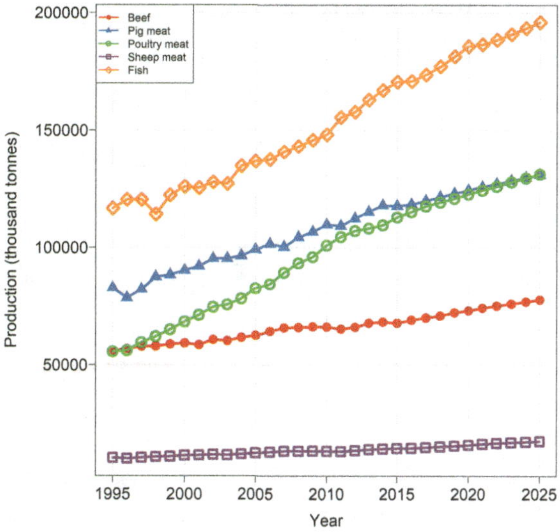

Fig. 4.1 World meat production. *Data Source* OECD/FAO (2016)

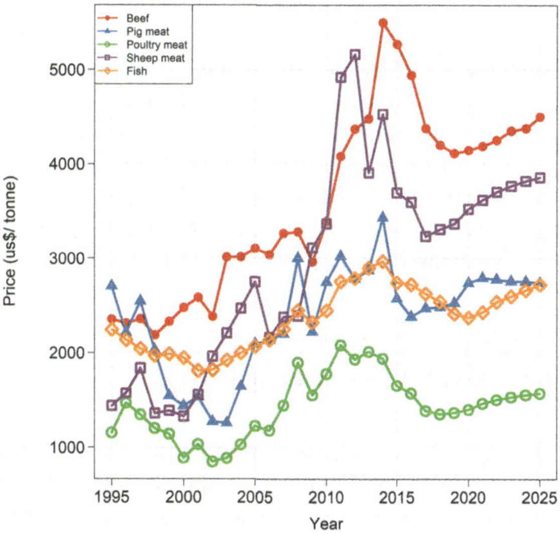

Fig. 4.2 Trend in price of world meat production. *Data Source* OECD/FAO (2016)

followed by poultry and fish will grow at 23.35, 19.02 and 17.39%, respectively during the same period.

The sluggish growth in global meat production can be correlated with the falling price of meat products (Fig. 4.2). Since the record hike in price of meat products observed in 2014, it is expected that that price will come down and stabilize around 2018–2019. Thereafter the price will again start to rise albeit in a slow pace. Beef, poultry and fish prices are projected to show negative average annual growth rates (−1.47, −0.42 and −0.03, respectively) in the next decade (2015–2025) compared

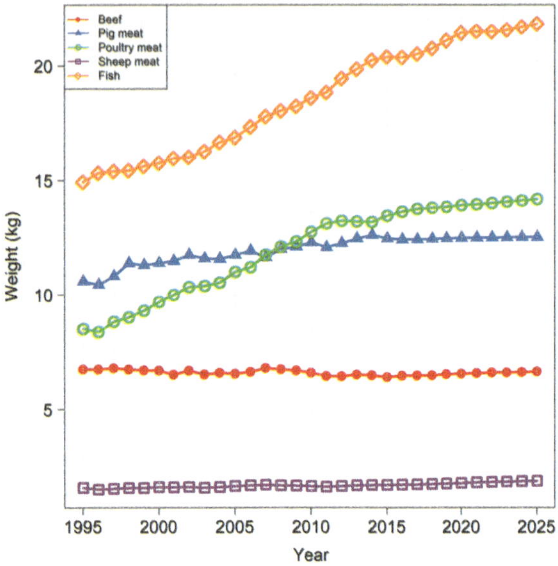

Fig. 4.3 World per capita meat consumption. *Data Source* OECD/FAO (2016)

to the 2005–2015 period. This softening of price, particularly because of strengthening of the dollar, El Niño effect, and slow down of emerging markets will positively affect the per capita consumption of meat products (Fig. 4.3) in the coming years (OECD/FAO 2016). Further lowering of price expected for meat products in the next decade may be due to a lower demand for meat products that can be attributed to economic difficulties in Russia, Brazil, China and Japan. However, strengthening USA and European economies will have a positive impact on the demand for meat products (OECD/FAO 2016).

4.3 Health Benefits of Fish

The human consumption of meat products in the next decade will be influenced by their price, the potential health benefits and risks as well as their perceived sustainability. The primary benefit of consuming meat lies in its nutritional composition. Animal meat continues to be the best source of dietary protein, fats, cholesterol, vitamins and minerals. However, there is still widespread discussion on the actual health benefits of many meat products. Several recent studies have shown a positive association between the consumption of red meat or processed meat and chronic diseases such as cancer, heart diseases and diabetes (De Smet and Vossen 2016; Kushi et al. 2006). In this context, fish is regarded as one of the most beneficial and safe animal meat products for human consumption (Photo 4.1).

The health benefits of fish are many. Fish is well known for its balanced composition of essential amino acids. Fish is a good source of vitamins especially

Photo 4.1 Seafood forms an essential part of a healthy diet. Photo by K.R. Salin

Photo 4.2 Appealing seafood display at a restaurant in Wuhan, China. Photo by K.R. Salin

vitamin D, A and B. Fish provides minerals such as calcium, iodine, zinc, iron and selenium. It is suggested that fish rich in ω-3 fatty acids may reduce the chances of cardiovascular diseases (Kushi et al. 2006). Studies conducted in animals indicate that fish oil suppresses the incidences of cancer (Kushi et al. 2006). Fish oil also helps in controlling obesity in human beings. (Wong et al. 2013). These positive health attributes are expected to lead to a surge in fish consumption in the next decade and place fish at an advantage, encouraging a substantial shift in consumer preference towards fish over the relatively inexpensive poultry products (Photo 4.2).

4.4 Paradigm Shift in Aquaculture Feed Sector

Boundless opportunities exist for the seafood sector as the world draws itself into a 'global village' with rapid urbanization, and the emergence of middle class population as major consumer segment in many parts of the world. Capture fisheries production has been stagnated at around 90 million tonnes since the 1990s (Fig. 4.4). However, the total fisheries sector (capture and aquaculture) has been growing at 3.2% since 1961 and has outpaced the global population growth with per capita consumption of fishery products reaching 20 kg in 2015 (SOFIA 2016). This remarkable growth has been achieved primarily through the contribution of aquaculture, one of the fastest growing food production sectors in the world. In the last decade (2006–2015), aquaculture has grown at an average annual rate of 5.53% but it is expected to slow to around 3% in the next decade. It is estimated that aquaculture has to grow by 33% in terms of production volume from 75.9 million tonnes (OECD/FAO 2016) in the next ten years to meet the projected additional output of nearly 25 million tonnes to reach a total production of 101 million tonnes by 2025 (Fig. 4.4).

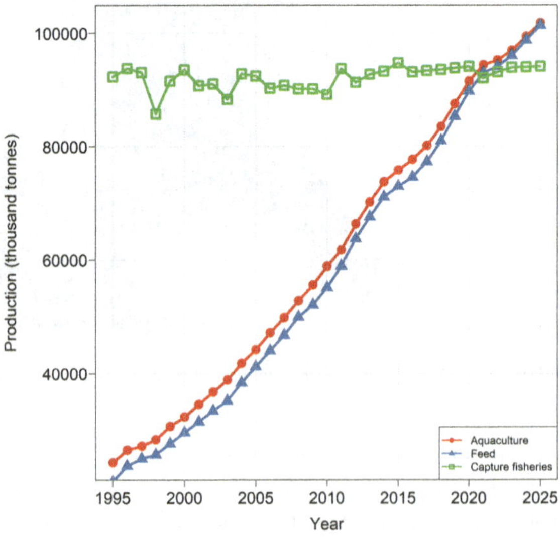

Fig. 4.4 Global production of fish from capture fisheries and aquaculture, and requirement of feed. *Data Source* OECD/FAO (2016)

However, key challenges that arise in the context of a sustainable annual growth rate of aquaculture include:

- The fishmeal challenge: how sustainable is to catch wild fish to feed the farmed fish?
- How to address the issue of carnivorous fish that are produced at a higher environmental and economic cost compared to the herbivores?
- How sustainable is the transition from fish meal to plant based ingredients in aquafeed?
- How to ensure and maintain nutritional superiority of aquafeeds with plant based ingredients in place of fish meal?
- How biotechnology can be applied to ensure sustainability of the aquafeed industry?

One of the major segments of the aquaculture production system is supplementary feed which accounts for over 60% of the total cost of production. Supplementary feeds provide the required macronutrients (protein and lipids) while relying on natural foods from the culture system (usually ponds) to supply expensive micronutrients (vitamins and minerals). Availability of good quality of feed at adequate volumes is essential to achieve the targeted aquaculture production. Over the past few decades the aquafeed industry has transformed from the traditional feeding using trash fish and rice bran/oil cake mixture to the high quality compounded pelleted feeds (Photos 4.3 and 4.4). Current level of feed technology addresses the nutritional requirement for farmed aquatic animals to the level of individual amino acids, vitamins and mineral requirements (complete feeds). In

Photo 4.3 Intensive raceway farms like this in Shanghai, China are based on high feeding rates of high quality feed. Photo by K.R. Salin

Photo 4.4 Pellet feed used in aquaculture. Photo by K.R. Salin

view of the aqua feed production over the previous decade (2006–2015) that has recorded an average annual growth rate of 5.9% and its projected growth rate of 3.3% in the next decade (2016–2025), the total feed requirement in 2025 for aquaculture sector is estimated to be 101 million tonnes (Fig. 4.4).

4.5 Feed Ingredients—The Fish Meal and Fish Oil Dilemma

One of the major constraints for aquafeed production is the limited availability of feed ingredients and their booming prices. The inclusion of fish meal and fish oil in aquafeeds and their positive impact on the composition of the final product are primarily responsible for the health benefits of aquaculture products compared to other meat products (Henriques et al. 2014). Rational use of fish meal and fish oil plays a major role in maintaining efficient feed conversion ratios (FCR) and optimum growth in aquatic organisms (Photo 4.5).

Fish meal is prepared by cooking, pressing, drying and milling of low value marine fishes, particularly the small pelagic fish that are not suitable for human consumption, and fish processing waste. Fish oil is prepared by centrifuging the press liquor obtained after fish meal production. Fish meal generally contains 60–72% protein depending upon the raw material used (Shepherd and Jackson 2013) and is a high quality protein with a uniquely balanced amino acid

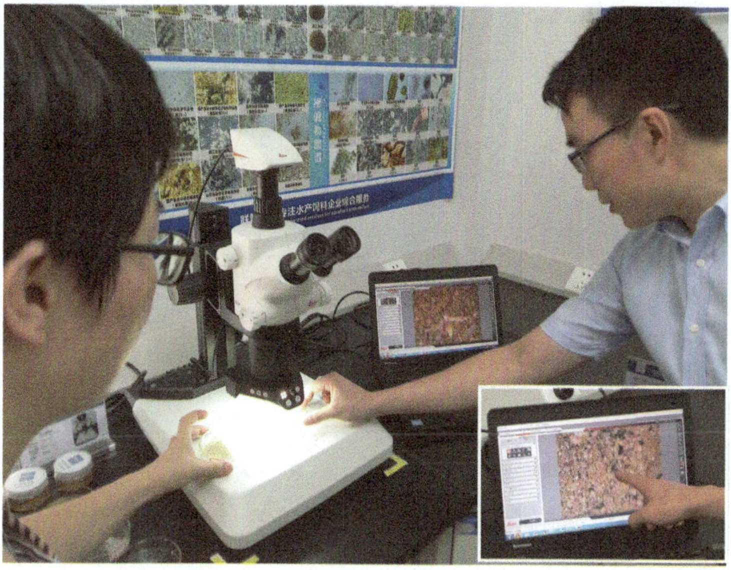

Photo 4.5 The quality of fish meal used in aquafeeds is critical. Microscopic examination can help to select the best quality fish meal for feed manufacturing process. Photo by K.R. Salin

composition, including all the essential amino acids. This property makes fish meal an ideal ingredient in aquatic and terrestrial animal feeds, which ensures the best growth, survival and reproduction in animals, compared to most plant based protein sources. Fish meal is a very good source of nucleotides, essential fatty acids and phospholipids. It also contains minerals like calcium, phosphorus, magnesium, zinc, manganese, selenium, iodine, molybdenum and chromium, in addition to the water soluble and fat soluble vitamins. Fish oil is a natural source of essential polyunsaturated fatty acids like eicosapentaenoic acid (EPA) and docosahexaenoic acid (DHA).

In 1960, the poultry and pig industries consumed as much as 98% of the global fish meal supply. However, as the salmon farming techniques became popular in temperate countries the aquaculture industry was using about 10% of the global supply of the fish meal by 1980 (Shepherd and Jackson 2013). This trend continued with the rapid expansion of aquaculture sector, while the terrestrial animal feeds moved closer to plant based ingredients. In 2012, aquafeed industry consumed almost 68 and 74% of the global fish meal and fish oil produced, respectively (Mallison 2013).

Global fish meal production increased over time until 1994 with a peak production of 7.5 million tonnes. Fish oil production also reached its peak of 1.5 million tonnes by 1994 (Fig. 4.5). The fish meal and fish oil production declined thereafter, although with inter-annual fluctuations. The lowest production levels of fish meal were observed during 1998, 2003, 2010 and 2014 with volumes of 5.3, 5.5, 4.4 and 4.1 million tonnes, respectively. Consequently, fish oil production was

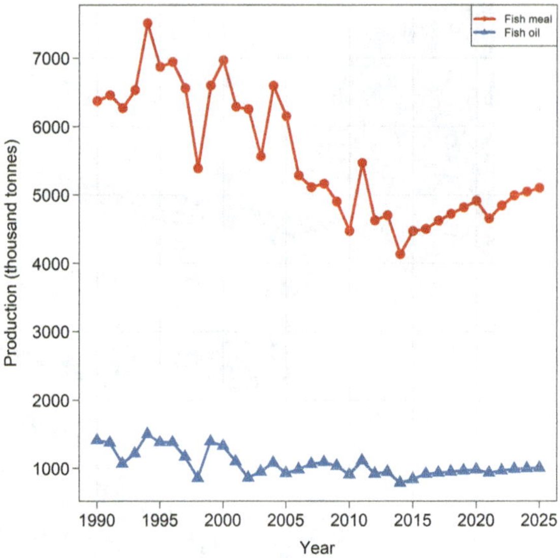

Fig. 4.5 Global fishmeal and fish oil production. *Data Source* OECD/FAO (2016)

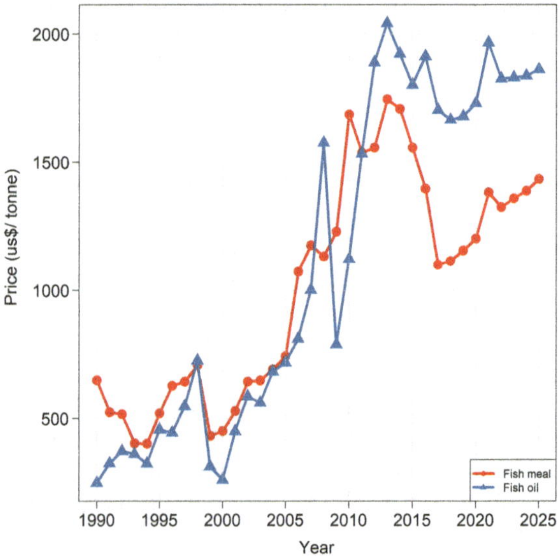

Fig. 4.6 Global fishmeal and fish oil price. *Data Source* OECD/FAO (2016)

also low during 1998, 2003, 2010 and 2014 with 0.85, 0.86, 0.9 and 0.78 million tonnes, respectively. Over the last decade (2006–2015) the total fish meal and fish oil production have shown negative annual growth rates of −2.57 and −0.29%, respectively.

4.6 Impact of Fish Meal and Fish Oil on Feed Cost

The cost of aquafeed depends to a great extent on the price of fish meal and fish oil. The average annual price of fish meal was the lowest in 1994 and 1999 at 403 and 433 US$/tonne, respectively (Fig. 4.6). Since 1999 the price had continued to increase reaching 1230 US$/tonne in 2009. However, fish meal prices surged steeply thereafter with 1687 US$/tonne in 2010 and a peak of 1747 US$/tonne in 2013. In the past decade (2006–2015) fish meal price increased at an average annual growth rate of 8.94%.

A similar trend is also observed in fish oil price where the lowest price levels of 325 and 262 US$/tonne were observed during 1994, 2000, respectively. The price of fish oil increased at a slow pace until 2002 and gained momentum thereafter. However, after 2010 there was a sudden spike reaching a peak in 2014 (1923 US$/tonne). It is also interesting to note that the annual average fish meal prices were higher than fish oil until 2010, when fish oil became more expensive than fish meal.

4.7 Status of the Major Fish Stocks Supporting Fish Meal and Oil

During 2013–2015 period nearly 17% of the total capture fisheries was utilized for fish meal production (SOFIA 2016). Fish meal is mainly derived from small pelagic fish with high oil content. These fish stocks are often characterized by early maturation and high fecundity. The species used for reduction to fish meal depend on the region of production. For example, the European fish meal production is mainly from fish such as capelin (*Mallotus villosus*), blue whiting (*Micromesistius poutassou*), small sand eel (*Ammodytes tobianus*) and Norway pout (*Trisopterus esmarki*). These species comprised 35% of the total fish meal requirement of the European feed industry with the rest imported from South America (20%), derived from Antarctic krill or, from food fish processing waste (Huntington and Hasan 2009).

A major global producer of fish meal is South America. Peruvian anchovy (*Engraulis ringens*) and Chilean jack mackerel (*Trachurus murphyi*) dominate the South American fish meal industry. Alaska Pollock (*Theragra chalcogramma*), Argentine hake (*Merluccius hubbsi*) and Southern blue whiting (*Micromesistius australis*) are also reduced to fish meal in South America. In Africa small pelagic fishes are mainly used for direct human consumption. South Africa has a fish meal industry dominated by European pilchard (*Sardina pilchardus*), other sardines and European anchovy (*Engraulis encrasicolus*) (Huntington and Hasan 2009).

A time series analysis of the major pelagic fish stocks which contribute to fish oil and fish meal production shows wide inter-annual fluctuations (Fig. 4.7). Over the past quarter of a century (1990–2014) small pelagic fish capture production was dominated by three species, namely Peruvian anchovy, Alaska pollock, and Chilean jack mackerel with shares of 45, 23 and 13%, respectively. Together these three species contributed 81% of the small pelagic fish captured for fish meal and fish oil. It is evident that Peruvian anchovy production account for much of the variation in small pelagic fish production. During the past 25 year, Peruvian anchovy peaked at 12.5 million tonnes in 1994, with its lowest point in 1998 (1.7 million tonnes), followed by a continuous decline. The catch of Alaska Pollock had shown a declining trend till 2000 thereafter maintaining around 2.8 million tonnes till 2010. However, this catch has improved thereafter showing some signs of recovery. The Chilean jack mackerel had also shown a strong decreasing trend. After registering a peak of 4.95 million tonnes in 1995, it reached a record low of 0.35 million tonnes in 2013 (data compiled from FAO database).

Major reasons for the declining trend in the harvests of small pelagic fish are believed to be over-exploitation of their wild stocks for direct human consumption and production of fish meal or fish oil, and climate change, particularly along the Pacific coast, which is rich in anchovy stocks (Shepherd et al. 2017). Global climate change combined with the El Niño phenomenon has escalated sea surface temperatures thereby shifting the anchovy stocks towards deeper areas making them hard to harvest (Shepherd et al. 2015). Reduced food (plankton) availability in deeper water also leads to emaciation and poor survival of the anchovy stocks (Pike and Tocher 2016). A sizeable El Niño effect was reported during 1998 and 2014, while the weaker ones were observed during 1992, 2003 and 2010 (Pike and Tocher 2016). After the strong El Niño in 2014, signs of recovery in anchovy stocks were seen in 2015 and in early 2016 thereby offering some relief on the prices of fish meal and fish oil (OECD/FAO 2016). It is projected that by 2018 the prices of fish

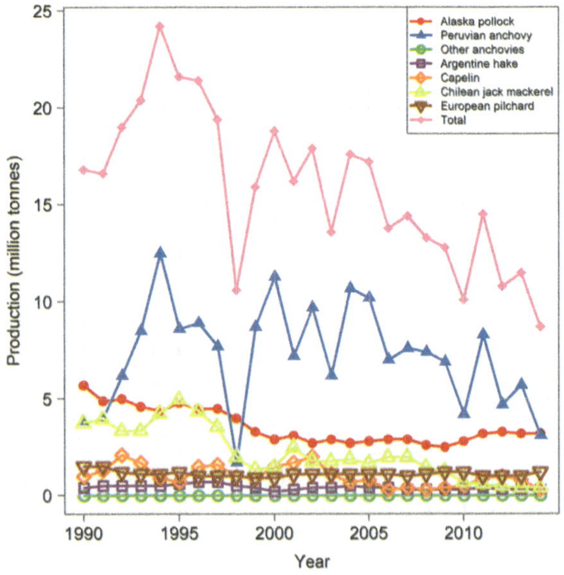

Fig. 4.7 Global pelagic fish production used for fishmeal and fish oil production (Total production includes Norway pout and Southern blue whiting in addition to the fishes mentioned in the graph). *Data Source* FAO fisheries and aquaculture data base

meal and fish oil will reach their levels of 2008, but the following decade is expected to witness a hike in their price (Fig. 4.6).

4.8 Non-aquafeed Uses of Fish Meal

Fish meal and fish oil are also consumed by livestock industries apart from aquafeed manufacturing. The global animal feed production in 2015 was 980 million tonnes contributed by poultry (45%), pig (27%), ruminant (20%), aquaculture (4%), pet (2%), and horse (1%) industries (Alltech 2015). While aquafeed accounts for only 4% of the total livestock feed production, a majority of the fish meal produced is consumed by aquaculture (68%) followed by pig (23%) and chicken (7%) feed industries. Aquafeed industry consumes 74% of the total fish oil produced and the remaining is mainly used for human consumption (22%) (Mallison 2013). The industries that process fish oil for human consumption can support higher prices than the aquafeed industry.

Of late it has been demonstrated that waste fish oil could be a raw material for production of biodiesel, which may in future result in much greater demand for fish oil globally (Behçet 2011; Lin and Li 2009). However, this demand could be offset by the prevailing lower crude oil prices and would depend on further fluctuations in their global prices to favor some interest on biodiesel production.

The static supply of fish meal from the wild harvest of small pelagic fish, combined with a rapid growth of the aquaculture industry means that alternatives to fish meal of a scale to adequately support the aquafeed manufacturing industry for a projected global production of 101 million tonnes of fish from aquaculture by 2025 must be identified and developed.

4.9 Alternative Sources of Feed Ingredients to Fish Meal

There have been a number of options available to supplement or replace fish meal as a protein source in aquafeeds. Important among them are fish processing waste and plant-based ingredients.

4.9.1 Fish Processing Waste

One of the earliest strategies for realizing sustainable aquafeeds involved the utilization of fish processing waste as raw material for the production of fish meal and fish oil. Improved per capita income and purchasing power of consumers, particularly in developing countries, have shifted the demand from whole fish to processed products like fish fillets. This generates huge amount of processing waste, which can present many environmental challenges related to their disposal. Waste

Table 4.1 Waste generation during fish processing (Arvanitoyannis and Kassaveti 2008)

Mode of processing	Fish (kg)	Solid waste (kg)
White fish filleting	1000	Skin: 40–50 Heads: 210–250 Bones: 240–340
Oily fish filleting	1000	400–450
Scaling of white fish	1000	Scales: 20–40
Deheading of white fish	1000	Head and debris: 270–320
Filleting of deheaded white fish	1000	Frames and off cuts: 200–300
Filleting of ungutted oily fish	1000	Entrails, tails, heads and frames: 400
Skinning white fish	1000	Skin: 40
Skinning oily fish	1000	Skin: 40

Table 4.2 Fish waste generated during filleting process (Waterman 1979)

Component	Average weight (%)
Head	21
Gut	7
Liver	5
Roe	4
Backbone	14
Fins and lungs	10
Skin	3
Fillet, skinned	36

Table 4.3 Proximate composition of fish waste (Esteban et al. 2007)

Nutrient	Fish waste
Crude protein (%)	57.92 ± 5.26
Fat (%)	19.10 ± 6.06
Crude fiber (%)	1.19 ± 1.21
Ash (%)	21.79 ± 3.52
Calcium (%)	5.80 ± 1.35
Phosphorous (%)	2.04 ± 0.64
Potassium (%)	0.68 ± 0.11
Sodium (%)	0.61 ± 0.08
Magnesium (%)	0.17 ± 0.04
Iron (ppm)	100 ± 42
Zinc (ppm)	62.00 ± 12
Manganese (ppm)	6 ± 7
Copper (ppm)	1 ± 1

products include fish frames, offal, trimmings and offcuts, all of which are nutritionally rich and could be used as a source of fish meal and fish oil (Ghaly et al. 2013) (Table 4.1).

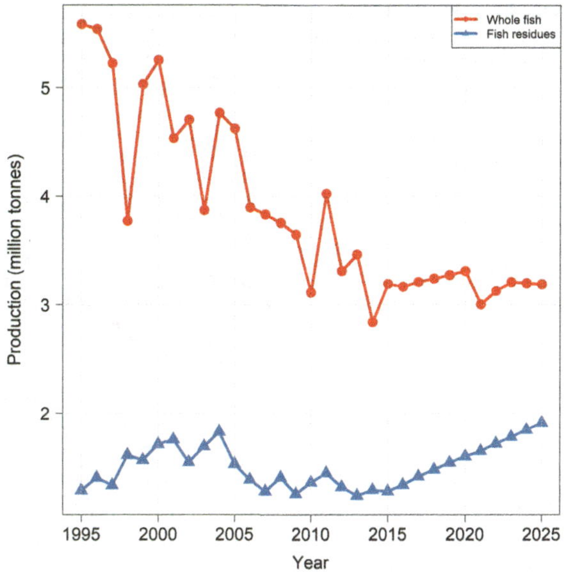

Fig. 4.8 Global fish production by resources. *Data Source* OECD/FAO (2016)

The fish processing waste generated depends on the final product prepared (Tables 4.2 and 4.3). Contribution of the fish meal produced from waste products is expected to reach 38% in the next decade (2015–2025) from the current 29% (Fig. 4.8). Even though recycling of fish products to fish meal can have a positive impact on sustainability, it may have an effect on the quality of the fish meal produced which is reported to have less protein, more ash content and an imbalanced amino acid composition compared to fish meal from capture fisheries (FAO 2016; SOFIA 2016).

Similar to fish processing waste, terrestrial animal protein meals and oils can also be used as alternatives for fish meal and fish oil. Meat by-product meals and fats reduced from the slaughtered animals like cattle, pig, sheep and poultry are some of the potential ingredients. Blood meals produced from farmed livestock can also substitute fish meal (Tacon et al. 2011). However, the inclusion levels of animal protein meals and lipids are limited by their nutrient imbalances, palatability issues, and the deficiency of certain nutrients.

4.9.2 Plant Based Ingredients—Protein Meals and Vegetable Oils

Over the past decade a major focus of the feed industry was to evaluate the inclusion of plant based ingredients as a replacement for fish meal and oil. The plant based protein meals (example: soybean meal, rapeseed meal, sunflower meal, groundnut meal, coconut meal, cotton seed meal and palm kernel meal) and

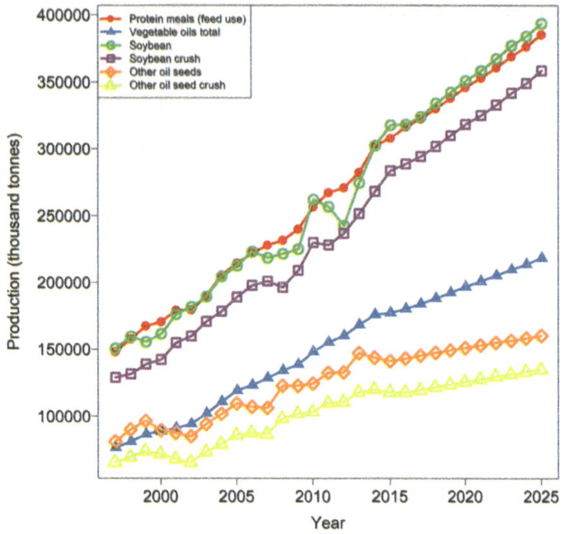

Fig. 4.9 Global oil seed and protein meal production. *Data Source* OECD/FAO (2016)

vegetable oils (soybean oil, rapeseed oil, cotton seed oil, coconut oil, palm oil, groundnut oil, sunflower oil and palm kernel oil) are increasingly included in aquafeeds to circumvent the shortage of fishmeal and fish oil.

Soybeans account for 69% of the total world production of oilseeds. Soybean production in 2015 was 318 million tonnes, of which 89% (284 million tonnes) was converted into soy crush to be further reduced as oil and protein meal (Fig. 4.9). The average annual growth rate of soybean crush production is expected to slow to 2.8% in the coming decade (2015–2025) compared to the previous decade (4.2%, 2005–2025). Soybean crush is expected to grow by 26% in volume, with an additional production of 75 million tonnes, in the next ten years.

Slow average annual growth is expected in other oilseed crush production sectors (1.7%) over the next decade compared over the previous decade (3.3%). The amount of protein meal consumed in feed production is projected to expand by 25% and reach 385 million tonnes by 2025. The production of protein meals used in feed is expected to grow at slower average annual growth rate of 2.6% during 2016–2025 compared to the period 2006–2015 (3.7%).

Vegetable oil production is projected to reach 219 million tonnes in 2025 from the current production of 178 million tonnes in 2015 with an annual growth rate of 23%. Average annual growth rate is expected to slow to 2.5% over the next decade (2016–2025) from 4% in the previous decade (2005–2015). The prices of protein meal, oilseeds and vegetable oils, which are often interdependent, are projected to be lower and stabilized in the coming decade (2015–2025) after reaching a record peak during the period from 2007 to 2014 (Fig. 4.10). Soybean price has reached a maximum of 562 US$/tonne in 2011. It is expected that the price of soybean, which is currently a major ingredient of the feed industry will stabilize around 397 US$/tonne during the next decade which is a positive sign for the feed industry.

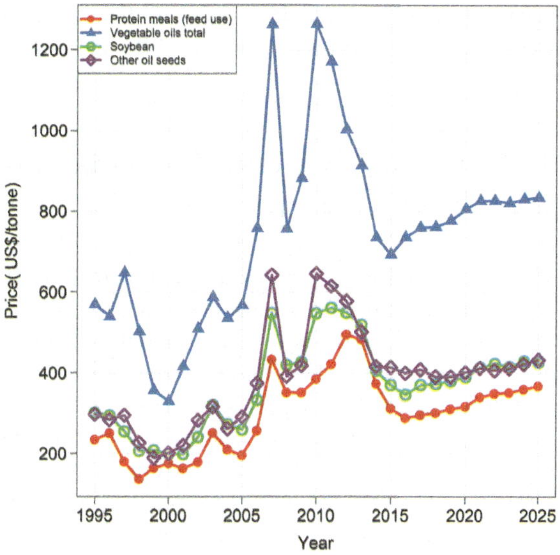

Fig. 4.10 Global oil seed and protein meal price. *Data Source* OECD/FAO (2016)

Average annual growth rate of the soybean price should fall to 1.2% in the next decade (2016–2025) compared to the previous decade's growth rate of 6.3% (2006–2015). Similarly, prices of other oilseeds should also come down during this period.

The average price of protein meal is projected to undergo a hike of 17% between the period from 2015 to 2025, and is expected to remain around 328US$/tonne during this period (Fig. 4.10). However, the average annual growth rate in protein meal price will significantly come down to 1.4% in the projected period compared to the previous 7.6% (2006–2015). Considerable reduction in the average vegetable oil price can be expected in the next decade. Average price will come down to 798 US$/tonne during 2016–2025 from the 945 US$/tonne in the last decade (2006–2015). Average annual growth rate of the vegetable oil price in the coming years is very minimal (1.8%, 2016–2025) compared to the previous decade's growth of 6.4%. The production of protein meal and vegetable oil is expected to grow in the next decade ensuring their greater availability for the feed industry. Their prices are expected to remain lower than their previous peak values making them a sustainable resource for animal feed production, particularly in aquafeeds (Fig. 4.10).

The inclusion of a wide range of plant based ingredients in aquafeed, in place of fish meal and fish oil is expected to enhance the sustainability of the industry. By reducing their utilization for manufacturing fish meal and fish oil, the low value, nutrient rich fish can be diverted for direct human consumption to boost the food, nutritional and livelihood security to alleviate the malnutrition problems, particularly in Low Income Food Deficit countries (Photo 4.6). However, major changes might be required in the preservation and transportation of such low valued fish in developing countries. This strategy is also an environmentally sustainable solution as it would also help to reduce overexploitation of fish stocks. In other words, an aquafeed industry that is less reliant on fishmeal and fish oil as a major ingredient would be more environmentally, socially and economically sustainable.

4.10 Journey to a True 'Vegetarian Fish'—The Salmon Story

The aquaculture feed industry has transformed itself over the past 15 years with a significant shift from a fishmeal based approach to formulations based on plant ingredients. One of the best examples of this change is the case of salmon feeds that are well known for their higher inclusion levels of fishmeal. While there were variations reported among different countries, the average inclusion level of fishmeal in salmon feeds was nearly 29% during the period from 1995 to 2008 (Fig. 4.11). Post 2008, there was a drastic reduction in fishmeal inclusion level observed that coincided with the surge in fishmeal price around that period and the following years. In 2015 the fishmeal inclusion level remained at 17.5% and was projected to come down further to a level of 12% by 2025 (OECD/FAO 2016; SOFIA 2016). The inclusion level of oil seed meal in salmon feed, which was as low as 3.4% in 1995 has reached 13% in 2015, and is expected to rise to 18.6% by 2025.

Another major consumer of fishmeal is the shrimp industry. Unlike in the salmon industry, the process of reduction in the levels of fishmeal inclusion in shrimp feed was gradual reaching 8.3% in 2015 from a peak of 19.3% in 1995 (Fig. 4.12). In

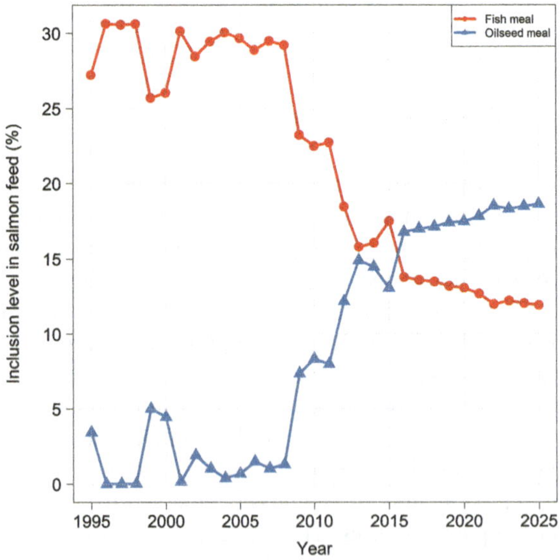

Fig. 4.11 Inclusion level of fishmeal and fish oil in Salmon feed. *Data Source* OECD/FAO (2016)

Fig. 4.12 Inclusion level of fishmeal and fish oil in Shrimp feed. *Data Source* OECD/FAO (2016)

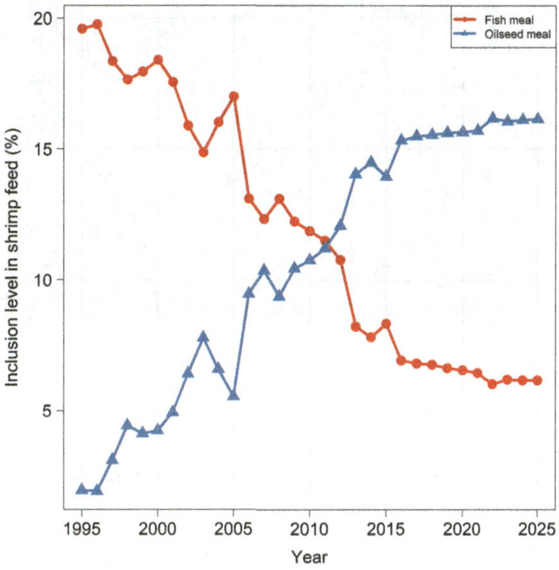

the next decade, the inclusion level would stabilize around 6% without significant reduction any further (OECD/FAO 2016; SOFIA 2016).

'Vegetarian fish' refers to fish fed with a feed containing no animal based ingredient. This concept is gaining momentum, particularly because of the surging prices of fishmeal and fish oil, and an increasing realization of the unsustainability of the animal based ingredients in aquafeeds. The question whether vegetarian fish is an option for the sustainability of aqua feed industries is still debatable. Perhaps it is one of the feasible solutions to address the emerging concerns of sustainability in the aqua feed industry, given the fact that in future, fishmeal would be confined as a strategic ingredient in larval feeds, while a lion's share of the growout feeds for majority species would be based on ingredients derived from plants.

4.11 Challenges to a Vegetarian Feed for Fish

There are several challenges to develop a 100% vegetarian fish, particularly in the case of salmon. Presently salmon is regarded as one of the most efficient livestock in terms of edible yield, FCR, energy retention and protein retention (Bjørkli 2002). Feed Conversion Ratio (FCR) realized in the case of Atlantic salmon is 1.15 compared to the 2.63, 1.79 and 6.3 for pig, chicken and lamb, respectively. It would be difficult to maintain the present lowest FCR and higher growth rates of salmon using a 100% vegetarian feed.

Photo 4.6 Small pelagic fish (left) and shrimp (right) harvested from Lake Victoria, East Africa used as a major food resource for the local population. Photo by K.R. Salin

4.11.1 Replacement of Fish Meal by Protein Meals

The most widely used fish meal replacement in the aquafeed industry is soybean. Soybean could replace up to 33% of fishmeal as protein source in salmon diets (Carter and Hauler 2000). However, at higher inclusion levels soybean meal was found to induce a condition called non-infectious sub-acute enteritis in salmon characterized by pathological changes in the mucosal lining of the distal intestine (Baeverfjord and Krogdahl 1996). Growth performance and feed efficiency were also reported to be affected negatively by the higher inclusion levels of soybean meal. This problem could however, be solved to a certain extent by using soy protein concentrate (Drew et al. 2007; Kaushik et al. 1995; Murai et al. 1987; Olli and Krogdahi 1994). The efficacy of using soybean meal as a single ingredient replacement of fishmeal versus a combination of plant and animal based diets was also studied. It was found that the combination could result in growth comparable to diets having fishmeal as the sole source of protein (Burr et al. 2012; Davidson et al. 2016; Øverland et al. 2009; Torstensen et al. 2008). In the largemouth bass, which is a strict predator, Tidwell et al. (2005) evaluated a series of plant and animal protein sources at different rates of inclusion and fishmeal replacement. They found that poultry by-product meal could fully replace fishmeal, but that the combination of blood meal and corn gluten meal (previously identified as the best combination protein) could not.

In a study conducted on post smolt Atlantic salmon, researchers compared the growth performance of fishmeal based diet and fishmeal free diets in a recirculating aquaculture system (Davidson et al. 2016). The fishmeal free diets included mixed nut meal, poultry meal, wheat flour, and corn protein concentrate, while the fishmeal based diets contained menhaden meal, poultry meal, soy protein concentrate, and blood meal proteins. Equal growth response was achieved in both the fish meal based diet and fish meal free diets. The combination of different plant based ingredients would ensure proper balance of all essential amino acids, vitamins and minerals (Davidson et al. 2016). This is a very promising result that supports the concept of vegetarian fish and would help aquaculture to maintain its preeminent position over other livestock in terms of FCR and growth.

4.11.2 Replacement of Fish Meal and Oil by Microalgal Products

Successful reports of the replacement of fish oil by using marine microalgae, particularly *Schizochytrium* spp. have been reported in the case of salmon. In the diets for Atlantic salmon parr, fish oil was completely replaced by the *Schizochytrium* oil. There was no significant difference in growth and FCR between fish fed algae oil and those fed fish oil. The DHA content of the muscles of algae oil fed fish were higher than the muscles of fish fed with fish oil diet (Miller et al. 2007), which is promising.

In channel catfish, addition of 2% dried *Schizochytrium* resulted in higher weight gain, feed efficiency and higher level of DHA (Li et al. 2009). In grouper, it was reported that a combination of soybean meal, soyprotein concentrate and *Schizochytrium* algae meal could replace fishmeal up to 40%. *Schizochytrium* algae oil could also completely replace fish oil in grouper diets without affecting growth performance and health (García-Ortega et al. 2016). In Nile tilapia (*Oreochromis niloticus*), significantly higher weight gain, lower FCR and high protein efficiency ratio (PER) were observed in the fishes fed with 100% *Schizochytrium* algae oil compared to the fishes fed with fish oil. It was also found that fish oil replacement with algae oil resulted in better deposition of polyunsaturated fatty acids (PUFA) in the fillets of Nile tilapia (Sarker et al. 2016). In another study in longfin yellowtail *Seriola rivoliana*, 80% of the fishmeal in the feed was replaced with a combination of soybean protein concentrate, squid and algal meals without compromising growth. The diet was supplemented by amino acids methionine, lysine, and taurine. It was also demonstrated that blends of fish oil, *Schizochytrium limacinum* meal, and canola oil could be used without affecting the growth of fish (Kissinger et al. 2016).

Schizochytrium can also be used as a dietary supplement for fish or shrimp. In larval microdiets of *Litopenaeus (Penaeus) vannamei*, 4% inclusion of the *Schizochytrium* meal significantly improved their growth performance without affecting PUFA deposition in the muscles (Wang et al. 2016). This study may change our perception that dietary inclusion of fishmeal is essential for ensuring good larval survival of shrimps. While fishmeal is generally regarded as an essential ingredient in larval diets, it is now evident that fishmeal could be replaced successfully by algae meal up to a certain extent. Future research would explore partial or complete replacement of fishmeal and fish oil in larval diets.

Other algae that are used as fishmeal and oil replacement in aquafeeds include *Desmodesmus*, freeze-dried *Isochrysis*, *Chlorella* meal, *Phaeodactylum tricornutum*, *Nannochloropsis* and Spirulina. In the feeds for European seabass

Dicentrarchus labrax, it was found that 20% of protein and 36% of lipid could be replaced using the freeze-dried *Isochrysis* without affecting growth performance (Tibaldi et al. 2015). In Atlantic salmon diets, 6% of fishmeal could be replaced with *Phaeodactylum tricornutum* meal without compromising growth and FCR (Sørensen et al. 2016). In juvenile European seabass diet, a 50% fish oil replacement by *Nannochloropsis* meal was done without any negative impact on growth performance (Haas et al. 2016). Juvenile Nile tilapia fed with 30% spirulina as replacement for fishmeal was shown to improve the growth and feed utilization efficiency (Velasquez et al. 2016). In Atlantic salmon it was possible to replace 20% of fishmeal with defatted *Desmodesmus* algae (Kiron et al. 2016). In juvenile channel catfish, *Ictalurus punctatus*, inclusion of up to 40% of *Chlorella* meal with or without supplemented lysine (2%) could significantly increase feed consumption and weight gain (Kupchinsky et al. 2015).

Compared to *Schizochytrium* which was tested for complete replacement of fish oil in the diets of fish, other algae were generally used as partial replacement. Most of these studies were conducted on marine fish, particularly a few carnivores. In a study on crucian carp, *Carassius auratus* it was demonstrated that Chlorella meal could completely replace fishmeal in diets when they were supplemented with cellulases at the level of 2 g kg^{-1} (Shi et al. 2017). Camelina (*Camelina sativa*) oil is another fish oil replacement that was described to have the potential to completely replace fish oil in Atlantic salmon without affecting their growth performance and health (Hixson et al. 2014). It can be concluded that the tremendous effort invested by scientists across the world over the past decade has completely transformed the non-vegetarian fish that were fed animal based diets into partial or complete vegetarians.

4.11.3 Nutritional Quality of Vegetarian Fish Meat

One of the major concerns about the vegetarian fish is centered on its taste and nutritional quality. It is important to evaluate the nutritional quality of vegetarian fish compared to the non-vegetarian counterparts, particularly with regard to their content of long chain polyunsaturated fatty acids like EPA and DHA which are important for human health.

Nutritional quality of the salmon products collected from retailers was analyzed to determine the effect of feed ingredients on meat quality (Henriques et al. 2014). The ω-3 PUFA present in salmon fillets are mainly derived from their feed because salmon has very limited capacity for the endogenous production of ω-3 PUFA. Three types of fillets from salmon, namely farmed salmon fillets (dominant of vegetable oil markers), farmed salmon fillets (dominant of fish oil markers) and wild salmon. The farmed salmon fillets which were fed with vegetable oil had predominantly the $18{:}1n-9$ and $18{:}2n-6$ fatty acids. In contrast the fillets produced from the fish fed with fish oil rich feed had greater ω-3 PUFA, especially EPA and DHA. The farmed fish had higher total lipid content than the wild fishes.

Farmed salmon had a minimum EPA + DHA content of ≥ 1 g/100 g flesh irrespective of the lipid source. The recommended dietary intake of EPA + DHA for humans is 500 mg/day or 3.5 g/week. Consumption of cultured salmon fillet with a portion of 150 g (two meals each with 75 g of flesh in a week) is sufficient to provide the recommended ration of EPA + DHA for good cardiac health. However, wild salmon fillet tested had <0.5 g of EPA + DHA in 100 g of flesh, and so their products would have to be consumed 4–5 times in a week to ensure the recommended dietary intake (Henriques et al. 2014).

The level of inclusion of fish oil in Norwegian salmon industry was 11% in 2013. This results in 2.5 g of EPA + DHA in 100 g of salmon fillet. Two servings of salmon fillets each with 75 g in a week are sufficient to have the recommended EPA + DHA content in the diet. If the fish oil inclusion was reduced from the current 11 to 5%, correspondingly EPA + DHA in the fillet would come down to 1.3 g/100 g of fillet. It would then be necessary to consume 3.6 servings of fillet each with 75 g in a week to meet the recommended dose of EPA + DHA (Bell et al. 2001; Pike and Tocher 2016; Ytrestøyl et al. 2015). The salmon fillet is an expensive commodity compared to other fish products. Doubling the recommended dietary intake resulting from a 5% inclusion level of fish oil is obviously not economically viable.

An extensive study conducted on 3500 farmed Scottish Atlantic salmon during 2006–2015 found that an increased incorporation of fatty acids of vegetable oil origin in feed had resulted in substantially lower levels of EPA and DHA, compromising the nutritional quality of salmon (Sprague et al. 2016). By 2010 the salmon feed industry had started replacing fish oil with vegetable oil because of the increased fish oil prices. Inclusion levels of fatty acids of vegetable oil origin such as 18:1n − 9, 18:2n − 6 and 18:3n − 3 had increased from 15, 5 and 2 to 30, 10 and 5%, respectively in 2015. As a consequence, the EPA and DHA content had reduced by nearly 50%, a significant reduction from 2.74 g/100 g in 2006 to 1.36 g/100 g flesh in 2015 (Sprague et al. 2016). Corroborating the earlier studies, this too warranted doubling the fillet intake to meet the recommended dietary dose of fish for maintaining cardiac health.

It is interesting to note that although the EPA + DHA content in the Atlantic salmon fillets fed a predominant vegetable oil feed (100 mg/100 g) is apparently low it could still deliver a higher dose of EPA + DHA than compared to poultry meat. In broiler chicken the average EPA + DHA content was only 34 mg/100 g (Dalziel et al. 2015). Researchers are looking at alternative ways to increase EPA and DHA concentrations in vegetarian fish.

4.12 Overcoming the Challenges—The GMO Approach

One of the approaches adopted to improve the nutritional composition of vegetarian fish is the development of genetically modified (GM) organisms for incorporation in salmon feed which could induce sufficient amounts of EPA and DHA in the fish flesh. Three genetically modified organisms (GMO) were developed, namely GM Camelina (*Camelina sativa*; false flax), GM yeast (*Yarrowia lipolytica*) and metabolically engineered diatom, *Phaeodactylum tricornutum*.

Two interesting studies on the use of genetically modified Camelina as a source of EPA and DHA have been published. In the first study, researchers produced two variants of the Camelina namely RRes_EPA (that only produces EPA) in which the seeds contained EPA levels of up to 31% (mean 24%); and RRes_DHA (that could produce EPA and DHA), in which the seeds accumulated up to 12% EPA and 14% DHA (mean of 11% EPA and 8% DHA). These levels were comparable to the EPA and DHA content in fish oils. However, low levels of undesirable C18 biosynthetic intermediates were present in the GM Camelina (Ruiz-Lopez et al. 2014). In other research, comparable levels of DHA (up to 12.4%) were obtained in GM Camelina but the EPA content was very poor (maximum of 3.2%) (Petrie et al. 2014). While both these studies used the same GM plant, variation in the yield of EPA and DHA might be due to the variations in the promoters, constructs and integration sites used in the genetic engineering process (Napier et al. 2015).

Studies comparing complete replacement of fish oil with GM and wild Camelina oils have been reported. The GM Camelina oil with 20% EPA was substituted for fish oil in the Atlantic salmon feed. In comparison with the fish fed with fish oil feed and wild Camelina oil feed, the inclusion of GM Camelina oil did not affect growth performance, feed efficiency or fish health. The fatty acid profile of the salmon flesh had sufficient EPA and DHA to meet the currently recommended nutritional requirement for humans (Betancor et al. 2015). In a later study a different source of GM Camelina was used to evaluate the growth performance as well as fatty acid composition of Atlantic salmon. In this case the GM oil that had 15% of total ω-3 LC-PUFA with equal EPA and DHA profile was compared with fish oil and wild Camelina oil. The growth performance and health was optimum in GM Camelina oil fed fishes. Similar to the previous study the fatty acid composition with respect to EPA and DHA was ideal for human consumption (Betancor et al. 2016).

In a different approach GM yeast was used to produce the ω-3 fatty acids. Scientists at DuPont collected nearly 40 strains of *Yarrowia lipolytica* and screened for their performance in fermentation and ability to accumulate ω-3 fatty acids. Finally, a strain American Type Culture Collection (ATCC) #20362 was selected for further studies (Xie et al. 2015). They produced three genetically modified strains, Gen I strain Y4305 that produced EPA at more than 15% of its dry cell weight (DCW); the Gen II strain Z1978 that produced EPA at more than 20% of its DCW, and the Gen III HP strain Z5567 that produced EPA at more than 25% of its DCW (Hong et al. 2014; Xue et al. 2013). At present two commercial products are available; the New Harvest™ EPA oil, for a human nutritional supplement; and Verlasso®, farmed salmon fed with GM oil produced from the yeast (Xie et al. 2015).

The Diatom *Phaeodactylum tricornutum* has also been genetically engineered to accumulate DHA in their cells. This GM algae could be used as a replacement for fish oil in fish feeds (Hamilton et al. 2014). However, further studies are required to evaluate its potential as fish oil replacement.

Genetically modified Camelina, yeast and algae have the potential to provide feasible alternatives for fish oil for the aquaculture feed sector. Among these the GM yeast needs special mention as it could be more environmentally sustainable unlike the oil crops which are known for their negative environmental impacts like deforestation, carbon footprint, consumption of water, use of fertilizers, use of pesticides, and impacts on biodiversity. Further, the major concerns on the ethical, economic and environmental sustainability of GM products are still debated and would need a general consensus on their use in commercial aquaculture enterprises. It is therefore necessary to evaluate the various facets of their sustainability, balancing the potential harmful effects as against their benefits before any commercialization.

4.13 Concluding Remarks

Replacement of fishmeal and fish oil in aquafeed by alternative ingredients has long been a topic of research. Partial success to replace fishmeal with ingredients such as soybean meal, poultry meal and blood meals have been reported. The use of plant-based oils, microalgae, and genetically modified (GM) yeast in place of fish oil shows varying results. One of the major challenges facing replacement of fish oil is a reduced nutritional quality, particularly the lower polyunsaturated fatty acids (PUFA) content in the fish produced. The sustainability of GM crops is still a contentious issue and would need further research aimed at consumer safety and acceptability to promote their application in commercial aquaculture. Future aquafeed research priorities may also include improved techniques for mass production of algae, exploring the potential of newer ingredients, and development of precision nutrition strategies to estimate and supplement the essential micronutrients using multiple ingredient combinations of vegetarian origin for optimum survival, growth, and reproduction of fish. The recent innovations in aquafeed technology seem to have brought us closer to the production of a 'vegetarian fish' fed exclusively by plant-based ingredients. It is promising to note that we are striding closer to this reality.

Acknowledgements The graphs used in this chapter have been drawn by Dr. Deepak George Pazhayamadom, Department of Primary Industry and Resources Northern Territory Government, Australia, which is gratefully acknowledged. The photo credits go to KRS.

References

Alltech (2015) Alltech global feed survey summary, 8p. Alltech, Nicholasville, Kentucky, USA. Available from http://www.alltech.com/sites/default/files/alltechglobalfeedsummary2014.pdf

Arvanitoyannis IS, Kassaveti A (2008) Fish industry waste: treatments, environmental impacts, current and potential uses. Int J Food Sci Technol 43(4):726–745

Baeverfjord G, Krogdahl Å (1996) Development and regression of soybean meal induced enteritis in Atlantic salmon, *Salmo salar* L., distal intestine: a comparison with the intestines of fasted fish. J Fish Dis 19(5):375–387

Behçet R (2011) Performance and emission study of waste anchovy fish biodiesel in a diesel engine. Fuel Process Technol 92(6):1187–1194

Bell JG, McEvoy J, Tocher DR, McGhee F, Campbell PJ, Sargent JR (2001) Replacement of fish oil with rapeseed oil in diets of Atlantic salmon (*Salmo salar*) affects tissue lipid compositions and hepatocyte fatty acid metabolism. J Nutr 131(5):1535–1543

Betancor M, Sprague M, Usher S, Sayanova O, Campbell P, Napier JA, Tocher DR (2015) A nutritionally-enhanced oil from transgenic *Camelina sativa* effectively replaces fish oil as a source of eicosapentaenoic acid for fish. Sci Rep 5

Betancor MB, Sprague M, Sayanova O, Usher S, Metochis C, Campbell PJ, Tocher DR (2016) Nutritional evaluation of an EPA-DHA oil from transgenic *Camelina sativa* in feeds for post-smolt Atlantic salmon (*Salmo salar* L.). PLoS One 11(7):e0159934

Bjørkli J (2002) Protein og energiregnskap hos laks, kylling, gris og lam (Protein and energy account in salmon, chicken, pig and lamb). M.Sc. Thesis, Norwegian University of Life Sciences (UMB), Norway

Burr GS, Wolters WR, Barrows FT, Hardy RW (2012) Replacing fishmeal with blends of alternative proteins on growth performance of rainbow trout (*Oncorhynchus mykiss*), and early or late stage juvenile Atlantic salmon (*Salmo salar*). Aquaculture 334:110–116

Carter C, Hauler R (2000) Fish meal replacement by plant meals in extruded feeds for Atlantic salmon, *Salmo salar* L. Aquaculture 185(3):299–311

Dalziel CJ, Kliem KE, Givens DI (2015) Fat and fatty acid composition of cooked meat from UK retail chickens labelled as from organic and non-organic production systems. Food Chem 179:103–108

Davidson J, Barrows FT, Kenney PB, Good C, Schroyer K, Summerfelt ST (2016) Effects of feeding a fishmeal-free versus a fishmeal-based diet on post-smolt Atlantic salmon *Salmo salar* performance, water quality, and waste production in recirculation aquaculture systems. Aquac Eng 74:38–51

De Smet S, Vossen E (2016) Meat: the balance between nutrition and health. A review. Meat Sci

Drew M, Borgeson T, Thiessen D (2007) A review of processing of feed ingredients to enhance diet digestibility in finfish. Anim Feed Sci Technol 138(2):118–136

Esteban M, Garcia A, Ramos P, Marquez M (2007) Evaluation of fruit–vegetable and fish wastes as alternative feedstuffs in pig diets. Waste Manag 27(2):193–200

FAO (2016) Food Outlook: Biannual report on global food markets, June 2016. Food and Agriculture Organization of the United Nations

García-Ortega A, Kissinger KR, Trushenski JT (2016) Evaluation of fish meal and fish oil replacement by soybean protein and algal meal from *Schizochytrium limacinum* in diets for giant grouper *Epinephelus lanceolatus*. Aquaculture 452:1–8

Ghaly A, Ramakrishnan V, Brooks M, Budge S, Dave D (2013) Fish processing wastes as a potential source of proteins, amino acids and oils: a critical review. J Microb Biochem Technol

Haas S, Bauer J, Adakli A, Meyer S, Lippemeier S, Schwarz K, Schulz C (2016) Marine microalgae *Pavlova viridis* and *Nannochloropsis* sp. as n-3 PUFA source in diets for juvenile European sea bass (*Dicentrarchus labrax* L.). J Appl Phycol 28(2):1011–1021

Hamilton ML, Haslam RP, Napier JA, Sayanova O (2014) Metabolic engineering of *Phaeodactylum tricornutum* for the enhanced accumulation of omega-3 long chain polyunsaturated fatty acids. Metab Eng 22:3–9

Henriques J, Dick JR, Tocher DR, Bell JG (2014) Nutritional quality of salmon products available from major retailers in the UK: content and composition of n-3 long-chain PUFA. Br J Nutr 112:964–975

Hixson SM, Parrish CC, Anderson DM (2014) Full substitution of fish oil with camelina (*Camelina sativa*) oil, with partial substitution of fish meal with camelina meal, in diets for farmed Atlantic salmon (*Salmo salar*) and its effect on tissue lipids and sensory quality. Food Chem 157:51–61

Hong SP, Sharpe PL, Xue Z, Yadav NS, Zhang H, Zhu QQ (2014) Recombinant microbial host cells for high eicosapentaenoic acid production: google patents

Huntington T, Hasan MR (2009) Fish as feed inputs for aquaculture–practices, sustainability and implications: a global synthesis fish as feed inputs for aquaculture: practices, sustainability and implications vol 518. FAO Rome, pp 1–61

Kaushik S, Cravedi J, Lalles J, Sumpter J, Fauconneau B, Laroche M (1995) Partial or total replacement of fish meal by soybean protein on growth, protein utilization, potential estrogenic or antigenic effects, cholesterolemia and flesh quality in rainbow trout, *Oncorhynchus mykiss*. Aquaculture 133(3):257–274

Kiron V, Sørensen M, Huntley M, Vasanth GK, Gong Y, Dahle D, Palihawadana AM (2016) Defatted biomass of the microalga, *Desmodesmus* sp., can replace fishmeal in the feeds for Atlantic salmon. Front Mar Sci 3:67

Kissinger KR, García-Ortega A, Trushenski JT (2016) Partial fish meal replacement by soy protein concentrate, squid and algal meals in low fish-oil diets containing *Schizochytrium limacinum* for longfin yellowtail *Seriola rivoliana*. Aquaculture 452:37–44

Kupchinsky ZA., Coyle SD, Bright LA, Tidwell JH (2015) Evaluation of heterotrophic algae meal as a diet ingredient for channel catfish, *Ictalurus punctatus*. J World Aquacult Soc 46(4): 445-452

Kushi LH, Byers T, Doyle C, Bandera EV, McCullough M, Gansler T, Thun MJ (2006) American cancer society guidelines on nutrition and physical activity for cancer prevention: reducing the risk of cancer with healthy food choices and physical activity. CA Cancer J Clin 56(5): 254–281

Li MH, Robinson EH, Tucker CS, Manning BB, Khoo L (2009) Effects of dried algae *Schizochytrium* sp., a rich source of docosahexaenoic acid, on growth, fatty acid composition, and sensory quality of channel catfish *Ictalurus punctatus*. Aquaculture 292(3):232–236

Lin C-Y, Li R-J (2009) Engine performance and emission characteristics of marine fish-oil biodiesel produced from the discarded parts of marine fish. Fuel Process Technol 90(7): 883–888

Mallison A (2013) Marine ingredients overview. Paper presented at the Intrafish Investment Forum, London. Available from http://www.intrafish.com/incoming/article1381588.ece/BINARY/8_Special+Pres_Mallison+Intrafish+London+19Nov13-2.pdf

Miller MR, Nichols PD, Carter CG (2007) Replacement of fish oil with thraustochytrid *Schizochytrium* sp. L oil in Atlantic salmon parr (*Salmo salar* L) diets. Comp Biochem Physiol A Mol Integr Physiol 148(2):382–392

Murai T, Ogata H, Hirasawa Y, Akiyama T, Nose T (1987) Portal absorption and hepatic uptake of amino acids in rainbow trout (*Onchorhynchus masou*) force-fed complete diets containing casein or crystalline amino acids. Bull Jpn Soc Sci Fish

Napier JA, Usher S, Haslam RP, Ruiz-Lopez N, Sayanova O (2015) Transgenic plants as a sustainable, terrestrial source of fish oils. Eur J Lipid Sci Technol 117(9):1317–1324

OECD/FAO (2016) OECD-FAO agricultural outlook. OECD agriculture statistics (database). http://dx.doi.org/10.1787/agr-data-en2025

Olli JJ, Krogdahi Å (1994) Nutritive value of four soybean products as protein sources in diets for rainbow trout (*Oncorhynchus mykiss*, Walbaum) reared in fresh water. Acta Agric Scand A Anim Sci 44(3):185–192

Øverland M, Sørensen M, Storebakken T, Penn M, Krogdahl Å, Skrede A (2009) Pea protein concentrate substituting fish meal or soybean meal in diets for Atlantic salmon (*Salmo salar*)—effect on growth performance, nutrient digestibility, carcass composition, gut health, and physical feed quality. Aquaculture 288(3):305–311

Petrie JR, Shrestha P, Belide S, Kennedy Y, Lester G, Liu Q, Nichols PD (2014) Metabolic engineering *Camelina sativa* with fish oil-like levels of DHA. PLoS One 9(1):e85061

Pike IH, Tocher DR (2016) Could an El Niño event put dietary supplies of n-3 long-chain polyunsaturated fatty acids (EPA and DHA) in jeopardy. Eur J Lipid Sci Technol

Ruiz-Lopez N, Haslam RP, Napier JA, Sayanova O (2014) Successful high-level accumulation of fish oil omega-3 long-chain polyunsaturated fatty acids in a transgenic oilseed crop. Plant J 77(2):198–208

Sarker PK, Kapuscinski AR, Lanois AJ, Livesey ED, Bernhard KP, Coley ML (2016) Towards sustainable aquafeeds: complete substitution of fish oil with marine *Microalga Schizochytrium* sp. Improves growth and fatty acid deposition in Juvenile Nile Tilapia (*Oreochromis niloticus*). PLoS One 11(6):e0156684

Shepherd C, Jackson A (2013) Global fishmeal and fish-oil supply: inputs, outputs and markets. J Fish Biol 83(4):1046–1066

Shepherd CJ, Monroig O, Tocher DR (2015) Production of high quality, healthy salmon from a changing raw material base, with special reference to a sustainable Scottish industry. In: A study commissioned by the Scottish aquaculture research forum (SARF)

Shepherd CJ, Monroig O, Tocher DR (2017) Future availability of raw materials for salmon feeds and supply chain implications: the case of Scottish farmed salmon. Aquaculture 467:49–62. http://dx.doi.org/10.1016/j.aquaculture.2016.08.021

Shi X, Luo Z, Chen F, Wei CC, Wu K, Zhu XM, Liu X (2017). Effect of fish meal replacement by Chlorella meal with dietary cellulase addition on growth performance, digestive enzymatic activities, histology and myogenic genes' expression for crucian carp *Carassius auratus*. Aquac Res 48:3244–3256. http://dx.doi.org/10.1111/are.13154

Sofia F (2016) State of world fisheries and aquaculture. FAO, Rome, Italy

Sørensen M, Berge GM, Reitan KI, Ruyter B (2016) Microalga *Phaeodactylum tricornutum* in feed for Atlantic salmon (*Salmo salar*)—effect on nutrient digestibility, growth and utilization of feed. Aquaculture 460:116–123

Sprague M, Dick JR, Tocher DR (2016) Impact of sustainable feeds on omega-3 long-chain fatty acid levels in farmed Atlantic salmon, 2006–2015. Sci Rep 6, 21892. http://doi.org/10.1038/srep21892

Tacon AG, Metian MR, Tacon MAG, Hasan MR, Metian M (2011) Demand and supply of feed ingredients for farmed fish and crustaceans: trends and prospects

Tibaldi E, Zittelli GC, Parisi G, Bruno M, Giorgi G, Tulli F, Poli B (2015) Growth performance and quality traits of European sea bass (*D. labrax*) fed diets including increasing levels of freeze-dried *Isochrysis* sp. (T-ISO) biomass as a source of protein and n-3 long chain PUFA in partial substitution of fish derivatives. Aquaculture 440:60–68

Tidwell JH, Coyle SD, Bright LA, Yasharian D (2005) Evaluation of plant and animal source proteins for replacement of fish meal in practical diets for the largemouth bass *Micropterus salmoides*. J World Aquac Soc 36(4):454–463

Torstensen B, Espe M, Sanden M, Stubhaug I, Waagbø R, Hemre G-I, Olsvik P (2008) Novel production of Atlantic salmon (*Salmo salar*) protein based on combined replacement of fish meal and fish oil with plant meal and vegetable oil blends. Aquaculture 285(1):193–200

UN (2015). Transforming our world: the 2030 agenda for sustainable development. Resolution adopted by the United Nations General Assembly, A/RES/70/1, 21 October 2015. United Nations

Velasquez SF, Chan MA, Abisado RG, Traifalgar RFM, Tayamen MM, Maliwat GCF, Ragaza JA (2016) Dietary Spirulina (*Arthrospira platensis*) replacement enhances performance of juvenile Nile tilapia (*Oreochromis niloticus*). J Appl Phycol 28(2):1023–1030

Wang Y, Li M, Filer K, Xue Y, Ai Q, Mai K (2016) Evaluation of *Schizochytrium* meal in microdiets of Pacific white shrimp (*Litopenaeus vannamei*) larvae. Aquacu Res

Waterman J (1979) Measures, stowage rates and yields of fishery products. Torry Research Station

Wong AT, Chan DC, Ooi EM, Ng TW, Watts GF, Barrett PHR (2013) Omega-3 fatty acid ethyl ester supplementation decreases very-low-density lipoprotein triacylglycerol secretion in obese men. Clin Sci 125(1):45–51

Xie D, Jackson EN, Zhu Q (2015) Sustainable source of omega-3 eicosapentaenoic acid from metabolically engineered *Yarrowia lipolytica*: from fundamental research to commercial production. Appl Microbiol Biotechnol 99(4):1599–1610

Xue Z, Sharpe PL, Hong S-P, Yadav NS, Xie D, Short DR, Wang J (2013) Production of omega-3 eicosapentaenoic acid by metabolic engineering of Yarrowia lipolytica. Nat Biotechnol 31(8): 734–740

Ytrestøyl T, Aas TS, Åsgård T (2015) Utilisation of feed resources in production of Atlantic salmon (Salmo salar) in Norway. Aquaculture 448:365–374

Chapter 5
Sustainable Production of Shrimp in Thailand

Pattira Pongtippatee, Krishna R. Salin, Gabriel Arome Ataguba and Boonsirm Withyachumnarnkul

Abstract World production of farmed shrimp is focused on a few species. At present, the Pacific whiteleg shrimp *Litopenaeus (Penaeus) vannamei* tops the list, followed by the black tiger shrimp *Penaeus monodon* and a few others. The former attains the market size between 15 and 25 g within three months in culture, while *P. monodon* requires at least four months to reach the marketable size of ~25 g and larger. It is therefore desirable to produce stocks of *P. monodon* with a fast-growth-rate trait, either through selective breeding or via other modes of scientific invention. Among others, chromosome manipulation of *P. monodon* that confers three sets of chromosomes (3n) to the shrimp, a condition called triploidy and a feat that was achieved through a non-GMO technique, could be the answer. Thermal chromosome set manipulation for triploid induction is considered as a safe and environmentally friendly technique to produce sterile offspring for genetic protection, and prevent genetic pollution from aquaculture stocks in the wild. The black tiger shrimp can be induced to yield a high percentage of triploid offspring with abnormal reproductive histology. Two studies applying different induction methods briefly at the very early stage of embryo formation—one from Australia (chemical shock) and another from Thailand (cold shock) have reported successful results. The triploid shrimps produced from chemical shock method had reduced growth rate, while that from cold shock displayed higher growth rate compared to

P. Pongtippatee
Faculty of Science and Industrial Technology, Aquatic Animal Biotechnology Research Center, Prince of Songkla University, Surat Thani Campus, Surat Thani, Thailand

B. Withyachumnarnkul (✉)
Department of Anatomy, Faculty of Science, Prince of Songkla University, Hat Yai, Songkhla, Thailand
e-mail: wboonsirm@yahoo.com

K. R. Salin · G. A. Ataguba
Aquaculture and Aquatic Resources Management (AARM), Asian Institute of Technology, Pathum Thani, Thailand

B. Withyachumnarnkul
Aqua Academy Farm, Tha Chana, Surat Thani, Thailand

F. I. Hai et al. (eds.), *Sustainable Aquaculture*, Applied Environmental Science and Engineering for a Sustainable Future,
https://doi.org/10.1007/978-3-319-73257-2_5

shrimp with the natural double set of chromosomes. Gender distribution of the triploid *P. monodon* induced by cold shock was skewed towards females, the larger size of the two sexes and hence is favored more in aquaculture. Because of the favorable results of the cold shock method, it is currently pursued to produce triploid *P. monodon* for commercial purposes. The production is accomplished through an automatic cold-shock induction system, consisting of spawning detection and cold shock treatment sections. This triploid induction project is an extension of the ongoing selective breeding program of a specific pathogen-free *P. monodon* that has already been commercially launched, and is part of an initiative to promote sustainable genetic stock improvement protocols for this shrimp species.

Keywords Triploid *Penaeus monodon* · Commercial production
Spawning detection device · Cold shock · Triploidy induction device

5.1 Introduction

Shrimp aquaculture dates back to the late twentieth century with efforts of the French, Chinese and North Americans in the development of culture techniques for various species of penaeid shrimp (Briggs et al. 2004). World shrimp production (penaeid shrimps and freshwater prawns) reached a peak of about 4 million metric tonnes in 2011 and grew by 6% between 2012 and 2014 with value increasing from 1.92 billion dollars in 2011 to 2.36 billion dollars in 2014 (Fig. 5.1) (FAO 2016). Thailand with half of its aquaculture production composed of crustaceans is a leading producer of shrimps globally at an estimated production level of 2.5×10^5 metric tonnes in 2013 (Ferdouse 2014). Total aquaculture production

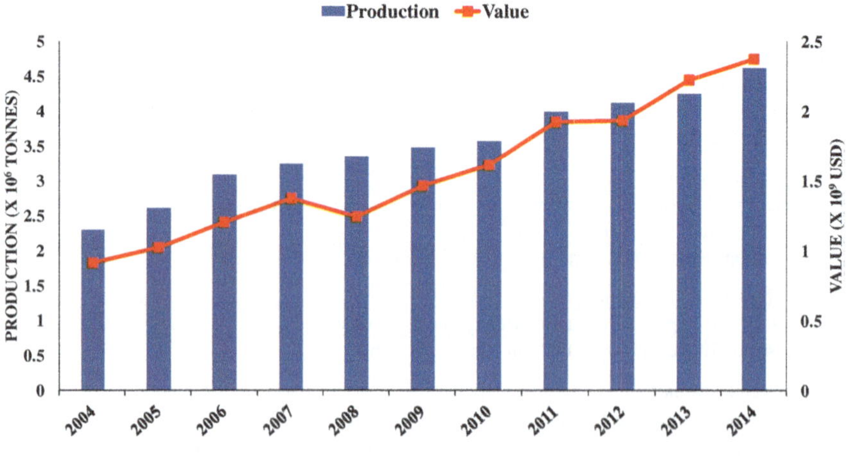

Fig. 5.1 Global production and value of shrimps from 2004–2014. *Source* FAO (2016)

from Thailand was estimated at 1.2 million metric tonnes between 2011 and 2012 with a drop of 0.2 million tonnes from the production volume in 2009 (FAO 2014).

Shrimp production has often come at a cost to the environment. Firstly, collection of wild postlarvae was the bane of the industry. Secondly, coastal shrimp culture has led to the destruction of mangrove forests in many places. Addressing sustainability will therefore involve technologies that ensure captive reproduction and development of high performing strains as well as culture techniques that de-emphasize mangrove clearing.

5.2 Impacts of Shrimp Aquaculture

Traditional shrimp farming practices that prevailed in Asia represented a low key, sustainable activity that had no serious impact on the environment. However, over the years, shrimp culture has evolved into a capital intensive industrial activity and massive corporate investments in the sector have led to considerable increase in production volumes and profits. The risks in shrimp farming have also consequently become greater and the impacts more visible. The impacts of shrimp aquaculture is often viewed in the context of its perceived effects on the surrounding environment, particularly in the changes on patterns of land use, eutrophication, and disease threats from farmed shrimps. A more detailed account of the general impacts of aquaculture is provided elsewhere in this book under the chapter on sustainability (see Chap. 1).

5.3 Land Cover Change

Shrimp aquaculture has expanded over the years and in Thailand, it blossomed in the early 1980s with tremendous impact on the coastal ecosystem. The effects of shrimp aquaculture can be either environmental or socio-economic (Bert 2007; Mitsch and Gosselink 2015; Patamasiriwat et al. 1999) but no matter the type of impact, its extent can be massive and perplexing (Bert 2007).

Shrimp aquaculture is the major reason behind land cover change in coastal areas through mangrove clearing to make way for aquaculture ponds. The mangrove cover in South and Southeast Asia dropped from 6.36×10^6 ha in 2000 to 6.02×10^6 ha in 2010 (FAO 2010). In Thailand, 50% of mangrove cover was lost between 1975 and 1993 principally due to shrimp farm expansion (EJF 2006) and reached a total area of 6.6×10^4 ha in 1996 (Barbier 2006). The total mangrove area coverage in Thailand had reduced from a level of 244,000 ha in 2010 to 240,000 ha in 2015 (FAO 2015).

5.4 Nutrient Loading and Pollution

In addition to the changes in land use, overstocking of shrimps in culture units has led to degradation of coastal areas by pollution, excessive sediment discharge, nutrient loading and eutrophication (Bert 2007; Patamasiriwat et al. 1999). Organic matter loads discharged from shrimp farms accounted for a huge proportion of nitrogen, phosphorous and suspended solids present in the receiving waters (Primavera 2006), while unpalatable changes in land structure have resulted from shrimp culture (Lakshmi and Rajagopalan 2000). Drastic losses in shrimp farms were reported in Asia within a decade of shrimp aquaculture development beginning 1980 due to water quality problems associated with overstocking (Flaherty and Karnjanakesorn 1995; Kautsky et al. 2000). Competing claims on resources as well as water channels used in shrimp aquaculture also create huge socio-economic issues. In Thailand, land ownership changes from shrimp aquaculture have been reported to favor the rich at the expense of the poor while the poor knowledge of management has not been clement to the small holders (Briggs 2001).

5.5 Health Impacts

Shrimp aquaculture also involved the use of chemicals, therapeutants and toxicants. According to Rico et al. (2012), a total of 36 different antibiotics were used in aquaculture activities in Asia. In Thailand too, the use of antibiotics in shrimp aquaculture had been reported (Holmstrom et al. 2003; Gräslund et al. 2003; Szuster 2006). The impacts of indiscriminate use are quite widespread and are often linked to human health concerns as well as the integrity of the ecosystem (Gräslund et al. 2003; Holmstrom et al. 2003). In a study of use of chemical and biological agents in shrimp aquaculture, Gräslund et al. (2003) reported that 74% of shrimp farmers in Thailand used one or several types of antibiotics including chloramphenicol and oxolinic acid, 67% used chlorine with a 58% using unidentified chlorine compounds, 43% used formalin/formaldehyde, 42% used quaternary ammonium compounds, 38% used Benzalkonium chloride (BKC), 37% used iodophors, and with organophosphates used by 22% of the farmers. Chloramphenicol poses a hazard of aplastic anaemia (CFS 2005) and development of antibiotic resistance in bacteria (Ng et al. 2014) with oxolinic acid also posing the latter hazard (Guardabassi et al. 2000) as well as persistence in the environment (Weston 1996). Organophosphates such as Trichlorfon are metabolic disruptors and derivatives of toxic compounds with possibility of carcinogenesis and mutagenesis (Kamrin 1997).

5.6 Mitigating the Impacts

The challenges posed by shrimp aquaculture to the environment demand a synergy among stakeholders to ensure their mitigation. Briggs (2001) and Patamasiriwat et al. (1999) presented several options for the remediation of issues relating to unsustainable shrimp aquaculture in Thailand. These include proper governance, site selection, control of effluent quality, management of introductions and transfers, proper pond design and management, disease management and reporting, control of chemical and drug use, improved aquaculture extension and training, control of feed quality, as well as enhanced research and development. Thailand has progressed well in most of these fronts with remarkable achievements over the past 20 years (since mid-1990s), although the sector has been under continued exposure to the challenges of controlling diseases, and increased efforts for enhancing the yield from shrimp aquaculture.

The much needed shift from the reliance of baby shrimp collected for aquaculture from the wild was brought about by the emergence of advanced hatchery techniques to produce good quality postlarvae, supported by successful broodstock domestication. A principal step in solving the issue of seed supply was development of domesticated broodstock of the black tiger shrimp, being a native species in many parts of Asia and in view of the issues related to the dependence on wild broodstock for hatchery production. The application of biotechnology to mitigate problems in shrimp aquaculture in Thailand started with the application of polymerase chain reaction (PCR) technology in shrimp aquaculture for improved growth and reproduction (Withyachumnarnkul et al. 2001). The use of PCR in shrimp aquaculture has further been expanded to the detection of pathogens in farmed shrimp and the culture environment.

5.7 Production of Black Tiger Shrimp in Thailand

The black tiger shrimp, *P. monodon* Fabricius (1798), is one of the largest penaeid shrimps in the world with considerable commercial importance in international markets. This shrimp species is indigenous to tropical oceans and its aquaculture has been successfully practiced in many tropical countries, including Thailand, for more than five decades, with significant contribution to the development of the sector. However, since 1989, *P. monodon* farming industry has been hard-hit by disease outbreaks especially from yellow-head disease (YHD), white-spot disease (WSD) and monodon slow-growth syndrome (MSGS) (Flegel 1997; Withyachumnarnkul et al. 2004), and these three diseases continue to affect shrimp farms in the region (Withyachumnarnkul, pers. comm.). The exotic Pacific whiteleg shrimp *Litopenaeus (Penaeus) vannamei* introduced to Thailand in 1998 for obvious economic reasons became popular soon after introduction and its culture methods were standardized by enterprising farmers. Following the outbreaks of

Table 5.1 Comparison of normal culture performance between the two most farmed shrimp species (before 2016)

Shrimp species	Marketable size (g)	Culture period (mo)	Stocking density (PL/m^2 or PL/m^3)[a]	Production (tonnes/ha)
Litopenaeus vannamei	12–15	2–3	80–100	15–25
Penaeus monodon	25 and higher	4 months and longer	25–40	5–15

[a]For *P. monodon*, the stocking density is of PL/m^2 and for *L. vannamei*, PL/m^3

MSGS in 2009, the majority of Thai shrimp farmers switched to *L. vannamei,* as their preferred species. Farming of *L. vannamei* flourished and had almost entirely wiped out the aquaculture of *P. monodon* until 2013 when a new disease, the acute hepatopancreatic necrosis disease (AHPND) caused more than 50% loss of the shrimp production in Thailand. This led to a fraction of the Thai farmers to switch back to *P. monodon* culture despite the fact that *P. monodon* was also affected by AHPND (FAO 2013). This augurs well for this shrimp being a native species, and its farming considered very important for sustainability of the shrimp farming industry in Thailand.

One of the major reasons that Thai farmers have switched back to *P. monodon* is the apparent difference between the two shrimps in size that fetches a better price, considering standard farming conditions for both species. At marketable size, *P. monodon* normally reaches 25–40 g body weight (BW) within a culture period of 4 months, from a stocking density of 25–40 postlarvae (PL)/m^2 (Table 5.1). In contrast, *L. vannamei* normally reaches only 12–15 g BW within 3 months, when stocked at 80–100 PL/m^2. In fact, *L. vannamei* can also grow up to 25–40 g BW within the same period and stocking density as *P. monodon*, but the price of large-sized *L. vannamei* in Thailand is usually lower than that of *P. monodon*. With improved farming technology, however, *L. vannamei* farming performance could be much higher than that depicted in Table 5.1. During 2016, *L. vannamei* production in Thailand has reported consistent production of 30–60 tonnes/ha following high stocking density and under innovative management.

5.8 Genetic Improvement Program for Black Tiger Shrimp

In 2006, the Shrimp Genetic Improvement Center (SGIC), a government organization under the supervision of the National Science and Technology Development Agency (NSTDA), Thailand, initiated a national program for domestication and selective breeding of a specific pathogen-free (SPF) *P. monodon*, which targeted fast growth and disease resistant traits. By 2016, the Center has successfully produced domesticated *P. monodon* up to the 7th generation, and disseminated the SPF

P. monodon seed to shrimp farmers. Performance trials in the field have revealed multiple advantages of the genetically improved SPF seed over the seed produced from wild broodstock. The demand for the SPF seed has been growing because of their superior performance in farm ponds.

A number of biotechnological tools have been used towards development of the SPF broodstock of *P. monodon*, in an attempt to improve growth characteristics of the genetically improved shrimp. Growth performance of black tiger shrimps in culture conditions vary widely based on environmental parameters, while it could also be a function of the sex-dependent size difference. It has been shown that females of *P. monodon* grow faster than males enabling them to attain harvestable size at least a month earlier than males (Hansford and Hewitt 1994; Hansford 1991) as evidenced in the case of several other penaeid shrimps (Campos-Ramos et al. 2006). Restricted breeding by producing sterile offspring is also vital to protect the wild stock from genetic introgression when captive bred shrimp are used for stock enhancement in the wild, as well as to provide some degree of genetic protection to the selectively bred stocks. Triploidy induction techniques were applied at SGIC with the aims to confer reproductive sterility for genetic protection of the stocks, and to skew the sex ratio towards faster-growing females. The outcome of this attempt was encouraging as the triploid (3n) *P. monodon* grew significantly faster than their diploid (2n) counterparts, and most 3n shrimp were females (Pongtippatee et al. 2012). These studies have given compelling evidence in support of rearing 3n over 2n *P. monodon*, demonstrating the feasibility of 3n *P. monodon* commercial production.

It is also important to note that animals produced by inducing triploidy are not regarded as genetically modified organisms (GMOs) in the context of food safety, environmental and ethical concerns because they contain no foreign gene. The 3n shrimp has its own DNA (deoxyribonucleic acid) except with a change in the ploidy (chromosome number), while triploidy is also a naturally occurring phenomenon in the case of many plants and animals including fish. GMOs are produced by genetic engineering (GE) methods often referred to as transgenesis, which involves the introduction of a foreign gene of interest to the target species by a series of GE protocols. According to the guidelines of the Food and Drug Administration (FDA) of USA genetically engineered organisms are those modified by recombinant DNA techniques, including their entire lineage of modified organisms (HHS-FDA 2009). Thus there is a clear demarcation between transgenic fish which are genetically engineered organisms and triploid fish that are produced by a non-GMO protocol for genetic improvement.

5.9 Principle of Triploidy Induction in Shrimps

Sellars et al. (2010) provided an elaborate review of triploidy induction on penaeid shrimp; therefore this review will concentrate chiefly on triploidy induction in *P. monodon*. Chromosome set manipulation is aimed at altering chromosome number in the resulting offspring to achieve gonadal sterilization, direct the gender, control the viability of hybrids, and to produce clones. This effectively checks any deleterious breeding interactions between feral and domesticated stocks. Production of all female populations of shrimp is one of the most useful outcomes of ploidy manipulation. This is also important in increasing productivity of the stock while conferring reproductive sterility on the females to prevent genetic interactions (Sellars et al. 2009). This in itself is subject to 100% success in induction of ploidy which is not achievable in most attempts (Harrell et al. 1998; Overturf 2009; Sellars et al. 2010; Vallero 2015).

It is the process of meiosis in sex cells and the sequence of events from post-fertilization to the first mitosis are the basis for understanding how triploidy is induced. Shortly after fertilization, the second meiotic division can be hindered by the application of a shock in the form of chemicals, mechanical pressure or thermal change to retain the second polar body in the fertilized egg, hence 2n from the egg in combination with 1n from the sperm results in the triploid condition. In the first meiosis, the first polar body is extruded and disintegrates. The second meiosis is completed after fertilization, during which the second polar body is extruded and disintegrates as well. If the extrusion of the second polar body is inhibited, then the set of chromatids that would have been normally extruded will instead remain inside the egg, so the fertilized egg ends up with three sets of chromosomes: two from the egg and one from the sperm (Fig. 5.2). This fertilized egg will pass on its

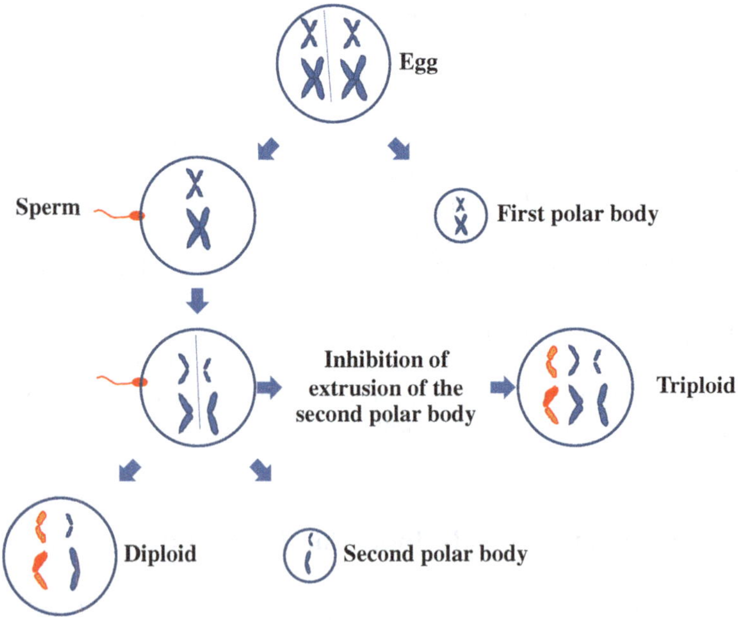

Fig. 5.2 Schematic illustration of triploidy induction by inhibiting extrusion of the second polar body

triploid condition in subsequent mitoses, and the animal will have triploid sets of chromosomes in every cell.

Inhibiting the extrusion of the second polar body is one way to induce the triploid condition, while there are also other means of induction. Inhibiting the extrusion of the first polar body appears to have the same effect as inhibiting the extrusion of the second polar body. Inhibition of cytokinesis (cytoplasmic division) of the first mitosis would result in four sets of chromosomes, or tetraploid chromosomes (4n); and mating 4n and 2n parents would result in the interploid, 3n offspring (Cassani et al. 1990). The choice between these methods to induce triploidy depends on their feasibility, which varies by the species and a need for the use of sustainable technology.

For instance, in *P. monodon*, the first polar body is extruded at 2–5 min and the second polar body at 8–14 min post-spawning (Hall et al. 1999; Pongtippatee-Taweepreda et al. 2004). It is possible to inhibit the extrusion of the first polar body, but because of the time constraint it is more convenient to inhibit the extrusion of the second polar body.

5.10 Current Methods in Triploidy Induction—Advantages and Disadvantages

The current methods for triploidy induction are based on thermal shock (heat or cold), pressure shock or chemical shock. Such shocks prevent the contraction of the spindles that pull the two sets of chromosomes apart during meiotic division, thus preventing the extrusion of the polar bodies.

Thermal and pressure shocks prevent the contraction of the spindles by physical means, while a chemical shock does so by disrupting microtubules of the spindle (Komen et al. 1991). Standardization of hydrostatic pressure for triploid induction varies according to species. An optimal pressure of 9000 psi applied for a duration of 2 min to the eggs of *Oreochromis niloticus* after 40 min of fertilization has been reported to induce triploidy in the species (Hussain et al. 1993). For coho salmon (*Oncorhynchus kisutch*), treatment of eggs for 4 min with a pressure of 10,000–12,000 psi, applied 20 min post-fertilization produced total triploidy (Teskeredzic et al. 1993). Similarly, a pressure of 10,000 psi applied for 5 min was also reported to induce triploidy in brown trout (*Salmo trutta*) (Preston et al. 2013). Shorter post-fertilization durations before hydrostatic pressure application have been reported for various species including the pearl oyster, *Pinctada martensii* (5–7 min) using a much lower pressure of 2800–3500 psi (Shen et al. 1993); the grass carp *Ctenopharyngodon idella* (4 min post-fertilization) and a pressure of 7000–8000 psi (Cassani and Caton 1986); and the European seabass *Dicentrarchus labrax* (6 min post-fertilization) with a pressure of 8500 psi (Peruzzi and Chatain 2000). Water temperature as well as species differences account for the variation in time intervals needed after fertilization as well as the duration of shock application.

Reports detailing the application of hydrostatic pressure in inducing triploidy in shrimp are rare. However, Foote et al. (2012) reported the use of hydrostatic pressure between 1000 and 3000 psi to induce triploidy in two penaeid shrimps *P. monodon* and *P. japonicus* using the spawning vessel as pressure chamber.

Currently, the chemicals used to induce triploidy are colchicine, cytochalasin B, nitric oxides, freon, 6-diaminomethylpurine (6-DMAP) and caffeine (Benfey 1999; Nell 2002; Tiwary et al. 2004; Maxime 2008; Piferrer et al. 2009; Sellars et al. 2010, 2006). Among these chemicals, 6-DMAP is the most popular one, probably because of its high success rate. Chemical method is carried out mainly as experimental activity in the laboratory (Rottmann et al. 1991), and not highly favored for commercial applications.

Each induction method has its advantages and disadvantages. Thermal shock is advantageous when large volumes of eggs need to be treated, and it is the least expensive method. Generally, it is believed that cold shock is more efficient than heat shock for triploidy induction in species from the tropical zone, while heat shock is more efficient with species living in the temperate zone (da Silva et al. 2007); this appears a plausible hypothesis but currently is not validated by sufficient scientific evidence. A thermal shock is comparatively more effective with smaller size of eggs, because a sudden temperature change penetrates smaller eggs faster than larger ones (Teskeredžić et al. 1993). This size difference is insignificant when a pressure shock is applied since pressure is transmitted instantly and uniformly throughout a confined fluid (Teskeredžić et al. 1993; Piferrer et al. 2009). However, an advantage of thermal shock over pressure shock is that no special equipment is needed even when large numbers of eggs are treated in one batch. Furthermore, pressure shocks would damage chromosomes causing fusions, bridges, fragments, gaps, and rings, and would lead to abnormal larval morphology, retarded development, or mortality (Yamazaki and Goodier 1993).

The major concerns on using chemical method of induction are its inconsistent results and low efficiency in producing triploids (Guo et al. 1996). There are also human safety concerns when animals aimed for human consumption are treated by chemical shock for triploid induction. There is evidence that 6-DMAP inhibits mitotic cell division (Simili et al. 1997; Rime et al. 1989). It has been reported to cause retarded growth of triploid *P. japonicus* in comparison with diploids (Coman et al. 2008). On the other hand, freon gas is indicted in ozone layer depletion (Larderel et al. 2001). Nitrogen oxide is also toxic to humans (WHO 2000). With the forgoing, chemical methods of triploidy induction are not environmentally friendly with adverse effects on both human and environmental wellbeing. Sustainable triploid production therefore lies in the use of thermal induction in comparison with other means of triploidy induction in shrimps.

5.11 Triploidy Induction for *P. monodon*

Polyploidy induction has been carried out in several shrimp species of economic significance, including Chinese shrimp *Fenneropenaeus chinensis*, Kuruma shrimp *Marsupenaeus japonicus*, red-legged banana shrimp *F. indicus*, red endeavor shrimp *Metapenaeus ensis*, banana shrimp *F. merguiensis,* and in *L. vannamei* (Sellars et al. 2006, 2010). All the induction methods mentioned have been tried in these species, and it was found that thermal and chemical shocks are more successful than pressure shocks.

The mechanism of triploidy induction is to consistently inhibit extrusion, either of the first or of the second polar body. In *P. monodon*, two induction methods have been employed successfully, namely by using the chemical 6-DMAP (Sellars et al. 2012b) or by cold shock (Pongtippatee et al. 2012). Both studies aimed at inhibiting extrusion of the second polar body. In the chemical shock treatment, 6-DMAP at 150–200 μM was introduced at 7 min post-spawning for treatment duration of 10 min. The cold shock induction was done using cold water of 8 °C applied to fertilized eggs for 10 min, starting from 8 min post-spawning.

The reason for setting a specific time and a treatment duration for triploid induction of *P. monodon* is to shock the majority of eggs when they are about to extrude their second polar bodies. The exact timing of this procedure is vital in view of the typical spawning and early embryonic development of *P. monodon* with a time gap between the first and last batch of eggs extruded from the gonopore during the spawning process (Fig. 5.3). The spawning duration for *P. monodon* broodstock

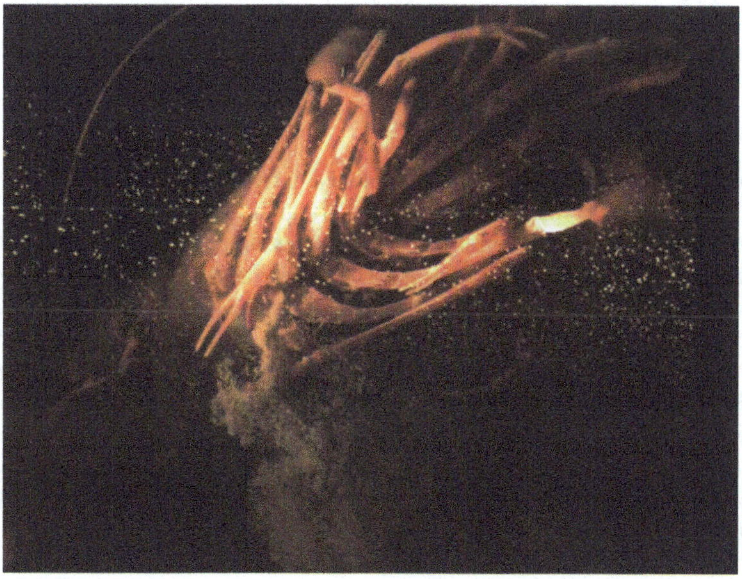

Fig. 5.3 Spawning of *Penaeus monodon*. Usually about 200,000 eggs are released from gonopore in each spawning, which lasts for 3–7 min

from the first to the last egg (200,000 eggs/spawn on average) is 3–7 min. Upon contact with seawater the eggs undergo cortical reaction, which is mainly the extrusion of cortical rods from the eggs, and this event is completed within the first minute post-spawning (Pongtippatee-Taweepreda et al. 2004). Therefore, the first egg that comes out from gonopore will undergo cortical reaction first and is more advanced in its embryogenesis than the eggs that are released subsequently. Obviously, this 3–7 min lag produces unequal staging of the eggs or their asynchronous development, particularly in cortical reaction, formation of hatching envelope, and extrusion of the first and the second polar body. The difference in extrusion times of the second polar body between the first and the last egg is the key factor that determines the specific timing of shock application to induce triploidy with a high yield or success rate.

The extrusions of the first and the second polar bodies in *P. monodon* are at 2–5 and 8–14 min post-spawning, respectively, and the first mitosis begins at about 1 h post-spawning (Pongtippatee-Taweepreda et al. 2004; Hall et al. 1999). Therefore, to inhibit the first polar body extrusion and to induce triploidy in this species, the induction must be within 2 min post-spawning. However, this time-constrained maneuver is difficult to perform in practical situations. Further, if the induction begins at the end of the spawning process (7 min after the first egg is spawned), the majority of the eggs would be in water for more than 2 min, and their first polar bodies would have been extruded before applying the shock. Therefore, the inhibition of the second polar body extrusion, which is at 8 min post-spawning of the first egg, is preferable for the induction of triploidy in *P. monodon*. At that time point, the majority of eggs would already have extruded the first but not yet the second polar body.

5.12 Ploidy Analysis

Various techniques for verification of ploidy have been developed (Kang et al. 2013). Currently, ploidy analysis of the shrimp is done by karyotyping or by fluorescence-activated cell sorting (FACS) flow cytometry (Fig. 5.4). Karyotyping is performed by staining chromosomes of regenerated cells in blastema of the pereiopod at metaphase, following colchicine treatment and staining with carbol fuchsin (Lakra et al. 1997). Because karyotyping is time-consuming and it is difficult to count the number of chromosomes, which is about 44 pairs in *P. monodon*, FACS is the preferred technique (Pongtippatee et al. 2012). In FACS, haemocytes are separated from the haemolymph drawn from shrimp, followed by treatment of the haemocytes with propidium iodide and ribonuclease A, and analyzed with a flow cytometer. Distinguishing between the triploid and the diploid haemocytes is based on cellular density that is higher for the triploid cells.

An attempt to develop a more convenient and more accurate method for determining triploidy of *P. monodon* has been made, using differential gene expression between the 3n and the 2n shrimp. By using a subtraction hybridization

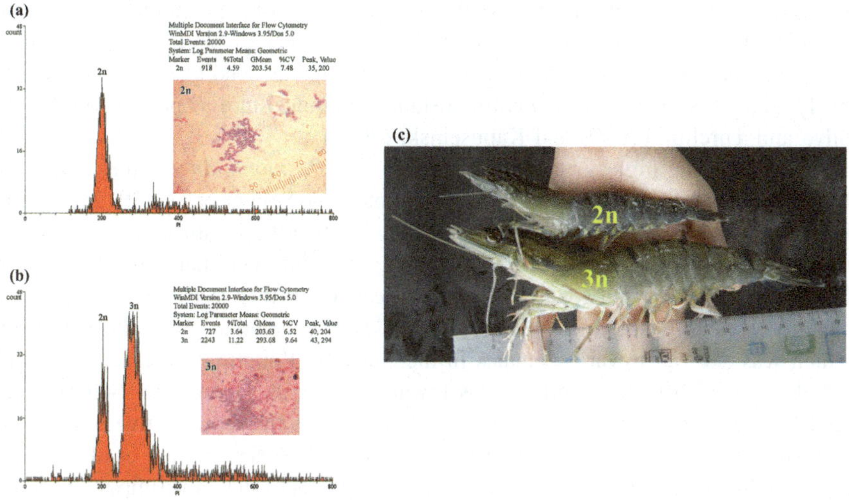

Fig. 5.4 Fluorescence-activated cell sorting (FACS) flow cytometry and karyotyping of diploid (2n) and triploid (3n) *P. monodon* **a** and **b** and the picture of 2n and 3n *P. monodon* **c** from the same batch and same age. From Pongtippatee et al. (2012)

technique, changes in certain genes' expression in the shrimp's muscle and hepatopancreas appear promising for distinguishing between triploid and diploid shrimp (Semchuchot et al., in preparation).

5.13 Growth Rate and Sex Ratio Advantages of Triploid Over Diploid *P. monodon*

Both Sellars et al. (2012b) and Pongtippatee et al. (2012) have reported similar triploid induction rates in *P. monodon*, despite different induction methods. Also, these studies reported similar survival rates of the triploid *P. monodon*, and the rates gradually declined in a grow-out culture. At 11 months of culture, the survival rate was below 10% (Pongtippatee, unpublished). One reason for the gradual loss of shrimp was the greater incidence of cannibalism, as dead carcasses with missing body parts were found more often in the 3n shrimp ponds than in those with 2n shrimp (Withyachumnarnkul, personal comm.). In general cannibalism occurs when newly molted shrimp are attacked by their cohorts. Whether the 3n shrimp are more aggressive than the 2n ones is an interesting research topic. However, at 4–5 months, the survival rates of these shrimps were comparable. The shrimp grew to 40–45 g BW under 25–30 postlarvae/m^2 stocking density within 4 months, with average daily growth (ADG) exceeding 0.3 g/d, which is about 25–30% faster than their 2n counterparts. Such growth performance could be a significant advantage to commercial farming of 3n *P. monodon*. The gains in growth using female 3n shrimp are also of significance to the sustainability of aquaculture since more production is achieved using diets of the same level of protein with shrimps attaining market size faster, and therefore reducing the feeding time. This effectively addresses the requirements for sustainable aquaculture as pointed out by De Silva and Turchini (2009) and Kapuscinski (2007).

However, the high growth rate of the 3n *P. monodon* in the study by Pongtippatee et al. (2012) was not corroborated by Sellars et al. (2012a). In the latter study, 3n *P. monodon* grew actually slower than the 2n shrimp. Although the reasons of this discrepancy are not known, the different methods of triploidy induction might account for it, and is an aspect for further research.

Another significant difference between these two studies is in the sex ratio, which was skewed towards females in the study by Pongtippatee et al. (2012) (2 females per each male), while it was towards males as reported by Sellars et al. (2012a) (1.625 male per each female). In both studies the reproductive function of triploid *P. monodon* was markedly reduced in both sexes. Sellars et al. (2012a) reported that the histological features of female ovaries of the 3n shrimp revealed no progressive stage beyond oogonia, with several cells undergoing apoptosis and with vacuolated spaces. In the male, a drastic reduction in sperm maturation and poor counts of sperm were observed in testis and vas deferens, and no mature spermatophore was present. This observation, however, was done at the age of 6 months, which is not a sexually mature age of *P. monodon*. A preliminary study on histology of the gonads of 3n *P. monodon* at 11 months in culture revealed that the ovary of the 3n female shrimp also contained mature oocytes (Pongtippatee et al., unpublished). It remains to be seen whether these 3n females spawn functionally normal eggs. In the male, similar histological features as reported by Sellars et al. (2012a) were observed. Some of these gonadal features were similar to those observed in triploids of other penaeid shrimp species (Li et al. 2003, 2006; Xiang et al. 2006; Sellars et al. 2009).

5.14 Commercial Triploid Induction Device for *P. Monodon*

Currently, the cold shock method for inducing 3n *P. monodon* is carried out manually. It is a labor-intensive and time-consuming procedure, which is a constraint for commercial production. An attempt to design and develop a device for inducing 3n shrimp is thus underway, and, although still at the preliminary stage, the initial results are promising.

This device functions in two steps; a spawning detection step and an induction step. In the first step spawning of the broodstock is detected within 22 s following its onset. This is accomplished through an electric signal that changes when eggs flow towards a filter (Mueangdee et al. 2013). This signal triggers the second step, which is composed of: (1) flow of spawning tank water to a reservoir, in order to concentrate the eggs; (2) flow of pre-cooled water to the concentrated eggs, to suddenly reduce the temperature to 8 °C; and (3) flow of pre-heated water to suddenly increase the temperature to 28 °C. The time from the spawning signal to the cooling stage is set at 8 min, and the cooling stage lasts for 10 min. Key functional characteristics of this device include the sudden temperature changes along with gentle mixing of water that avoids damage to the eggs.

5.15 Sustainability of the Stock Improvement Program for *P. monodon* in Thailand

While the introduction of exotic *L. vannamei* signified a drastic change in the way shrimp farming was carried out in Thailand, the role of *P. monodon* is still important because of its potential to be amenable to sustainable farming. The use of exotic species and escape of farmed strains had negative impacts on fitness and genetic diversity of shrimp stocks in the wild. Briggs et al. (2005) highlighted several perceived impacts of introductions of two shrimp species in Asia. Shrimp diseases have spread as a result of movements across nations and continents. A comprehensive list of activities that promote the spread of exotic aquaculture species as well as those that curtail excesses in spread have been identified (Minchin 2007). Furthermore, trade in shrimp products would also lead to transmission of shrimp pathogens across boundaries with possibility of infection of other crustaceans (Jones 2012). In terms of genetic diversity, it has been reported that cultured *L. vannamei* in Mexico had a lower variability occasioned by population bottlenecks that affected the effective population size (Vela-Avitúa et al. 2013). Although Klinbunga et al. (2006) had reported that the Thai tiger shrimps had less genetic differentiation, later Khamnamtong et al. (2009) observed a high level of genetic diversity among ecotypes of the tiger shrimp in Thailand. This conflicting information perhaps would have stemmed from the genetic markers used for the investigations; the latter used EST markers while the former used mtDNA markers (COI) which are very conservative. The high genetic diversity existing among the wild stocks of *P. monodon* is promising as it signifies the potential for greater genetic improvement that would benefit its aquaculture.

5.16 Concluding Remarks

The black tiger shrimp *P. monodon* is one of the most valuable species for aquaculture in Asia, particularly in Thailand with a key role in sustainable aquaculture. Although the Pacific whiteleg shrimp has dominated the shrimp farming industry in Thailand since the time of its introduction, tiger shrimp farming continues to be favored by many progressive farmers. The genetic stock improvement program currently underway in Thailand has revealed its potential to produce shrimp broodstock free of most of the known virus infections. The Specific Pathogen Free (SPF) broodstock of this species has been successfully produced up to the seventh generation, and various biotechnological interventions have been adopted to improve their growth performance in aquaculture. Induction of triploidy is one of the means of producing genetically sterile shrimps with a high proportion of females that have been proven to grow faster than the males. The triploid induction technology has been standardized with no significant impact on welfare of the animal or on the environment. The triploid induction device by the application of cold shock that is currently under the testing phase holds great promise to apply the triploidy induction procedure on a commercial scale. This simplified procedure using lowering of temperature and with no use of chemicals is expected to be a significant step in stock improvement of *P. monodon*, as triploidy is a major non-GMO technique for genetic improvement without any concern on fish welfare, and could be considered as a sustainable alternative to produce fast growing aquaculture stocks of shrimp.

Acknowledgements This article forms part of a study funded by the Thailand Research Fund, Research Career Development, grant no. RSA5780052; and by the Prince of Songkla University, grant no. S&T550429S. We thank Dr. Seppo Karrila from the Faculty of Science and Industrial Technology, Prince of Songkla University, Surat Thani for providing valuable suggestions.

References

Barbier EB (2006) Mangrove dependency and the livelihoods of coastal communities in Thailand. In: Hoanh CT, Tuong TP, Gowing JW, Hardy B (eds) Environment and livelihoods in tropical coastal zones: managing agriculture-fishery-aquaculture conflicts, vol 2. CABI. Retrieved from <Go to ISI>://000246405900010, pp 126–139

Benfey TJ (1999) The physiology and behavior of triploid fishes. Rev Fish Sci 7(1):39–67. https://doi.org/10.1080/10641269991319162

Bert TM (2007) Ecological and genetic implications of aquaculture, vol 6. Springer, Dordrecht

Briggs M, Funge-Smith S, Subasinghe RP, Phillips M (2004) Introductions and movement of two penaeid shrimp species in Asia and the Pacific. FAO fisheries technical paper 476. https://doi.org/10.1017/CBO9781107415324.004

Briggs M, Funge-Smith S, Subasinghe RP, Phillips M (2005) Introductions and movement of two penaeid shrimp species in Asia and the Pacific. FAO fisheries technical paper 476. FAO, Rome. https://doi.org/10.1017/CBO9781107415324.00

Briggs MRP (2001) Status, problems and solutions for a sustainable shrimp culture industry with special reference to Thailand. Retrieved from http://www.dfid.stir.ac.uk/afgrp/greylit/FR006.pdf

Campos-Ramos R, Garza-Torres R, Guerrero-Tortolero DA, Maeda-Martinez AM, Obregon-Barboza AH (2006) Environmental sex determination, external sex differentiation and structure of the androgenic gland in the Pacific white shrimp *Litopenaeus vannamei* (Boone). Aquac Res 37:1583–1593

Cassani JR, Caton WE (1986) Efficient production of triploid grass carp (*Ctenopharyngodon idella*) utilizing hydrostatic pressure. Aquaculture 55:43–50

Cassani JR, Maloney DR, Allaire HP, Kerby JH (1990) Problems associated with tetraploid induction and survival in grass carp, *Ctenopharyngodon idella*. Aquaculture 88:273–284

CFS (2005) The catch with seafood: human health impacts of drugs & chemicals used by the aquaculture industry. Center for Food Safety, Washington, DC

Coman FE, Sellars MJ, Norris BJ, Coman GJ, Preston NP (2008) The effects of triploidy on *Penaeus (Marsupenaeus) japonicus* (Bate) survival, growth and gender when compared to diploid siblings. Aquaculture 276(1–4):50–59. https://doi.org/10.1016/j.aquaculture.2008.01.031

da Silva FSD, Moreira RG, Orozco-Zapata CR, Hilsdorf AWS (2007) Triploidy induction by cold shock in the South American catfish, *Rhamdia quelen* (Siluriformes) (Quoy & Gaimard, 1824). Aquaculture 272(Suppl. 1):110–114. https://doi.org/10.1016/j.aquaculture.2007.08.006

De Silva SS, Turchini GM (2009) Use of wild fish and other aquatic organisms as feed in aquaculture—a review of practices and implications in the Asia-Pacific. In: Hasan MR, Halwart M (eds) Fish as feed inputs for aquaculture: practices, sustainability and implications. FAO, Rome, Italy, pp 63–127

EJF (2006) Mangroves : nature's defence against Tsunamis: a report on the impact of mangrove loss and shrimp farm development on coastal defences

FAO (2010) Global forest resources assessment 2010, vol 163. http://www.fao.org/docrep/013/i1757e/i1757e.pdf

FAO (2013) FAO/MARD technical workshop on early mortality syndrome (EMS) or acute hepatopancreatic necrosis syndrome (AHPNS) of cultured shrimp (under TCP/VIE/3304). 25–27 June, Hanoi, Vietnam

FAO (2014) The state of world fisheries and aquaculture: opportunities and challenges (trans: Aquaculture Fa). Food and Agriculture Organization of the United Nations, Rome

FAO (2015) Global forest resources assessment 2015. In: Desk reference

FAO (2016) Global aquaculture production. Retrieved on 6 June 2016, from http://www.fao.org/fishery/statistics/global-aquaculture-production/query/en

Ferdouse F (2014) The situation of farmed shrimp production in Asian countries. Infofish Retrieved from http://fenacam.com.br/pdf/fenacam2014/carcinicultura/5-situacao-da-producao-de-camarao-de-cultivo-em-paises-asiaticos-afetados-pela-sindrome-da-mortalidade-precoce-(ems)-_-fatima-ferdousepdf

Flaherty M, Karnjanakesorn C (1995) Marine shrimp aquaculture and natural resource degradation in Thailand. Environ Manage 19(1):27–37. https://doi.org/10.1007/BF02472001

Flegel TW (1997) Special topic review: major viral diseases of the black tiger prawn (Penaeus monodon) in Thailand. J Microbiol 13:433–442. https://doi.org/10.1023/A:1018580301578

Foote A, Wood A, Mair G, Sellars M (2012) Ploidy manipulation of shrimp through hydrostatic pressure: optimization and commercial potential. In: Merican Z, Juárez L, Tan R, Sage C (eds) AQUA 2012. World aquaculture society. Retrieved from https://www.was.org/Meetings/ShowAbstract.aspx?Id=2602

Gräslund S, Holmström K, Wahlström A (2003) A field survey of chemicals and biological products used in shrimp farming. Mar Pollut Bull 46(1):81–90. https://doi.org/10.1016/s0025-326x(02)00320-x

Guardabassi L, Dalsgaard A, Raffatellu M, Olsen JE (2000) Increase in the prevalence of oxolinic acid resistant Acinetobacter spp. observed in a stream receiving the effluent from a freshwater trout farm following the treatment with oxolinic acid-medicated feed. Aquaculture 188(3–4):205–218. https://doi.org/10.1016/S0044-8486(00)00340-9

Guo XM, DeBrosse GA, Allen SK (1996) All-triploid Pacific oysters (*Crassostrea gigas* Thunberg) produced by mating tetraploids and diploids. Aquaculture 142(3–4):149–161. https://doi.org/10.1016/0044-8486(95)01243-5

Hall MR, Young N, Kenway M (1999) Manual for the determination of egg fertility in *Penaeus monodon*. Australian Institute of Marine Science. Retrieved from http://epubs.aims.gov.au/handle/11068/1545

Hansford SW (1991) More females in prawn ponds could cut production costs. Austral Fish 50 (10):32–33

Hansford SW, Hewitt DR (1994) Growth and nutrient digestibility by male and female *Penaeus monodon*: evidence of sexual dimorphism. Aquaculture 125:147–154

Harrell RM, van Heukelem W, Kerby JH (1998) A comparison of triploid induction validation techniques. Progressive Fish-Culturist 60(3):221–226. https://doi.org/10.1577/1548-8640(1998)060<0221

HHS-FDA (2009) Guidance for industry regulation of genetically engineered animals containing heritable recombinant DNA constructs. United States Department of Health & Human Services, Rockville

Holmstrom K, Graslund S, Wahlstrom A, Poungshompoo S, Bengtsson B-E, Kautsky N (2003) Antibiotic use in shrimp farming and implications for environmental impacts and human health. Int J Food Sci Technol 38(3):255–266. https://doi.org/10.1046/j.1365-2621.2003.00671.x

Hussain MG, Penman DJ, Mcandrew BJ, Johnstone R (1993) Suppression of 1st cleavage in the Nile Tilapia, Oreochromis-niloticus L—a comparison of the relative effectiveness of pressure and heat shocks. Aquaculture 111(4):263–270

Jones B (2012) Transboundary movement of shrimp viruses in crustaceans and their products: a special risk? J Invertebr Pathol 110(2):196–200

Kamrin MA (1997) Pesticide profiles: toxicity, environmental impact, and fate. CRC Press, Boca Raton

Kang J, Lim HJ, Kang H, Lee J, Baby S, Kim J (2013) Development of genetic markers for triploid verification of the Pacific Oyster, *Crassostrea gigas*. Asian-Australas J Anim Sci 26(7):916–920. https://doi.org/10.5713/ajas.2013.13108

Kapuscinski AR (2007) Environmental risk assessment of genetically modified organisms. CABI. Retrieved from https://books.google.co.th/books?id=cQxMMaO3KOgC

Kautsky N, Ronnback P, Tedengren M, Troell M (2000) Ecosystem perspectives on management of disease in shrimp pond farming. Aquaculture 191(1–3):145–161. https://doi.org/10.1016/S0044-8486(00)00424-5

Khamnamtong B, Klinbunga S, Menasveta P (2009) Genetic diversity and geographic differentiation of the giant tiger shrimp (*Penaeus monodon*) in Thailand analyzed by mitochondrial COI sequences. Biochem Genet 47:42–55. https://doi.org/10.1007/s10528-008-9205-3

Klinbunga S, Preechaphol R, Thumrungtanakit S, Leelatanawit R, Aoki T, Jarayabhand P, Menasveta P (2006) Genetic diversity of the giant tiger shrimp (*Penaeus monodon*) in Thailand revealed by PCR-SSCP of polymorphic EST-derived markers. Biochem Genet 44:222–236. https://doi.org/10.1007/s10528-006-9027-0

Komen J, Bongers ABJ, Richter CJJ, van Muiswinkel WB, Huisman EA (1991) Gynogenesis in common carp (Cyprinus carpio L.). 2. The production of homozygous gynogenetic clones and F sub(1) hybrids. Aquaculture 92:127–142

Lakra WS, Kumar P, Das M, Goswami U (1997) Improved techniques of chromosome preparation from shrimps and prawn. Asian Fish Sci 10:117–121

Lakshmi A, Rajagopalan R (2000) Socio-economic implications of coastal zone degradation and their mitigation: a case study from coastal villages in India. Ocean Coast Manag 43:749–762. https://doi.org/10.1016/S0964-5691(00)00057-0

Larderel JAD, Shende RM, Mercado C, Robinson A (2001) Protecting the ozone layer, p 51

Li F, Xiang J, Zhang X, Zhou L, Zhang C, Wu C (2003) Gonad development characteristics and sex ratio in triploid Chinese shrimp (*Fenneropenaeus chinensis*). Mar Biotechnol 5:528–535. https://doi.org/10.1007/s10126-002-0103-0

Li F, Zhang C, Yu K, Liu X, Zhang X, Zhou L, Xiang J (2006) Larval metamorphosis and morphological characteristic analysis of triploid shrimp *Fenneropenaeus chinensis* (Osbeck 1765). Aquac Res 37:1180–1186. https://doi.org/10.1111/j.1365-2109.2006.01545.x

Maxime V (2008) The physiology of triploid fish: current knowledge and comparisons with diploid fish. Fish Fish 9:67–78. https://doi.org/10.1111/j.1467-2979.2007.00269.x

Minchin D (2007) Aquaculture and transport in a changing environment: overlap and links in the spread of alien biota. Mar Pollut Bull 55:302–313. https://doi.org/10.1016/j.marpolbul.2006.11.017

Mitsch W, Gosselink JG (2015) Wetlands, 5th edn. Wiley, New Jersey

Mueangdee N, Pongtippatee P, Promkun P, Bootterb M, Chuaymung U, Sunyaprark W, Phrommoon J, Kanjanasopa W, Withyachumnarnkul B (2013) Spawning detection device for the black tiger shrimp *Penaeus monodon* broodstock. Aquaculture 380–383:21–22. https://doi.org/10.1016/j.aquaculture.2012.11.029

Nell JA (2002) Farming triploid oysters. Aquaculture 210:69–88. https://doi.org/10.1016/S0044-8486(01)00861-4

Ng KH, Samuel L, Kathleen MM, Leong SS, Felecia C (2014) Distribution and prevalence of chloramphenicol-resistance gene in Escherichia coli isolated from aquaculture and other environment. Int Food Res J 21(4):1321–1325

Overturf K (2009) Molecular research in aquaculture. Wiley-Blackwell, Hoboken, 400 p

Patamasiriwat D, Kuik O, Pednekar SS (1999) The shrimp aquaculture sector in Thailand: a review of economic, environmental and trade issues. International Institute for Environment and Development. Retrieved from https://books.google.comng/books?id=bWfjjFusxbkC

Peruzzi S, Chatain B (2000) Pressure and cold shock induction of meiotic gynogenesis and triploidy in the European sea bass, *Dicentrarchus labrax* L.: relative efficiency of methods and parental variability. Aquaculture 189:23–37. https://doi.org/10.1016/S0044-8486(00)00355-0

Piferrer F, Beaumont A, Falguière JC, Flajšhans M, Haffray P, Colombo L (2009) Polyploid fish and shellfish: production, biology and applications to aquaculture for performance improvement and genetic containment. Aquaculture 293:125–156. https://doi.org/10.1016/j.aquaculture.2009.04.036

Pongtippatee-Taweepreda P, Chavadej J, Plodpai P, Pratoomchart B, Sobhon P, Weerachatyanukul W, Withyachumnarnkul B (2004) Egg activation in the black tiger shrimp *Penaeus monodon*. Aquaculture 234(1–4):183–198. https://doi.org/10.1016/j.aquaculture.2003.10.036

Pongtippatee P, Laburee K, Thaweethamsewee P, Hiranphan R, Asuvapongpatana S, Weerachatyanukul W, Srisawat T, Withyachumnarnkul B (2012) Triploid *Penaeus monodon*: sex ratio and growth rate. Aquaculture 356–357:7–13. https://doi.org/10.1016/j.aquaculture.2012.06.004

Preston AC, Taylor JF, Craig B, Bozzolla P, Penman DJ, Migaud H (2013) Optimisation of triploidy induction in brown trout (*Salmo trutta* L.). Aquaculture 414–415:160–166. https://doi.org/10.1016/j.aquaculture.2013.07.034

Primavera JH (2006) Overcoming the impacts of aquaculture on the coastal zone. Ocean Coast Manag 49:531–545. https://doi.org/10.1016/j.ocecoaman.2006.06.018

Rico A, Satapornvanit K, Haque MM, Min J, Nguyen PT, Telfer TC, van den Brink PJ (2012) Use of chemicals and biological products in Asian aquaculture and their potential environmental risks: a critical review. Rev Aquac 4:75–93. https://doi.org/10.1111/j.1753-5131.2012.01062.x

Rime H, Neant I, Guerrier P, Ozon R (1989) 6-Dimethylaminopurine (6-DMAP), a reversible inhibitor of the transition to metaphase during the first meiotic cell division of the mouse oocyte. Dev Biol 133:169–179. https://doi.org/10.1016/0012-1606(89)90308-4

Rottmann RW, Shireman JV, Chapman FA (1991) Induction and verification of triploidy in fish. SRAC Publication no. 427. Southern Regional Aquaculture Center

Sellars MJ, Arce SM, Hertzler PL (2012a) Triploidy Induction in the Pacific white shrimp *Litopenaeus vannamei*: an assessment of induction agents and parameters, Embryo Viability, and Early Larval Survival. Mar Biotechnol 14:740–751. https://doi.org/10.1007/s10126-012-9439-2

Sellars MJ, Degnan BM, Preston NP (2006) Production of triploid Kuruma shrimp, Marsupenaeus (Penaeus) japonicus (Bate) nauplii through inhibition of polar body I, or polar body I and II extrusion using 6-dimethylaminopurine. Aquaculture 256:337–345. https://doi.org/10.1016/j.aquaculture.2006.02.052

Sellars MJ, Li F, Preston NP, Xiang J (2010) Penaeid shrimp polyploidy: global status and future direction. Aquaculture 310:1–7. https://doi.org/10.1016/j.aquaculture.2010.10.002
Sellars MJ, Wood A, Murphy B, McCulloch RM, Preston NP (2012b) Triploid Black Tiger shrimp (*Penaeus monodon*) performance from egg to harvest age. Aquaculture 324–325:242–249. https://doi.org/10.1016/j.aquaculture.2011.10.038
Sellars MJ, Wood AT, Murphy B (2009) Challenges to commercializing shrimp triploidy: technique to produce all-female, genetically protected populations. Global Aquaculture Advocate
Shen YP, Zhang XY, He HP, Ma LJ (1993) Triploidy induction by hydrostatic pressure in the pearl oyster, Pinctada martensii Dunker. Aquaculture 110:221–227. https://doi.org/10.1016/0044-8486(93)90370-E
Simili M, Pellerano P, Pigullo S, Tavosanis G, Ottaggio L, De Saint-Georges L, Bonatti S (1997) 6DMAP inhibition of early cell cycle events and induction of mitotic abnormalities. Mutagenesis 12:313–319. https://doi.org/10.1093/mutage/12.5.313
Szuster B (2006) Coastal shrimp farming in Thailand: searching for sustainability. environment and livelihoods in tropical coastal zones: managing agriculture-fishery-aquaculture conflicts, vol 2. https://doi.org/10.1079/9781845931070.0086
Teskeredžić E, Donaldson E, Teskeredžić Z, Solar I, McLean E (1993) Comparison of hydrostatic pressure and thermal slhocks to induce triploidy in coho salmon (Oncorhynchus kisutch). Blood 117:47–55
Teskeredzic E, Teskeredzic Z, Donaldson EM, McLean E, Solar I (1993) Triploidization of coho salmon following application of heat and electric shocks. Aquaculture 116:287–294. https://doi.org/10.1016/0044-8486(93)90015-Q
Tiwary BK, Kirubagaran R, Ray AK (2004) The biology of triploid fish. Rev Fish Biol Fish 14:391–402. https://doi.org/10.1007/s11160-004-8361-8
Vallero D (2015) Environmental biotechnology: a biosystems approach, p 746
Vela-Avitúa S, Montaldo H, Márquez-Valdelamar L, Campos-Montes G, Castillo-Juárez H (2013) Decline of genetic variability in a captive population of Pacific white shrimp Penaeus (Litopenaeus) vannamei using microsatellite and pedigree information. Electron J Biotechnol 16. https://doi.org/10.2225/vol16-issue4-fulltext-11
Weston DP (1996) Environmental considerations in the use of antibacterial drugs in aquaculture. In: Baird DJ, Beveridge MCM, Kelly LA, Muir JF (eds) Aquaculture and water resource management. Fishing News Books, Oxford, pp 140–165
WHO (2000) Air quality guidelines for Europe. WHO Regional Publications, European Series no 91:288. https://doi.org/10.1007/BF02986808
Withyachumnarnkul B, Plodphai P, Nash GD (2001) Growth rate and reproductive performance of F4 domesticated Penaeus monodon broodstock. In: The 3rd national symposium of marine shrimp, Sirikit National Convention Center, Bangkok, Thailand, 8–9 Nov, 2001, pp 33–40
Withyachumnarnkul B, Boon-Nad A, Anantasomboon G, Chayaburakul K, Sriurairatana S, Flegel TW (2004) Lymphoid organ extracts of growth retarded *Penaeus monodon* contain a growth retardation agent, p 647. Book of Abstracts, Annual Meeting World Aquaculture Society (WAS). March 1–5, Honolulu, Hawaii
Xiang J, Li F, Zhang C, Zhang X, Yu K, Zhou L, Wu C (2006) Evaluation of induced triploid shrimp Penaeus (Fenneropenaeus) chinensis cultured under laboratory conditions. Aquaculture 259:108–115. https://doi.org/10.1016/j.aquaculture.2006.05.033
Yamazaki F, Goodier J (1993) Cytogenetic effects of hydrostatic pressure treatment to suppress the first cleavage of salmon embryos. Aquaculture 110:51–59. https://doi.org/10.1016/0044-8486(93)90433-Y

Chapter 6
Aquaponics: A Commercial Niche for Sustainable Modern Aquaculture

Paul Rye Kledal and Ragnheidur Thorarinsdottir

Abstract Aquaponics—the combination of recirculating aquaculture system (RAS) and horticulture—has received increasing interest globally as a way to introduce a more circular economy within aqua- and horticulture and hence secure a more resource productive growth with reduced pollution to the environment. However, aquaponics production have mainly been based on small scale low-tech and labor intensive systems built by hobbyist and research units, but during the last decade larger and more complex systems based on modern RAS and hydroponics techniques have been designed and constructed. This new development has mainly been driven forward by researchers and risk taking entrepreneurs worldwide, but commercial oriented production units are emerging with participation of industry partners from both the aqua- and horticultural sectors. The biological dependence is one of the major constraints for going large-scale and commercial. De-coupled aquaponics holds the prospect of reducing or even eliminating the biological dependence, but in the same time acquire the symbiotic benefits of combining fish and plants in a circular production system.

Keywords Aquaponics · Sustainable aquaculture · Hydroponics
Organic aquaculture · Decoupled aquaponics

6.1 Introduction

The word *aquaponics* originate from the Latin and the Greek words for water (aqua and hydro), applied respectively to the words *aquaculture* (fish farming), and *hydroponics* for growing plants in water without soil used today in modern

P. R. Kledal (✉) · R. Thorarinsdottir
Reykjavik, Iceland
e-mail: paul@igff.dk

F. I. Hai et al. (eds.), *Sustainable Aquaculture*, Applied Environmental Science and Engineering for a Sustainable Future,
https://doi.org/10.1007/978-3-319-73257-2_6

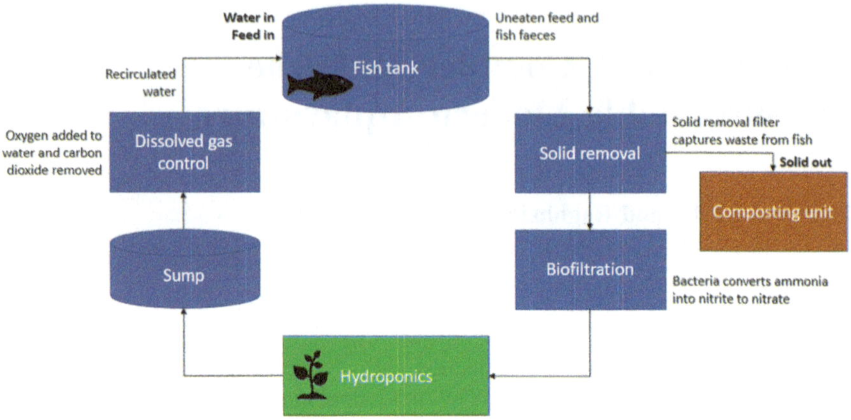

Fig. 6.1 The concept of aquaponics (Modified after Thorarinsdottir 2015)

horticulture production. Aquaponics itself represents a fusion of the two production systems combined into a singular food system producing terrestrial plants and aquatic organisms. In an aquaponics system water is kept in circulation as illustrated in Fig. 6.1.

Wastewater from the fish tanks is led to a solid removal system and further to a bio-filter, where bacteria converts ammonia to nitrate. These dissolved nutrients are circulated onwards to the horticultural part of the system where plants take up the nutrients, and hence cleanse the water before being returned to the fish. The solid removal filter captures uneaten fish and fish faeces. The solids contain a large part of the nutrients, but need to be dissolved in a composting unit or by aeration to be usable in the hydroponicsp or alternatively turned into fertilizer to be used in soil systems.

Likewise, the fish produce CO_2 valuable to the plants, and the fish tanks can obtain heat and act as a temperature buffer during night in the greenhouse saving energy costs (Körner et al. 2014). Besides these symbiotic effects, aquaponics offers to be a resource efficient closed loop food production system mimicking nature itself. Aquaponics relates to the 'cradle-to-cradle' design described in several articles (McDonough and Braungart 2002; Braungart et al. 2007; Kumar and Putnam 2008) presenting eco-effectiveness even moving beyond zero emissions and provides public goods and services when taking social, economic and environmental benefits into account.

The principles behind aquaponics—combining an aquatic feed supply to a terrestrial plant production—is not new in itself, but has been used effectively back in history by the Aztec Indians in the former valley of Mexico back in the fourteenth-century, as well as by Chinese farmers in the Pearl River Delta of South China around the same period.

6.2 The History of Aquaponics

6.2.1 The Chinampas of the Aztec

When first approaching the basin of Mexico and the Aztec capital of Tenochtitlan in 1519, Hernando Cortes and his men were amazed. They saw an astounding white city, anchored to the shores by three long causeways, floated on a glittering lake. This lake-borne city, Tenochtitlan, was the world largest at the time estimated to occupy around 3–350,000 inhabitants.

Cortes and his men also found the practice of a unique agricultural system. This method of farming, which still persists in limited intensity today, consists of land development through the construction of what is called *chinampas* in marshy areas and shallow lakes. The chinampa system is a network of raised fields on manmade low islands in the lakes and marshes. The method consists of piling lakebed clays and mud, aquatic plants and dry land crop silage, as well as silted muck and manures in precise layers between reed fences secured in the bottom of the lakes or marshlands. Once the ground was raised to its proper height, fast growing willow trees were planted on the edges, which prevented the erosion of the raised ground. These willow trees also provided shade and firewood, and restrained the crop-damaging pests (Aghajanian 2007).

The chinampas were between 5 and 10 m wide, up to 90 m long and around ½ m above the water level. In between the chinampa beds were canals of 1–1.3 m giving, not only life to an abundant wildlife and fish, but also providing an efficient transport system for canoes supplying labor and food.

This astonishing eco-effectiveness of combining an aquatic feed supply to a terrestrial plant production gave the chinampa agriculture a unique role in sustaining the population pressure in the Valley of Mexico during the Aztec period.

6.2.2 The Chinese Dike-Pond System

In the Pearl Delta of south China, a land-water farming system, also known as the dike-pond system, evolved during the mid-fourteenth century. The dike-pond system evolved as an important flood control measure in the delta. Water control measures were started in the lower-lying areas, where small watercourses were dammed and created to make fishponds. Ponds were dug to drain the marshes and natural ponds in order to create agricultural land, and the excavated was used to construct dykes. The fishponds were stocked with carp fry naturally occurring in the delta (Ruddle and Zhong 1988).

The first commercial crops to be grown on the dikes were Litchi and Longan followed later on by mulberry. The mulberry leaves provided an important feed for the cash crop: silkworms. Silkworm excrement was thrown into the pond, and accidentally gave way to the discovery that it could feed the fish. The mud at the bottom of the ponds was used to fertilize trees, when there was a shortage of animal manure.

Since then, several types of dike pond systems have been developed. The above example is called *the mulberry-dike-pond system*, others are called *the fruit-dike-pond system* and so on.

The pond is the heart of dike pond system. Most ponds are rectangular, 0.4–0.6 ha and 2–3 m deep. The dikes are usually 6–10 m wide and 0.5–1.0 m above the pond surface. Different species live in different water depths and have different feeding needs, which ensure full utilization of the water and pond ecology.

Pond mud contains many nutrients and can be used as fertilizer for crops on the dikes. The pond is drained two or three times a year, and mud at the bottom is retrieved up on the dikes, which then are repaired while the depth of the pond is restored.

Livestock are also part of a dike-pond system. Both small and large animals like pigs or ducks can be bred on the dykes and their manure can be thrown into the pond and thus promote growth of algae which the carp can feed on. Through photosynthesis the algae in the pond give off oxygen and produce glucose, added nutrients that benefit both fish and aquatic plants.

Fish fodder may also be cultivated on the dykes, for example Miscanthus, or fodder for animals that live on the dikes.

The idea of the dike pond system is that it is a circuit where the components are complementary, while no waste is produced, because everything is recycled and transformed. The energy input from outside is minimal and consists mainly of labor and solar energy. Solar energy has the advantage that it is renewable and free (Stenkjaer 2011).

Like in aquaponics the dike-pond system is an interrelated ecosystem that brings into full play the productive potential of humans and their environment and promotes the development of different branches of both aqua-, agri- and horticulture.

6.3 The Modern Paths of Aquaponics

In the late 1960s and beginning of the seventies 'Limits to growth' was a global discussion theme due to emerging and simultaneous crisis world-wide in food production, population growth, urban sprawl, pollution, energy and raw material supplies etc. Many experiments to find new and more sustainable and self-sufficient low-input solutions in production as well as consumption took place in this period, both from grass root movements as well as industry and university pioneers.

Aquaponics itself emerged from the aquaculture industry as fish farmers were exploring methods of raising fish while trying to decrease their dependence on the land, water and other resources. Traditionally, fish were raised in large ponds, or in netted pens off coastlines, but much progress has been made since then in Recirculating Aquaculture Systems (RAS) (Bradley 2014; Dalsgaard et al. 2013).

The advantage of RAS is that fish can be stocked much more densely, thus using only a fraction of the water and space to grow the same amount of fish as in pond or netting based systems. A major disadvantage with RAS is the large amount of concentrated waste water that quickly accumulates and the antibiotics needed to keep the fish healthy.

The term aquaponics is often attributed to the various works of the New Alchemy Institute and the works of Dr. Mark McMurtry at the North Carolina State University. In 1969, John and Nancy Todd and William McLarney founded the New Alchemy Institute.[1] The culmination of their efforts was the construction of a prototype Bioshelter, the "Ark". The Ark was a solar-powered, self-sufficient, bio-shelter designed to accommodate the year-round needs of a family of four using holistic methods to provide fish, vegetables and shelter (Bradley 2014).

At the same time in the 1970s, research on using plants as a natural filter within fish farm systems began, most notably by Dr. James Rakocy at the University of the Virgin Islands (UVI) (Rakocy et al. 2007). In the late nineties they developed the still much applied *UVI- system*, which is described further below.

During the mid-1980s, Mark McMurtry and Professor Doug Sanders at North Carolina State University developed an aqua-vegeculture system based on Tilapia fish tanks sunken below the greenhouse floor. Effluent from the fish tanks was trickle-irrigated onto sand-cultured hydroponic vegetable beds located at ground level. The nutrients in the irrigation water fed tomato and cucumber crops, and the plants and sand beds served as a bio-filter. After draining from the beds, the water recirculated back into the fish tanks. The only fertility input to the system was fish feed (32% protein) (Diver 2000).

The first larger scale commercial aquaponics facility, Bioshelters in Amherst, MA, was established in the mid-1980s. Then in the early 1990s, Missouri farmers Tom and Paula Speraneo inspired by Mark McMurtry, introduced their *Bioponics* concept. They grew herbs and vegetables in 'ebb and flow gravel grow beds' irrigated by the nutrient rich water from a 2200 L tank in which they raised Tilapia (Bradley 2014).

While gravel grow beds had been used for decades by hydroponics growers, the Speraneos were the first to make effective use of them in Aquaponics—remembering prior to this, sand was the main growing medium used in emerging aquaponics systems. Their system was practical and has been widely duplicated, and many present day DIY (Do-It-Yourself) aquaponics owes its origin to the Speraneos. They wrote a 'how-to manual' that became a springboard for many home based or school educational systems built throughout the world.

However, the Speraneos system of substituting sand for gravel in ebb and flow beds only works well if the system is fitted with dedicated mechanical and biological filtration. If not, the system will bear the risk of an eventual 'collapse', due

[1]The New Alchemy Institute evolved in 1991 to the Green Center Inc., which is a non-profit educational institute, and the custodian and distributor of publications of New Alchemy's ecological research conducted from 1971 to 1991. www.thegreencenter.net.

to the accumulation of organic matter using up oxygen in the system needed for the fish and furthermore reduced aeration of media bacteria and the plant root zone.

By 1997, Rakocy and his colleagues developed the use of deep-water culture hydroponic grow beds in a large-scale aquaponics system.

The UVI system has been the inspiring layout of several minor commercial systems in the US, Canada and Europe, and also applied by university researchers due to its proven reliability over several decades. The University of Virgin Islands has also been active in aquaponics research for more than thirty years and has a globally recognized aquaponics education program. The system developed at UVI is a raft hydroponic system and the aquaculture part focus is on Tilapia production (Rakocy et al. 1997, 2007).

A continuous operation was run at UVI for 2.5 years (1995–1997) with red Tilapia and leaf lettuce production (Rakocy et al. 1997, 2007). The system (Fig. 6.2) was based on four fish rearing tanks, each with 7.8 m^3 water volume (total 31.2 m^3), two cylindro-conical clarifiers (3.8 m^3 each), four rectangular filter tanks (0.7 m^3 each) containing orchard netting, six hydroponic tanks (11.5 m^3 each) and a sump (0.6 m^3). The hydroponic tanks were 30.5 m long by 1.2 m wide by 0.4 m deep and had a combined surface area of 214 m^2. Thus, the surface area to fish tank volume was 6.85 m^2/m^3. The water volume was 110 m^3. A 0.5 hp in-line pump moved water at an average rate of 378 L/min from the sump to the fish rearing tanks (mean retention time of water 1.5 h), from which effluent flowed with gravity through the system. Air diffusers were used both in fish and hydroponic tanks through air stones supplied by air from a 1.5 hp blower for fish and 1 hp blower for plants (Rakocy et al. 2007).

The daily fish feed input averaged 12 kg equivalent to 56 g/m^2 plant growing area. The waste water from the fish was only supplemented with potassium (K), calcium (Ca) and iron (Fe) to provide sufficient amounts of the essential nutrients for normal plant growth. These additions were equivalent to 16.1 g KOH, 3.3 g CaO, 13.7 g $Ca(OH)^2$ (more economical than CaO) and 6.0 g iron chelate (10%) per kg of fish feed. The annual production of tilapia was 3096 kg and the lettuce production was projected to 1694 cases (appr. 11 tons), or appr. 3.5 tons lettuce per ton Tilapia produced and the land use was 0.04 ha, which can be considered being a small-scale system (Rakocy et al. 1997).

6.3.1 Current Status

At present, the interest in aquaponics is increasing globally (Goddek et al. 2015). In Europe several strong collaboration networks have been established e.g. the COST FA1305 Aquaponics hub running from 2014–18 with 26 participating countries. Several ongoing projects in semi-commercial scale aquaponics and research units are delivering results to support upscaling of aquaponics systems capable of contributing to a new integrated and sustainable food production methodology (Thorarinsdottir 2015).

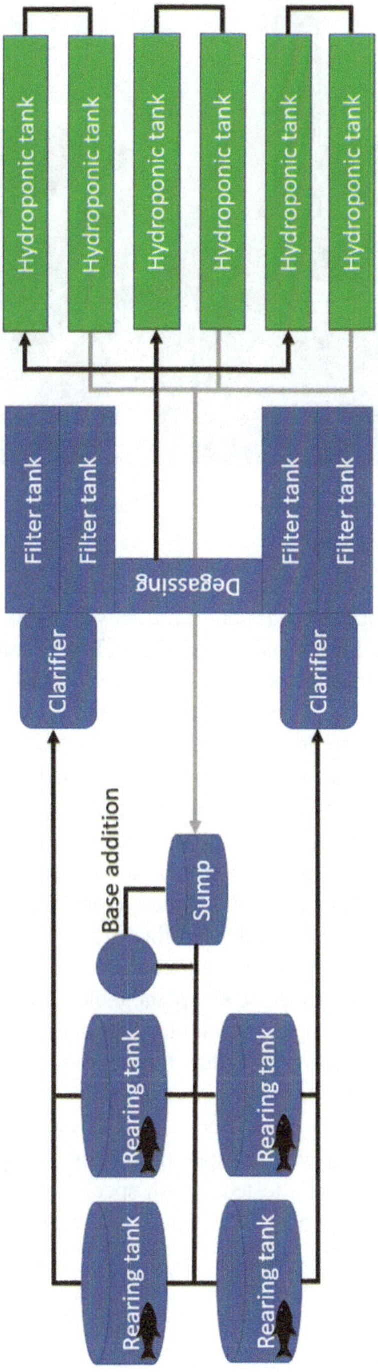

Fig. 6.2 UVI aquaponics system diagram (Modified after Rakocy et al. 1997)

Fig. 6.3 IGFF aquaponics pilot unit, Denmark (Photo, Paul Rye Kledal/IGFF)

The renewed growing interest in aquaponics can also be seen by the incorporation of technological improvements and productivity gains made within modern RAS driven by high capital costs and increasing technical complexity putting large demands on RAS system management and productivity (Dalsgaard et al. 2013) and hydroponics over the last two decades (Stickney and Granvil 2012; Resh 2013). In RAS this goes in regards to drum-filters, bio-filters, low-energy pumps, low-cost measurement equipment, data logging, broad range of fish feed etc. In horticulture recent developments have resulted in improved fertilizer, grow media, hydroponic pipe optimization, accessible bio-pest management, IT climate control etc.

In general the *UVI-system* is still the foundation for the renewed interest in aquaponics, which is understandable. It is reliable and have been tested over two decades, but new innovations are made on other areas.

IGFF[2] in Denmark has developed a decoupled aquaponics unit of 70 m^2 (Fig. 6.3). The IGFF unit consists of six plant tables arranged in three pairs of 1.45×7.50 m on the top of three rectangular fish tanks ($3 \times 1 \times 0.8$ m) with a usable volume of 2 m^3 each. Moveable plant tables produce horticulture products in pots with compost to open up for the prospect of getting an organic certification for the aquaponics system. Soil is used because to obtain an organic certification requires plants to be grown in various specified types of soil. Silver Tilapia, Red Tilapia and Pike perch have been tested as fish species and various plants such as lettuce, basil, tomatoes and peppers have been grown successfully on the plant

[2]www.igff.dk.

Fig. 6.4 The semi-commercial pilot unit of Svinna-verkfraedi Ltd. in Iceland (Photo, Ragnheidur Thorarinsdottir)

tables. Water to the plants is supplied by the 'flood and ebb' principle, and the plant tables are placed above the tanks to provide shade as well as optimize 'economies of space' in the greenhouse.

In Iceland Svinna-verkfraedi Ltd. (www.svinna.is) has constructed a semi-commercial pilot unit in a greenhouse farm in South Iceland, see Fig. 6.4. The design is based on decoupling the RAS and the plant system to obtain optimum growing conditions for both the aquaculture and plant production parts. The RAS unit consists of three 4 m^3 fish tanks, a drum filter, a bio-filter and a sump tank. A sedimentation tank and second sump tank is used for collecting water from the drum filter for plant irrigation and the water is not recirculated to the RAS again. During the first development phase a relatively small 50 m^2 hydroponics unit has been included in the circulation to stabilize the dissolved nitrogen level in the RAS.

NER-Breen in Hondarribia, Basque Country is developing a 6000 m^2 commercial aquaponics based on the development by the innovation company Breen.[3] The farm is being constructed and is designed partly as a decoupled farm. However, parts of the plants are used to control the dissolved nutrient level in the RAS. The farm is under construction, see Fig. 6.5 and the plan is to start production in 2016.

Other interesting steps are planned for decoupling the aquaculture and the plant production units as for example suggested by Aqua4C fish farm in Kruishoutem Belgium. The plan is to use residual energy from an adjacent tomato farm for the

[3]www.breen.es.

Fig. 6.5 The Breen aquaponics plant of 6000 m^2 planning to be starting production in 2016 (Photo, Ragnheidur Thorarinsdottir)

fish farm and to link the water systems together as well so the rainwater collected in the greenhouses could be used for the aquaculture before it is used for tomatoes (Hortidaily.com 2015).

Another interesting development is the new urban aquaponics farm in St. Paul Minnesota designed and constructed by the RAS company Pentair plc in collaboration with Urban Organics LCC (Pump Industry Analyst 2015). The companies are setting up an 87,000 ft^2 indoor aquaponics facility with the potential to produce 275,000 lbs. of fresh salmonid and 400,000 lbs. of leafy greens.

6.4 Aquaponics and Economic Organization Typology

Overall the type of aquaponics production systems prevailing today can roughly be divided according to its choice of economic organization. In Fig. 6.6, the aquaponics systems are divided whether they are operating in the *non-market area*, pure *market oriented* or a *hybrid* (a mix of the two others). The type of economic organization will to a large extend also determine the choice and level of technology, capital and labor input.

The *non-market* oriented aquaponics system will typically have a low input of technology and capital, but require a higher level of labor on surveillance and maintenance in relation to production output. Homemade or various types of DIY systems as well as small educational kits will normally belong to this group. Likewise, research units on universities will also belong to this category.

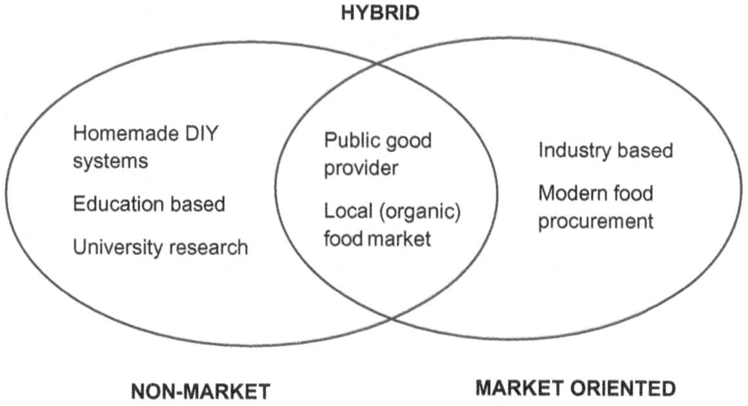

Fig. 6.6 Type of aquaponics system according to its economic organization

The pure *market oriented* aquaponics system will normally require a high input of technology and capital in relation to production output, but labor cost in comparison will be relatively low. Markets will be modern food procurement systems targeting grocers, restaurants and supermarkets.

Market oriented aquaponics systems will still be very small in terms of production output, and in comparison with present day specialized RAS or horticultural systems. However, with the societal demand for universities to take more part in 'blue-green' bio-circular innovations, and the advent of commercial oriented aquaponics systems, we foresee both a growing and closer cooperation between research entities and the infant aquaponics industry. This type of research would be welcome, because a major part of the aquaponics research done at universities are limited in its value for commercial oriented entrepreneurs targeting larger scale risk markets. A more 'dynamic' and 'hands-on' research approach in close cooperation with producers focusing on a variety of fish:plant relations, and their potential production flexibility to cope with fluctuating markets, would be an important research step to support a growth in the market oriented aquaponics industry.

Hybrids will often be an economic organization chosen when a certain scale of food production is being provided to a neighborhood or a community—for example an urban farm. The food system is in the same time also providing various types of social and/or environmental goods to the community in terms of job creation, social inclusion and environmental education. Typically, a public or a non-profit entity will support such an economic organization in return, hence making it a hybrid between a pure market and non-market economic organization.

However, despite the many technological improvements in both RAS and hydroponics being incorporated into aquaponics, one of the major constraints for aquaponics production to move into a larger market-oriented scale, is the higher economic risk associated compared to a specialized RAS or horticulture production. The reason for this is firstly the biological dependency built in the system.

Secondly, the knowledge complexity rises exponential since a producer needs to be specialized, in not only fish and plants, but also understand their interaction with the life cycle of bacteria in bio-filters.

The biological dependency as a risk factor is constantly present with most aquaponics systems operating today. If the plant production declines due to diseases or a pest attack, its ability to function as an efficient cleansing bio-filter is reduced dramatically, and will affect the fish production negatively. Similarly, if the fish production fails because of disease or problems with the bacteria in the mechanical bio-filter, the quality and output of the plant production will be reduced significantly due to fertilizer deficiencies. Therefore, the larger a production, the more technology and capital inputs are required, and the larger economic risk will be the consequence if a system failure occurs.

For aquaponics production to move into larger scale commercial markets the circular dependency in the system has to be decoupled in such a way, that both the biological and economical risks are minimized to a degree where the symbiotic benefits are much greater.

6.5 Future Developments and Research Foci

6.5.1 Decoupled Aquaponics

In today's aquaponics systems the water circulates from fish to plants and via bio-filtering back to the fish as originally shown in Fig. 6.1. The water quality is specifically managed to fit the requirements of the fish species being cultured, and suitable plants are normally chosen to fit the fish environment. It is not always guaranteed that the fish preferences are completely aligned with the optimum requirements of the plants. This calls for compromising of the plant's needs, and as a result they may not achieve their full growth capacity, hence reducing a full optimization of the production and its investments. Likewise, the biological dependency built in aquaponics is a major risk factor hindering large-scale market orientation.

Focus is therefore oriented towards dividing the water flow into two independent subsystems that can occasionally communicate whenever plants need a boost in nutrients or the fish require reclaimed water from plants to dilute the wastes accumulating in the fish sub-unit. This solution, which is referred to as a "decoupled" system (Fig. 6.7) would not only better secure optimal environmental conditions for both the plant and fish production units, but also eliminate the biological dependency in aquaponics hence minimize the economic risk substantially.

The processing of sludge from the aquaponics system has recently received increased interest. Not only is prompt removal from the system helping to maintain healthier and more resilient systems for fish, but also it improves the productivity by better capitalizing of by-products through their reintroduction into the production

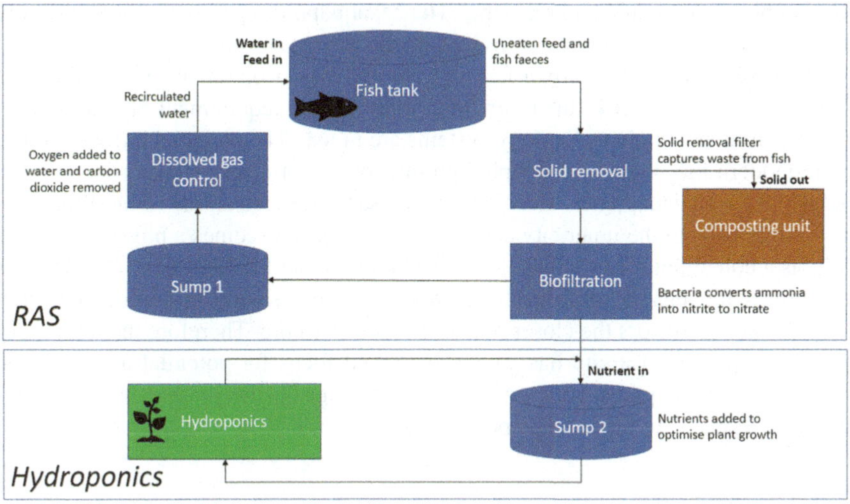

Fig. 6.7 The principles of a decoupled aquaponics system (Thorarinsdottir 2015)

system. Several ideas are being tested aiming for zero waste solutions, using the sludge as e.g. feed for crayfish, farming of worms and/or black soldier flies, or making fertilizer through aerobic or anaerobic digestion.

6.5.2 Aquaponics and Urban Farming

It is important that cities reclaim foods that were once characteristic of their community because these foods instill a sense of place, strengthening the communal bonds between the residents. In addition, these foods can generate jobs and revenue for the community (Yang 2012).

Likewise, there has been a fast growing interest in urban farming in both developing as well as developed countries since the world food crisis of 2007 (FAO 2010; de Zeeuw et al. 2011). The vulnerability and dependency of food for the world urban community, already reaching 60%, and a reliance based on a very concentrated food system, made it clear for a growing number of city councils as well as a varied range of non-governmental organizations (NGO's), that change in the food system from 'farm to fork' had to change (http://www.urbanagricultureeurope.la.rwth-aachen.de/)

Aquaponics is becoming very popular among these various urban farm initiatives around the world. This is especially true for community building in areas of unemployment, food security issues and social problems related to inclusiveness; other factors include educational awareness through giving a simple overview of the complexity of nature and recycling. Hence, aquaponics production will typically be organized as a *non-market* or *hybrid* economic organization.

Aquaponics related to urban farming are still based on small production units (Sommerville et al. 2014) due to first and foremost the requirement of a large plant area when more simple technology systems are in use. The potential introduction of more modern bio-filters or decoupled production systems opens up for larger scale industrial based aquaponics. However, the closer you move an urban farm from the peri-urban zone to the inner city center the more space becomes a physical issue as well as a constraint for establishing a financially viable food business. The latter is also true with regards to space becoming a scarce resource in competition with other economic sectors the closer one gets to the city core. Therefore the 'empty' or 'free' roof spaces of a city has gained increased focus for potential areas of food production, but requires often an expensive change in the present building construction to carry an urban (aquaponics) roof-top farm.

6.5.3 Aquaponics and Organic Certification

Organic certification appears to be a natural step for an aquaponics producer since the whole system is based on a holistic thinking in terms of recycling, lowering the resource intake and securing zero pollution. However, the present organic regulatory regime does *not* have any standards or regulations for certifying organic aquaponics. The RAS technology is even *forbidden* under the present organic regulation, which seems to be more of an economic protection to the extensive open pond systems prevalent in organic fish production rather than having anything to do with fish welfare or the aquatic environment.

It is only possible to have an aquaponics production system completely certified organic if a *non-holistic* approach is made, meaning a certification towards the organic fish- and horticulture regulation is made separately. Firstly, the plants must be grown in soil. Secondly, the fish produced must be fed with organic certified fish feed, and thirdly the fish can only be produced in a RAS system if the fish are sold as fingerlings for further growth in open-air pond systems certified organic. However, if aquaponic produce gains markets and moves into larger scale production systems, it seems indisputably; that the organic farm movement will need to revise its present regulation focusing on specific technologies rather than having a more principle based approach allowing for new resource productive and holistic production systems such as aquaponics.

The present regulatory framework for organic fish and horticultural production in the EU is regulated by the Council Regulation (EC) No. 834/2007 whereas more detailed rules are regulated by the Commission Regulations (EC) No. 889/2008, and (EC) no. 710/2009.

6.5.3.1 Horticultural Produce

For organic horticultural production current Commission Regulation (2008), implementing Council Regulation (2007), contains only one element specific to greenhouse production:

art. 4 which bans hydroponic production and allows organic cultivation only in soil.

Since most aquaponics production systems are based on a soilless hydroponic technology, the plants produced under such a system cannot be certified as organic. This leaves with the only option to adopt culturing practices applying soil through decoupled aquaponics/RAS wastewater.

6.5.3.2 Aquacultural Produce

For organic aquaculture the production is regulated by Commission Regulations of (2008) and (2009). In parr. 11. Commission Regulation (2009) recirculating systems are clearly prohibited in organic aquaculture, except for the specific production in hatcheries and nurseries making and selling fingerlings for further growth in open-air pond systems.

Parr. 11.
Recent technical development has led to increasing use of closed recirculation systems for aquaculture production, such systems depend on external input and high energy but permit reduction of waste discharges and prevention of escapes. Due to the principle that organic production should be as close as possible to Nature, the use of such systems should not be allowed for organic production until further knowledge is available. Exceptional use should be possible only for the specific production situation of hatcheries and nurseries.

Since recirculating technology is at the core of the aquaponics production system it is at present not possible to get a complete organic certification on an aquaponics system, if all of the finishing produce is to be sold for the consumer market.

6.5.3.3 Future of Organic Aquaponics

The crux for aquaponics producers to get an organic certification in the future lies in the acceptance of the recirculating technology within organic regulation itself, as well as presenting aquaponics as an ideal closed loop, non-pollute and holistic food production system.

Short-term strategies in this regard could be to:

(1) View aquaponics as a farm based on a necessary harmony and biomass ratio between husbandry (the fish), and a soil-based horticultural production as the field turning waste into valuable resources and providing a food production with no discharges to the environment.
(2) Work towards a specific regulation on aquaponics. This would imply allowing recirculating technology such as RAS, and an intensification of the fish production. Fish intensification is already regulated by the organic regulation, but is based on an extensive open-air pond system and the question of discharge of fish manure to the aquatic environment.

Paragraph 24 in the Commission Regulation (2009) opens up for an interpretation that such steps for a revision in the organic rules could be allowed. Especially the last four lines in Parr. 24 implies that national initiatives could be taken with the aim of improving the common EU regulation on organic aquaculture. This would require a more dedicated willingness in the organic movement to commence a process in this direction, but unfortunately this dedication and willingness does not seem to exist at present.

Parr. 24
Organic aquaculture is a relatively new field of organic production compared to organic agriculture, where long experience exists at the farm level. Given consumers' growing interest in organic aquaculture products further growth in the conversion of aquaculture units to organic production is likely. This will soon lead to increased experience and technical knowledge. Moreover, planned research is expected to result in new knowledge in particular on containment systems, the need of non- organic feed ingredients, or stocking densities for certain species. New knowledge and technical development, which would lead to an improvement in organic aquaculture, should be reflected in the production rules. Therefore provision should be made to review the present legislation with a view to modifying it where appropriate.

6.6 Perspectives

Aquaponics is rapidly moving into new development phases and presenting industrial and commercial potential. The UVI-production system is still prevalent, but the technological innovations made in both the industrial aqua- and horticulture sectors within the last two decades has led to the emergence of new approaches.

De-coupled aquaponics holds the prospect of reducing or even eliminating the biological dependence, but at the same time acquires the symbiotic benefits of combining fish and plants in a circular production system. A more dynamic and hands-on relevant research targeting de-coupled aquaponics would be very valuable for the development of a *generation 2* within the aquaponics industry.

The large amounts of CO_2 produced from both the aqua- and horticulture sector could become a serious growth constraint for these sectors. Use of CO_2 produced by the fish in an aquaponics system by plants placed in a closed greenhouse environments could therefore provide a huge potential for the 'blue and green' sectors in the future.

Sufficient phosphorus production will also be a major concern in the near future. In current recirculating aquaculture systems and aquaponics setups, 30–65% of the phosphorus added to the system via fish feed is lost in the form of fish sludge that is filtered out by mechanical filtration. Since phosphorus is a major component of agricultural fertilizer, the development of phosphorus recycling production systems in aquaponics would be an important contribution for future food production.

With the emergence of a *generation 2* de-coupled aquaponics system, the next constraint required to be addressed within aquaponics would be the huge plant area required to maintain a sustainable fish production both economically and environmentally. A viable commercial and technological development in bio-gassing the sludge from the fish manure would be a major breakthrough for introducing large industrial scale aquaponics. A viable bio-gassing of the fish manure would mean a tremendous reduction in the required plant area to uphold the land-based aquaculture production systems known today. It would see the emergence of a *generation 3* within aquaponics presented as a flexible environmental production module that could be applied to almost any existing aqua- or horticultural production of today.

If the major constraints listed above are addressed and viable solutions found, the aquaponics industry holds the prospect of being an important niche within the aquaculture sector itself just as the organic sector is within agriculture today.

References

Aghajanian A (2007) Chinampas. Their role in Aztec Empire-building & expansion. IndoEuropean Publishing.com

Bradley K (2014) Aquaponics: a brief history. https://www.milkwood.net/2014/01/20/aquaponics-a-brief-history/

Braungart M, McDonough W, Bollinger A (2007) Cradle-to-cradle design: creating healthy emissions—strategy for eco-effective product and system design. Int J Clean, 1337–1348

Commission Regulation (2008) (EC) No 889/2008 of 5 September 2008 laying down detailed rules for the implementation of Council Regulation (EC) No 834/2007 on organic production and labelling of organic products with regard to organic production, labelling and control

Commission Regulation (2009) (EC) No 710/2009 of 5 August 2009 amending Regulation (EC) No 889/2008 laying down detailed rules for the implementation of Council Regulation

(EC) No 834/2007, as regards laying down detailed rules on organic aquaculture animal and seaweed production
Council Regulation (2007) (EC) No 834/2007 of 28 June 2007 on organic production and labelling of organic products and repealing Regulation (EEC) No 2092/91
Dalsgaard J, Lund I, Thorarinsdottir R, Drengstig A, Arvonen K, Pedersen PB (2013) Farming different species in RAS in Nordic countries: current status and future perspectives. Aquac Eng 53:2–13
de Zeeuw H, van Veenhizen R, Dubbeling M (2011) The role of urban agriculture in building resilient cities in developing countries. J Agric Sci 149:153–163
Diver S (2000) Aquaponics—integration of hydroponics with aquaculture. http://backyardaquaponics.com/Travis/Attra%20Aqua.pdf
FAO (2010) Growing Greener Cities
Goddek S, Delaide B, Mankasingh U, Ragnarsdottir KV, Jijakli H, Thorarinsdottir RI (2015) Challenges of sustainable and commercial aquaponics. Sustainability 7(4):4199–4224. https://doi.org/10.3390/su7044199
Hortidaily.com (2015) Belgian tomato grower delivers heat and electricity to fish farm. Hortidaily.com Publication date: 11/30/2015. http://www.hortidaily.com/article/22414/Belgian-tomato-grower-delivers-heat-and-electricity-to-fish-farm
Kumar S, Putnam V (2008) Cradle-to-cradle: reverse logistics strategies and opportunities across three sectors. Product Econ, 305–315
Körner O, Gutzmann E, Kledal PR (2014) Modelling the symbiotic effects in aquaponics. In: Conference paper for the European Aquaculture Society, Adding Value, Donastia, San Sebastian, Spain, October 14–17, 2014
McDonough W, Braungart M (2002) Cradle to cradle: remaking the way we make things. North Point Press, New York
Pump Industry Analyst (2015) Pentair partners with urban organics to advance aquaponics. Volume 2015 (9), pp 12–13, September 2015. https://doi.org/10.1016/S1359-6128(15)30333-5
Rakocy JE, Bailey DS, Shultz KA, Cole WM (1997) Evaluation of a commercial-scale aquaponic unit for the production of tilapia and lettuce. In: Fitzsimmons K (ed) Tilapia aquaculture: proceedings of the fourth international symposium on tilapia in aquaculture, Orlando, Florida, pp 357–372
Rakocy JE, Bailey DS, Shultz RC, Danaher JJ (2007) Fish and vegetable production in a commercial aquaponic system: 25 years of research at the University of the Virgin Islands. In: Proceedings of the 2007 national Canadian aquaculture conference, Edmonton, Alberta, Canada
Resh HM (2013) Hydroponic food production—a definitive guidebook for the advanced home gardener and the commercial hydroponic grower, 7th edn. CRC Press, Taylor & Francis Group
Ruddle K, Zhong G (1988) Integrated agriculture-aquaculture in South China: the dike-pond system of the Zhujiang Delta. Cambridge University Press, Cambridge
Sommerville C, Cohen M, Pantanella E, Stankus A, Lovatelli A (2014) Small-scale aquaponics food production—integrated fish and plant farming. FAO Fisheries and Aquaculture Technical Paper No. 589, FAO, Rome 2014
Stenkjaer N (2011) Dike pond system. http://www.folkecenter.net/gb/rd/biogas/biomassbiogas-at-folkecenter/dike_pond/
Stickney RR, Granvil DT (2012) History of aquaculture (Chapter 2). In: Tidwell JH (ed) Aquaculture production systems. Willey-Blackwell
Thorarinsdottir R (ed) Aquaponics guidelines. University of Iceland, August 2015, Haskolaprent, Reykjavik, Iceland (ISBN: 978–9935-9283-1-3) https://skemman.is/handle/1946/23343
Yang J (2012) Food production. A paper of case studies, PLAC 5500

Chapter 7
Aquaponics Production, Practices and Opportunities

Edoardo Pantanella

Abstract Aquaponics is a plant production system that integrates soilless cultivation and recirculating aquaculture. Aquaponics is an environmentally-friendly system that makes full reuse of wastes that are used as fertilizers for plants. At the same time it is more productive than soil-based agriculture and has consistent water savings, which makes it the ideal technology to produce food in resource-limited and climate-change affected areas. The chapter seeks to provide an understanding of aquaponics by giving an overview of the state of the art of past and current research and by outlining advantages and disadvantages of aquaponics against traditional agriculture (soil, soilless) and aquaculture. A comprehensive description of the aquaponic components is given together with a summary of the different systems in use, providing keys for understanding their characteristics and suitability in different climatic and operating conditions. Beside the production of quality crops for both market and backyard consumption, aquaponics could be a tool to address food insecurity in developing countries. Furthermore, new opportunities for aquaponics are also seen in the use of saline waters to provide tool for bioremediation of brackish-water and marine aquaculture, but management systems need to be adapted to the range of salt-tolerant plants and seaweeds available for either food, feed or fuel, as well as the market demand.

Keywords Integrated aquaculture · Soilless cultivation · Hydroponics
Zero-waste · Sustainable farming

E. Pantanella (✉)
CIHEAM IAM Bari, Valenzano, BA, Italy
e-mail: edpantanella@gmail.com; pantanella@iamb.it

F. I. Hai et al. (eds.), *Sustainable Aquaculture*, Applied Environmental Science and Engineering for a Sustainable Future,
https://doi.org/10.1007/978-3-319-73257-2_7

7.1 Overview

There is a significance concern about how future generations will produce more in a sustainable way. Intensive food production forces agriculture to overexploit natural resources: the conversion of natural lands/forests into arable lands, the pollution from the massive use of fertilizers and chemicals, the reduction of soil fertility and carbon stocks are indeed some of the main raising issues on farming sustainability (Tillman et al. 2002). In the last twenty years the nitrogen content in the oceans has increased by twenty times due to the indiscriminate use of fertilization (Downing et al. 1999) causing severe eutrophication to water bodies. Therefore, the closing of the loop between inputs and wastes, such as the re-reuse of crops and animals by-products, is one of the few possibilities to improve the water and nutrient efficiency and to reclaim organic wastes back to useful productions. The pace in achieving higher crop productivity, as it was during the Green Revolution in the sixties, raises questions whether higher outputs from agriculture are still possible by technical or scientific breakthroughs (Brown and Kane 1994; Waggoner 1994). There are robust evidences that natural resources, such as water and fossil energy, are over exploited (UN-Water 2012) and cannot easily guarantee further agriculture expansion. The conversion of wild lands to agriculture, such are forests, not necessarily has brought long lasting advantages for food production due to the low fertility or the fast degradation of the soil, which eventually has caused irreversible losses of land. At the same time the excessive intensification of agriculture in fertile productive areas if from one side has disrupted natural ecosystems through the massive use of fertilizers and pesticides, from the other side has progressively reduced the fertility and productivity. Therefore the increased pressure on production makes the exploitation of natural resources inevitable unless new strategies and production techniques are adopted to make farming systems more self-sufficient and resilient.

In the case of horticulture the adoption of soilless agriculture, more commonly known as hydroponics, shows undoubting advantages for its higher nutrient and water use efficiency compared to soil based agriculture (Resh 2004; Leoni 2003). Hydroponics shows in fact yields that are 2–3 fold higher and water consumptions patterns that can be up to ten time lower than traditional farming. This is due to the improved water distribution and the better growing conditions of plants that receive punctual fertilization of nutrients with no competition from weeds as well as limited risks of pests and soil-borne pathogens.

Likewise, the farming of animals with the lowest footprint and the highest feed conversion efficiency, would eventually reduce the impact on water and land and increase the overall food output (Verdegem et al. 2006). In the case of fish the adoption of recirculating aquaculture systems (RAS), in which the fish rearing water is almost completely recycled after a filtration stage, can reduce the water footprint of traditional aquaculture by hundreds of times and avoid any discharge of organic wastes (fish excrements, uneaten feed) into the environment.

Aquaponics combines the benefits from both soilless culture and RAS. The build-up of nutrients in a closed aquaculture system can in fact reach concentrations that are ideal for the commercial production of plants. At the same time plants take up nutrients and reclaim water back to fish by also adding a profit from the costs for water treatment normally occurring in RAS. Aquaponics allows intensive and high-quality production of vegetables without any impact on the environment for either pollution from chemical fertilizers (agriculture) or animal wastes (aquaculture).

The present chapter aims at exploring the potentials of the integration of aquaculture with soilless culture for the sustainable development of agriculture in rural areas and wherever traditional agriculture could not be efficiently developed due to disturbed soil or adverse environmental conditions. Aquaponics is analysed in its components and compared against aquaculture, hydroponics and traditional agriculture practices, with the objective to unveil its advantages and disadvantages in the context of sustainable food production and food security.

7.1.1 What Is Hydroponics

Hydroponics is a combined word that joins two Greek terms: water (hydro) and work (ponos). Plants grow by means of a nutritive solution in which adequate quantities of macro and micronutrients are dissolved to support their growth. Hydroponics is also called *soilless culture*, as plants do not grow on soil but rather on inorganic substrates (sand, gravel, perlite, rockwool slabs) (Fig. 7.1), organic substrates (peat, sawdust, rice husk) or even with bare roots within an aqueous media (floating system) (Fig. 7.2). Substrates in soilless culture provide only mechanical support to the plants.

Hydroponics moved its first steps in commercial production in the first half of the 20th century following the intensification of agriculture and the need to overcome the problems of soil-borne diseases caused by the continuous monoculture practices in the greenhouses (Leoni 2003). Further expansion occurred from the fifties, when the adoption of plastic lowered the production costs and made the investment on greenhouses and climate control affordable by many (Resh 2004). The initial use of bulk substrates was successively substituted by the nutrient film technique (NFT) and the adoption of rockwool in the seventies, which opened up new horizons in commercial-scale horticulture.

Nevertheless hydroponics has a long history that witnesses the constant research of farmers for more productive and cost-effective solutions under different designs and plant nutrient sources, even with low-tech approaches.

In the Middle East the hanging gardens of Babylon, built more than 25 centuries ago, were the first example of soilless roof-top agriculture that used sludge and ash as plant nutrients (Leoni 2003).

In Mexico organic matter was the growing media and fertilizer used for the chinampas, a type of integrated aquaculture-agriculture system in shape of terrains surrounded by canals. The production of food with this technique was one of the most intensive agricultural system in the pre-Colombian era (Sutton and Anderson

Fig. 7.1 Hydroponic tomato on Rockwool media

Fig. 7.2 Small floating system with a plant bed and nutrient tank

2004) to the extent that chinampas could support the food needs of 10–18 people per hectare (Adams 2005), which is far above the current productivity of modern soil-based agriculture. The common factor that favoured the diffusion of all these agricultural systems was the lack of cultivable land and the need to increase the acreage for crops, a common problem still existing in many flooded areas of developing countries.

In S.E. Asia floating agriculture is still in use as a low-tech type of hydroponic. Floating rafts made of aquatic macrophites (i.e. water hyacinths) provide both support and nutrients to plants through the release of minerals released from the decaying organic matter. In the Inle Lake in Myanmar such type of system is still the backbone of local horticultural productions that supply vegetables to the domestic markets.

Nevertheless the solutions provided by floating agriculture can well integrate traditional aquaculture systems, such as ponds, in which quality vegetables can grow on water with yields that are similar or higher than soil crops (Pantanella et al. 2011d). At the same time floating rafts made with decaying organic matter can release nutrients in ponds to promote microalgae blooms that constitute the food of planktonic fishes. On the other hand floating pots filled with inert media that are suspended on pond water can provide tools for bioremediation for intensive farming by simply stripping nutrients from water.

7.1.1.1 Advantages of Hydroponics Against Soil Production

Soilless cultivation addresses many issues of traditional farming. The presence of an inert media in lieu of soil allows plants to grow with very limited incidence of soil-borne pathogens and pests. At the same time the hydroponics' real-time delivery of nutrients, which are monitored and distributed according to the growth stage of the plants, maximizes the productivity and quality traits of the produce. The lack of soil also avoids any need for weed control, which directly helps the crops to grow without the competition for nutrients and space by invasive plants. The delivery of water and nutrients is also engineered in such a way to avoid any leakage or spill outside of the system, thus minimizing any pollution risks. Such controlled management let hydroponics be up to ten times more efficient in its water use efficiency than traditional agriculture.

Hydroponics is at least 20–25% more productive than soil-based intensive greenhouse farming, which makes massive use of fertilizers and soil sterilization (Resh 2004). On the other hand for outdoor crops hydroponics shows 4–10 times higher yields than soil (Table 7.1).

Soilless cultivation is ubiquitous, as it allows to produce food even in places where traditional agriculture cannot be developed due to unsuitable soil or water scarcity: deserts, salinated or unproductive lands, roof tops in urban areas, contaminated land under reclamation.

The advantages of hydroponics against conventional agriculture can be summarized in the improved adaptability to farm in unfavourable areas, in the better efficiency of inputs' uses, in the higher yields and qualitative traits, in the reduced use of chemicals to overcome plants' soil-borne diseases (Jensen 1981, 1997; Tesi, 2002; Resh 2004) (Table 7.2).

Table 7.1 Comparative yields per hectare in soil and soilless culture (Resh 2004—modified)

Crop	Soil ton ha^{-1}	Soilless ton ha^{-1}
Beans	12.5	52.5
Beets	10.0	30.0
Cabbage	14.7	20.3
Cucumber	7.9	31.6
Lettuce	10.2	23.7
Peas	2.5	22.5
Potatoes	20.0	175.0
Tomatoes	12.5–25	150–750
Wheat	0.7	4.6

Table 7.2 Soil versus soilless production

	Soil	Soilless
Farming in new areas	Not always possible. Depends on the type of soil, fertility, salinity	Agriculture possible in any condition
Cultivation	Constant preparation of soil, need of machines, fuel intensive	No needed, substrates preparation or positioning on troughs/ground
Intensification of production	Limited. Monoculture brings "soil tiredness" and already decreases yields after two successive crops Soil tiredness requires crop rotation, fallow or soil sterilization, which is time consuming and interrupts crop cycles for 2–3 weeks	Monoculture is possible with no decadence of performances Substrates could be sterilized with simple means and no crop interruptions Inert media or water do not face risk of any fertility losses due to their characteristics
Plant nutrition	Variable delivery. The release depends on soil characteristics. Some deficiencies are possible. The precise delivery of nutrients according to the plant growth stage is not possible	Real time distribution of nutrients and pH according to the growth stage of the plants. Real-time control of the levels of nutrients required by plants
Nutrient use efficiency	Fertilizers broadcasted broadly, High dispersal through leaching and runoff in outdoor conditions	Minimal amount required due to microirrigation and containment of media. Water and nutrients monitoring avoid the loss of nutrients
Water use efficiency	Efficiency affected by soil texture and irrigation system	Optimal delivery trough microirrigation supported by sensors
Weed control	Need continuous control	No need of any control
Diseases and pests	Affected by soil-borne diseases and pests. Needs sterilization, crop rotation	Not affected because of no use of soil
Quality	Product characteristics depends on of the type of soil and management	Standardized production with full control of nutrients. Optimized growth

(continued)

Table 7.2 (continued)

	Soil	Soilless
Production costs	Normal, but use of machinery necessary for soil cultivation and higher use of inputs (water). Higher costs if greenhouses/nethouses are used	Higher costs due to more expensive setting in greenhouses/nethouses and the presence of a monitoring system,
Farm management	Standard level	Expert level. Needs higher knowledge for the higher technology used

7.1.1.2 Hydroponic Systems in Use

Hydroponic systems are classified into two main categories depending on whether plants grow with their bare roots in an aqueous media or if they benefit from the mechanical support given by substrates (Resh 2004). The first type, also called water culture, is the most used especially for leafy vegetables. Three main designs are used: nutrient film technique (NFT), deep water culture or floating system (DWC) and aeroponics.

Nutrient Film Technique (NFT) consists of flat-bottomed plastic pipes with holes on their top in which plants are positioned (Figs. 7.19 and 7.20). The plants' roots develop inside the pipes. A thin layer of nutritive solution wets the bottom of the pipes at a very low water flow (1–2 L min^{-1}) supplying plants with nutrients and water. In general the water flow can be continuous or intermittent, in the latter case roots take some additional oxygen from the air. This type of system is mainly closed, with water continuously recirculating between the troughs and the tank containing the nutritive solution, where water get oxygenated.

Floating system/deep water culture (DWC) consists of tanks of variable depth (from 7 to 30 cm) on which plants grow supported by floating polystyrene rafts (Figs. 7.15 and 7.16). Plants have bare roots into the nutritive solution, which is kept aerated by air stones. The volume of water allows for multiple production cycles, with nutrients being re-integrated from tanks containing concentrated stock solutions. The system is quite resilient against black outs, as oxygenated water is always in contact with the roots. DWC is largely used for leafy greens, culinary herbs and a variety of fruity plants (Leoni 2003).

Aeroponics has plants roots suspended in the air continuously wetted by nozzles spraying the nutritive solution. Plants are positioned on oblique trays to optimize the space into greenhouses and to create a volume for the spraying systems. Like the NFT the nutritive solution is continuously collected into a sump and minerals are reintegrated from tanks containing stock solution. Aeroponics needs the uniform distribution of the nutritive solution to the roots to guarantee a uniform growth, and a tailored management of the spraying cycles, which vary according to the type plants and their growth stage (Leoni 2003). Aeroponics takes advantage of the great root oxygenation, but it is prone to wilt in case of any disruption in the water distribution system.

Substrate culture has plants growing in pots, beds or bags filled with a growing media. There is a wide range of organic or inorganic media available, each with different characteristics, prices and availability: peat moss, sawdust, coconut, rice husk, rock wool, sand, gravel, perlite, clay balls, polyurethane, and polystyrene. The media is always kept separated from the ground underneath to prevent any risk of contamination with the soil. The nutritive solution can be delivered through micro-irrigators, sub-irrigation or ebb & flow (cyclic flood and drain of the media with the nutritive solution). The media systems can be either open (flow-through), with the nutritive solution used only once, or closed with a continuous recycling of the water (Tesi 2002; Resh 2004).

7.1.2 What Is Aquaponics

Aquaponics is an integrated system that combines hydroponics and recirculating aquaculture. Plants grow using the nutrients dissolved in the aquaculture effluents and provide tools for bioremediation by reclaiming the wastewater back to fish.

There have been many types of integrations between aquaculture and agriculture in the past, with fish water mainly delivered to the plants in open systems. Examples can be seen in the irrigation of crops with pond water or in ditch-dyke systems, where narrow strips of land are surrounded by a network of small water canals stocked with fish. The uptake of water for irrigation from the ponds, which are then refilled with new aqueous sources, is undoubtedly a good practice to maintain good water quality for healthy fish growth and to reduce the impact of aquaculture pollution (Barnabé 1990; Diana et al. 1997).

The feeding of fish in fact increases the levels of excreted ammonia into the water, whose build up is toxic for the aquatic animals. Therefore, the progressive intensification of aquaculture production, with higher fish densities, needs increased water exchange to avoid toxicity and deaths (Barnabé 1990). Although these integrated systems cover the water needs of the plants, the concentrations of nutrients available to the plants are still not sufficient to reach yields and sizes of commercial value, unless plants are further fertilized. The reason stands in the still low densities of fish, the competition for nutrients with microalgae growing in the pond and eventually the continuous dilution of nutrients by large volumes of new water used to refill the ponds.

Aquaponics has been developed mainly within recirculating aquaculture systems (RAS) where waste water is continuously recycled and reclaimed back to fish after a biofiltration stage (Rakocy 1989; Rakocy and Hargreaves 1993; Lennard 2004) (Fig. 7.3). RAS technology was developed to overcome all the problems linked with water use and pollution from traditional aquaculture systems (open systems) in which large volumes of water are discharged to avoid the build-up of wastes and toxic metabolites (Barnabé 1990; Diana et al. 1997). Traditional aquaculture has in

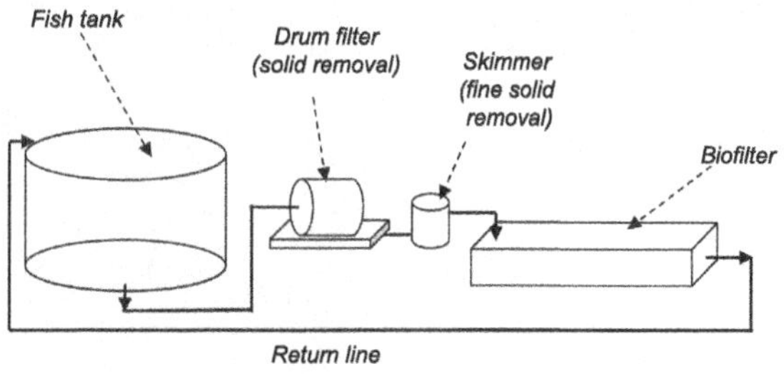

Fig. 7.3 Design of a generic recirculating aquaculture system—RAS

fact a consistent impact on the environment (Piedrahita 2003; Verdegem et al. 1999, 2006) due to the big water footprint. In addition the pollution from open aquaculture systems raises concerns about the sustainability of intensive fish farming (Costa Pierce 1996).

RAS has a very limited use of water and discharges very little amounts of wastewater (Verdegem et al. 2006). The core management of RAS focuses on the continuous reuse of water, which is possible through mechanical waste removal (uneaten feed, fish solids, dead fish) and the oxidation of nitrogen wastes operated by biological filtration (van Rijn 1996) to convert ammonia into no-toxic nitrate (nitrification).

An aquaponic system is a RAS in which the biological filtration is partly operated by plants (Fig. 7.4), whose roots host the bacteria responsible for the nitrification and directly uptake nutrients for plant growth. Aquaponics takes advantage from the nutrient build-up normally occurring in closed system due to the higher fish stocking densities and the much lower water exchange needed to get rid of fish excreta than traditional aquaculture. The increasing levels of nutrients allow aquaponics to achieve concentrations similar to chemical hydroponics and to obtain consistent productions of plants of commercial value.

The higher fish densities than traditional aquaculture is allowed by the continuous mechanical filtration to remove solids (Fig. 7.5) and the nitrification of ammonia, which prevents toxicity to fish and allow plants to take up nitrate, the most assimilable form of nitrogen. Likewise, the presence of other beneficial microorganisms such as fungi, microplankton, mineralizing bacteria, rhizobacteria help not only the system to increase the pool of essential nutrients available to plants, but also improves the resilience of the system against plant pathogens (Savidov 2005).

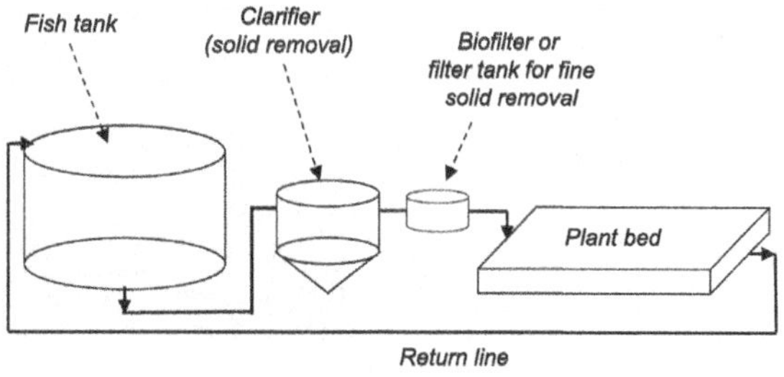

Fig. 7.4 Design of a generic aquaponic system

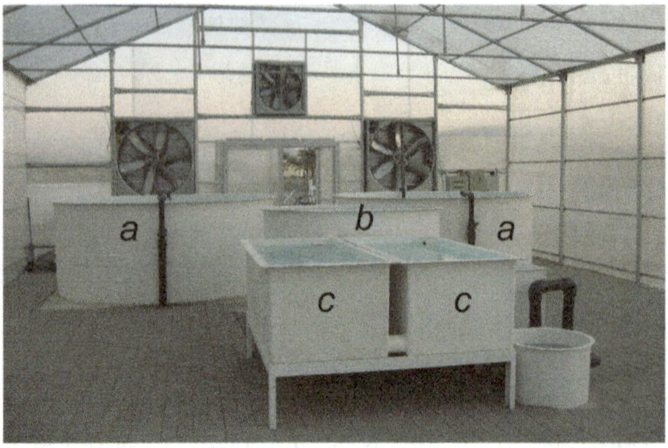

Fig. 7.5 Aquaponic system. *a* fish tanks; *b* clarifier; *c* filter tank for fine solid removal

Differences between aquaponics and RAS can be found in the simpler solid waste management, of the former, since a lower efficiency in solid removal improves the opportunities to obtain plant nutrients from the mineralization of fine suspended wastes. On the contrary RAS water must have the lowest concentrations of fine wastes to avoid any clogging of the biofilter, which eventually reduce the efficiency of the nitrifying bacteria.

Aquaponics does not differ from hydroponics for what concerns the types of systems in use: DWC (Fig. 7.6), NFT and media beds. However there are some differences in the management, since the combined presence of fish and plants requires some compromises in the setting of the environmental conditions and water parameters, which should always meet the optimum for both. This, in some cases, limits the choice of crops due to suboptimal environmental conditions of certain plants.

One constrain in aquaponics is the limited choices available in crop protection, due to the presence of fish. In aquaponics in fact no chemical pesticides can be used and even many of the remedies in use in organic agriculture may result toxic to

Fig. 7.6 Plant tanks in an aquaponic system where plants grow with bare roots

fishes. This, if on one side pull farmers to adopt mainly preventive strategies against pathogens and pests, on the other side allows for safer productions that are highly appreciated by a large number of consumers.

One recent advance in aquaponics is the use of hybrid or decoupled systems (Figs. 7.24 and 7.25), in which the fish and plant subsystems work as standalone units (RAS + hydroponics). In the former the separation between fish and plants is temporarily void to let the hydroponic unit be refilled with nutrients from fish wastewater and the RAS unit to use reclaimed water from plants; in the latter the water only goes from the fish to the plant unit. Such systems keep optimal growing conditions for both animals and plants and would give more freedom in using the traditional remedies in organic crop protection without any risks of fish toxicity.

Another difference in aquaponics stands in its complex ecosystem, in which the presence of microorganisms is not prevented but rather encouraged to improve the conversion of suspended wastes into nutrients for plants and to create a highly competitive and resilient environment where pathogens have difficulties to thrive.

In plant nutrition aquaponics supplies most of the nutrients to the plants, with the only exceptions of calcium, potassium and iron that need to be integrated through the regular addition of buffers to control the water pH into the systems. Following the natural conversion of ammonia into nitrate operated by nitrifying bacteria both aquaponics and RAS water tend in fact to acidify and need to be re-balanced by adding alkali to maintain optimal operating conditions for both bacteria, fish and plants.

Aquaponics has lower concentrations of circulating nutrients than hydroponics (Table 7.3). Despite the low levels of minerals aquaponics shows same yields of hydroponics firstly because nutrients are continuously supplied by fish to plants,

Table 7.3 Concentrations of nutrients in hydroponics compared against aquaponics at the Agriculture Experimental Station of the University of Virgin Islands (Massantini 1968; Rakocy et al. 1992, 2004a, b, 2006)

	Hydroponics			Aquaponics
	Minimum ($mg\ L^{-1}$)	Optimal ($mg\ L^{-1}$)	Maximum ($mg\ L^{-1}$)	Average concentrations ($mg\ L^{-1}$)
Nitrate	40	60–160	200	26.3–42
Ammonia		0–40	100	0.95–2.2
Phosphorus	15	30–90	130	8.2–16.4
Potassium	100	200–400	600	44–63.5
Calcium	75	150–400	600	11.9–24.2
Magnesium	25	25–75	150	6.0–6.5
Sulphur	50	75–300	600	18.3
Iron		2–4	10	1.3 -2.5
Boron		0.2–1	5	0.09–0.19
Manganese		0.2–2	15	0.06–0.8
Copper		0.01–1	5	0.03–0.05
Zinc		0.01–1	20	0.34–0.44
Chloride			600	11.5

and secondly because the water movement into the system enhances the flow of nutrients at root level and the consequent plant uptake.

7.2 Past and Present Research

Aquaponics research started to move its first steps by looking at bioremediation through the use of plants and the integration of systems for alternative productions. Initial trials focused on finding optimal component designs and assessing the most efficient fish/plants nutrient ratios and yields. Successive researches focused on the optimization of the plant nutrients in the systems and in the improvement of waste management for the full use of minerals for plant nutrition.

More recently the interest has been centred on the upgrade of the systems for commercial productions, in particular on the use of lights, decoupled systems as well as in exploring the beneficial interactions between plants and microorganism thriving into the systems.

The research on aquaponics started during the seventies. Several universities across North America and Europe started testing fish with plants in closed systems (Naegel 1977), but with designs of components still at their primordial stage. Since the eighties many researchers developed extensive studies on aquaponics with the focus on the component ratio. Prof. James E. Rakocy dedicated three decades on aquaponics at the University of the Virgin Islands (UVI), USA where he built and managed a commercial-scale system (Fig. 7.7). He was among the first to identify the optimal balance between fish and plants and to optimize the components for floating systems. In Australia Dr. Wilson Lennard carried out trials with an extensive number of plant species, optimized fish/plant ratios and compared the performances from different system designs. In Canada Dr. Nick Savidov pioneered research on improved mineralization for UVI systems in greenhouses in cold climates.

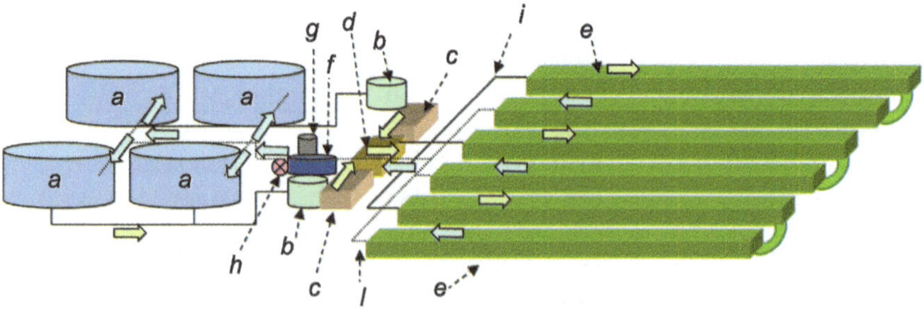

Fig. 7.7 Diagram of the UVI system: ***a*** fish tank, ***b*** clarifier, ***c*** filter tank, ***d*** degassing unit, ***e*** plant tanks (floating system), ***f*** sump, ***g*** base addition tank, ***h*** water pump, ***i*** influent water from the degassing tank to the plant tanks, ***l*** effluent water from the plant system to the sump

7.2.1 The UVI System

The University of the Virgin Islands (UVI) has been the first academy to test a pilot system for both research and commercial production (Fig. 7.7). The system, still in full activity, has four fish tanks accounting for a total volume of 31.2 m^3, two conical clarifiers of 3.8 m^3 each, four filter tanks of 0.7 m^3 each, which are used to remove fine particles through orchard nets, and a degassing tank where the intense aeration occurring removes carbon dioxide from the water before it goes to the plant tanks. The fish unit supply fertilized water to six hydroponic tanks 30.5 m long 1.2 m large and 0.4 m deep that account for a total surface of 214 m^2. Water from the plant tanks is collected into a sump where is pumped back to the fish tanks after being buffered by a small base addition tank. The total water volume of the system is 110 m^3.

Water is circulated by means of a ½ hp water pump that guarantee for a mean retention time of 1.5 h in the fish tanks. The system is supplied by two blowers, one of 1.5 hp for the fish tanks, and another of 1 hp for the plant troughs.

The UVI system can produce approximately 3 tons of tilapia and 11 tons of lettuce a year from a total surface of 400 m^2.

7.2.2 Fish Species

Aquaponics was tested with several fish species, which testimonies the large versatility of this system to different climatic conditions. The species cultured were tilapia (*Sarotherodon aurea*, *Oreochromis niloticus*, *Tilapia rendalli*) (Watten and Butsh 1984; Rakocy and Hargreaves 1993; Rakocy et al. 1999a, b; Seawright et al. 1998), African catfish (*Clariar gariepinus*) (Endut et al. 2010; Pantanella et al. 2011a) Murray cod (*Maccullochella peelii peelii*) (Lennard and Leonard 2004), rainbow trout (*Oncorhynchus mykiss*) (Adler et al. 2003), common carp (*Cyprinus carpio*) (Naegel 1977) Asian Barramundi (*Lates calcarifer*) (Rakocy 2007), mullet (*Mugil cephalus*) (Pantanella et al. 2011b) Eurasian perch (*Perca fluviatilis*) (Graber and Junge 2009), largemouth bass (*Micropterus salmoides*) (Pantanella, unpublished data) and bester sturgeon (*Acipenser ruthenus* × *Huso huso*) (Dediu et al. 2012).

At commercial level tilapia is the most reared fish in aquaponics systems, followed by ornamental fish, perch, bluegill, trout, bass and, for a certain number of farms, barramundi, carp, Pangasius and crayfish (Love et al. 2015).

7.2.3 Feed/Plant Ratios

As previously mentioned one of the primary objectives of research was to find optimal feed/plant ratios to determine the essential management criteria of the

Table 7.4 Daily nitrogen and phosphorus sink per plant according to different studies

Type of crop	Nitrogen sink ($g\ m^{-2}\ day^{-1}$)	Phosphorus sink ($g\ m^{-2}\ day^{-1}$)	Author
Lettuce	0.83	0.17	Gloger (1995)
Lettuce	0.94	0.1	Alder (2003)
Lettuce	1.0–1.1	na	Dediu et al. (2012)
Lettuce Sweet basil	0.13–0.32 0.34–0.51	na na	Pantanella et al. (2012b)
Aubergine Tomato Cucumber	3.3 0.6 0.4	0.4 na 0.1	Graber and Junge (2009)
Salsola	0.2–0.4	na	Pantanella et al. (2011b)

Table 7.5 Feed to plant ratio determined from past researches

Type of crop	Daily amount of feed ($g\ m^{-2}$)	Feed crude protein (%)	Fish species	Author
Lettuce	56	32	Tilapia	Rakocy et al. (1997)
Sweet basil	81–100	32	Tilapia	Rakocy et al. (2004a)
Lettuce	33	43	Murray cod	Lennard (2004)
Water spinach	15–42	32	African catfish	Endut et al. (2010)

systems from its many variables: feed quantity, feed protein content (percentage of crude protein, %CP), type of plant cropped, environmental conditions and type of aquaponics system. At the University of Virgin Island (UVI) Rakocy et al. (1992) determined that 2.4 g day^{-1} of feed with 36% protein to tilapia could supply nutrients to one plant of lettuce grown on floating system.

An approach in mass balance used the daily plant nutrient uptake (Table 6.4), mainly with a focus on the sink needed for wastewater treatment, although it is worth reminding that the nutrient sink in plants sensitively vary among species/varieties and environmental conditions, which eventually affect the growth rate and yields of the plants.

On the other hand a reference on the amount of nutrients released by feed was also given by Graber and Junge (2009) who calculated that 1 kg fish feed at 45% CP eaten by tilapia could supply 46 g of nitrogen, 6.0 g phosphorus and 1.0 g potassium (Table 7.4).

Trials carried out during the years gave more practical feed ratios, which are based on the amount of feed needed to supply nutrients to plants growing at standard densities (i.e. 20–30 plants for leaf vegetables) (Table 7.5).

As said earlier the feed ratios depend on the crop being cultivated, the type of system, and the environmental and climatic conditions. These parameters can sensitively affect the nutrient availability for the plants: from 60 to 100 g m^{-2} day^{-1} (Rakocy et al. 2006) for the standard UVI-type systems down to 16 g m^{-2} day^{-1} for lettuce growing in systems with minimum denitrification and additional waste mineralization provided by offline tanks where solids are kept continuously oxygenated (Lennard 2013).

Based on projects' experiences FAO delivered rule of thumb feed to plant ratios: 40–50 g m^{-2} day^{-1} for leaf vegetables and 50–80 g m^{-2} day^{-1} for fruiting vegetables (FAO 2014).

Plant uptake can be increased by the type of system in use (Lennard and Leonard 2006), since deeper contact of roots in the nutritive solution eases the absorption of nutrients. In trials using different beds the growth of plants was higher in deep water culture (DWC) and nutrient film technique (NFT) because of the wider root exposure to the nutritive solution (Pantanella et al. 2012c). Nevertheless the resulting higher nitrogen sink was not due to higher concentrations of nutrients accumulated in the plant tissues, which were constant, but rather from the increased biomass obtained by the plants. Good nutrient sink is therefore the result of yield maximization in which optimal vegetable biomass growth is obtained.

7.2.4 Nutrient Concentrations in Aquaponics

In aquaponics the levels of nutrients are in general much lower than hydroponics. At the University of the Virgin Island Experimental Station (UVI) (Rakocy et al. 1992, 2004a, b, 2006) aquaponic vegetables were cropped with only a small percentage of the optimal hydroponic concentrations (Table 6.3): nitrate ($N\text{-}NO_3$) 16–70%, ammonia ($N\text{–}NH_4$) 0–5.5%, phosphorus (P) 9–55%, potassium (K) 11–32%, calcium (Ca) 0.4–16%, magnesium (Mg) 8–25%, sulphur (S) 6–24%, iron (Fe) 32–100%, manganese (Mn) 3–40%. Higher concentrations of nitrogen than those measured at UVI are possible and can easily reach levels similar to hydroponics. Nevertheless, yields and quality of lettuce heads from aquaponics are similar to hydroponics even at concentrations ten times lower than those in use in hydroponics (Pantanella et al. 2012a). Nitrate concentration in RAS can vary from 100 to 1000 mg L^{-1} (Van Rijn 2010), thus resulting in more than sufficient levels of nutrients for commercial production of plants. However, some attention must be also put in the optimal nitrate concentrations for fish since different growth rates or toxicity responses are seen depending on the fish species. Losordo et al. (1998) noted that fish can tolerate nitrate levels of 200 mg L^{-1}, but concentrations above 300 mg L^{-1} appear to bring some toxicity (Masser et al. 1999). Likewise, trials done with African catfish demonstrated a decrease of growth and increase of nitrate in the plasma for concentrations in the water above 140 mg L^{-1} (Schram et al. 2012). Marine fishes are also less tolerant to high nitrate concentrations than freshwater fishes.

In aquaponics there are some plant limiting nutrients, which are not adequately supplemented by the fish feed. Main deficiencies are found in iron, potassium and calcium (Rakocy et al. 1993). However, calcium and potassium can be supplemented to aquaponic systems in the form of calcium carbonate, potassium bicarbonate, calcium hydroxide and potassium hydroxide in order to raise the pH or to increase the alkalinity in water (Rakocy et al. 1993), which is consumed by the nitrification process.

Resh (2004), Leoni (2003) and Sonneveld and Straver (1989) indicated optimal nutrient concentrations for plants. Nevertheless, the nutrients' needs vary according to the growth stage of each plant (Leoni 2003), which can be a challenging factor in aquaponic systems because of the difficulties in quickly adjusting the pool of minerals in systems containing big volumes of water or without stressing the fish. Van Anrooy (2002) remarked that high concentration of nitrates favours vegetative growth but lower levels of nitrogen are required during the fruiting stage.

Equal concentrations of nitrogen and potassium (N:K ratio of 1:1) bring cucumber productions to same or higher yields than hydroponics (Savidov 2005; Pantanella, unpublished data), while higher N:K ratios apparently reduce fruit yields (Graber and Junge 2009). The low concentrations of potassium is a limiting factor in fruity plants, as this element is essential in fruit setting, ripening and sweetness. Reductions of N:K ratios can be obtained by either buffering the system with alkali (potassium hydroxide, potassium bicarbonate) or by increasing the denitrification in dedicated tanks of the system, thus letting nitrogen in the water to be eliminated into the atmosphere (Rakocy, personal communication 2008).

7.2.5 Water Parameters

In aquaponics the nutrient availability is affected by pH (Rakocy et al. 2006; Losordo et al. 1998, Tyson et al. 2004). Values of pH of 7–8.5 favour nitrifying bacteria and thus improve the efficiency in the elimination of ammonia from the water (Tyson et al. 2008), however such higher levels may affect macro and micronutrient availability outside the pH range of pH 5.5–6 that is considered optimal for plants (Jones 1997; Resh 2004; Tyson et al. 2008).

Other relevant water parameters are found in electrical conductivity, which should range between 2.00 and 4.00 dS m^{-1} or less to avoid plant/leaf phytotoxicity (Resh 2004; Rakocy et al. 1992, 2006); alkalinity above 100 mg L^{-1} for optimal nitrification buffering (Rakocy 1997); biologic oxygen demand (BOD) below 20 mg L^{-1}, dissolved oxygen (DO) above 5 mg L^{-1} both for optimal fish, plant growth and development of nitrifying bacteria. Low BOD and high DO is needed to avoid oxygen depletion by aerobic bacteria and the creation of anaerobic conditions that could harm fish due to the production of hydrogen sulphide, an extremely toxic gas for fish (Rakocy 1997).

7.2.6 Water Use in Aquaponics and Recirculating Systems

The water consumption in aquaponics includes both fish and plant management. Replacement takes into account the discharge of sludge from fish faeces and uneaten feed, the plants evapotranspiration, the accumulation of water in the plant tissues, the water evaporation from fish tanks as well as other variable losses depending on the system design (evaporation from airstones bubbles, water splashes from pipes, etc). Researches have assessed in 0.5–4.6% of the total aquaponic system volume the daily amount of water consumed that needs to be added (Naegel 1977; Watten and Buschs 1984; Mc Murty et al. 1997; Rakocy et al. 1997, 2004a; Savidov 2005; Al-Hafedh et al. 2008). Nevertheless such water consumption is in the lower part of the above range in greenhouses or in systems with advanced aeration systems that keep the water evaporative losses to a minimum.

More practical information refer to the amount of water required to grow one kilogram of fish, which has been determined in 0.5–1.4 m^3 kg^{-1} in intensive RAS, while increasing consumption patterns in traditional aquaculture are observed depending on the management intensification: from 4.7 to 7.8 m^3 in aerated ponds, 11.5 m^3 in extensive ponds, up to 30 m^3 in aerated pond with water exchange (Verdegem et al. 2006).

As already mentioned water use in aquaponics is affected by the additional evaporative losses from the plant system. However a correct account of water consumption in aquaponics must consider the break-even of nutrients in which minerals are maintained at constant concentrations as the result of the equilibrium between the nutrients released by the feed/fish, the losses of nutrients from solids removed or denitrification, and the minerals directly used by plants and roots.

An assessment carried out in UVI-type systems (Pantanella et al. 2012b) showed that for one kilogram of tilapia (*Oreochromis niloticus*) body weight gain (equivalent to 1.0–1.2 kg of feed with 40–43% CP) the water consumption at nitrogen break-even point is 637–1373 L, and is balanced by 11–25 kg of lettuce sink. On the other hand the growth of one kilogram of African catfish (*Clarias gariepinus*), equivalent to 1.0–1.3 kg of feed with 40% CP, requires 243–395 L of water when growing 5–6.5 kg of sweet basil (Pantanella et al. 2012b). Therefore, the water consumption in aquaponics is affected by the type of crop, because plants capable of stocking more nitrogen in their tissues would eventually require less biomass to keep the nutrients into the aquaponic systems at a steady level, which eventually results in lower water volumes needed to grow plants.

However, the water consumption between aquaponics and hydroponics is 70–130% higher in UVI-type systems due to the presence of extended water surfaces from the fish tanks and the intense aeration occurring to keep fish with sufficient dissolved oxygen. Such increments suggest for the adoption of specific design solutions to further reduce the evaporative losses (e.g. bubbling) wherever water supply is an issue (Pantanella et al. 2012b).

7.2.7 Aquaponic Yields

Aquaponic/hydroponic plant production show higher yields than conventional soil crops. Resh (2004) stated that soilless cultivation could at least double the yields of conventional horticultural plants: from 1 kg m^{-2} in soil to 2.3 kg m^{-2} in soilless for lettuce; from 1.2–2.4 kg m^{-2} in soil to 14–74 kg m^{-2} in soilless for tomato. Rakocy et al. (1992) showed yields of 4.35 kg m^{-2} from lettuce grown at 25 plants m^{-2} in 21 days. On the advantages of aquaponics against soil production Mc Murty et al. (1990) showed higher yields with cucumber (7.3 vs. 4.6 kg m^{-2}) but lower production with tomato (4.6 vs. 6.1 kg m^{-2}). Rakocy et al. (2004a, b) showed higher productivity in basil with yields of 1.8–2.0 kg m^{-2} (0.6–1.0 kg m^{-2} in soil) and okra with 2.5–2.9 kg m^{-2} (0.15 kg m^{-2} in soil). The use of the aquaponic concept, with higher nutrients in the water and its recirculation between a fish tank/basin and rice could enhance rice production by 66% against traditionally fertilized rice, with yields per crop of 8.5 ton ha^{-1} against 5.1 ton ha^{-1} (Pantanella et al. 2011c).

Aquaponics shows higher productivity than hydroponics in mature systems for either tomato (31–59 vs. 41–45 kg m^{-2}) and cucumber (42–80 vs. 50 kg m^{-2}) and whenever the N:K ratio is close to 1 (Savidov 2005). For higher N:K ratios certain fruity plants can still perform well against hydroponics, as in the case of aubergine (7.7 kg vs. 8.0 kg m^{-2}) and tomato (23.7 vs. 26.3 kg m^{-2}), but cucumber seems to shows reduced performance (3.3 vs. 5.2 kg m^{-2}) (Graber and Junge 2009). Nevertheless N:K ratio at 1 even with lower nutrient concentrations than hydroponics (up to three time lower) can provide similar yields (7.6 vs. 7.5 kg m^{-2}) and quality of fruits (sweetness, vitamin C) in cucumber (Pantanella, unpublished data). For leaf vegetables there are no differences between aquaponics and hydroponics for both lettuce and basil productivity and quality (Pantanella et al. 2010, 2011a, 2012a). However concentrations of nitrates above 20 mg L^{-1} should be maintained to secure good growth and greenness in leaves. In saline crops, such as salsola aquaponics shows sensitive advantages than hydroponics even at lower nutrient concentrations (Pantanella 2011b). However, the best growth responses for both plants and fish should take into account the most favourable nutrients balances and climatic conditions, which must be adequate for the species being produced into the systems.

Aquaponic sub-systems design (floating system, gravel, NFT) could also help to raise plant yields and to increase water quality. Lennard and Leonard (2004) outlined the enhanced nitrification obtainable from gravel systems and, at the same time, the potential buffering capacity of gravel. Rakocy et al. (2006) also confirmed that substrates in the form of sand and gravel are optimal, but care should be put in delivering solid free water to avoid clogging. Media however present some drawback because it requires more maintenance to grow plants (e.g. digging holes during transplant) or because it may bring some stem damages wherever aquaponics is developed in windy outdoor conditions (Rakocy et al. 2006).

7.2.8 Economics

Economic assessment from literature showed high profitability of aquaponics leaf vegetables than fruit productions. Mc Murty et al. (1997) projected annual yields of

41.5–54 kg m^{-3} of tilapia and 29.2–59.6 kg m^{-2} of tomato, which was respectively equivalent to 109–142 USD m^{-3} and 50–102 USD m^{-2} depending on the system design. However Rakocy outlined that the biomass harvestable from each crop of Nile tilapia is 61.5 kg m^{-3} and 70.7 kg m^{-3} for red tilapia (Rakocy et al. 2004b). Savidov (2005) estimated a gross return of 342 USD m^{-3} every 24 weeks from tilapia reared in tanks (sold at 5 USD kg^{-1}), while basil returns were 184–236 USD m^{-2} per year (price of basil 15.4 USD kg^{-1}). Rakocy et al. (2004a) estimated a gross revenue of 515–550 USD m^{-2} from aquaponic basil against a revenue of 172 USD m^{-2} from soil. The experience from researchers and commercial scale operators shows how the highest incomes come from the vegetable side of the aquaponic systems, with nearly 3-fold gains than fish (Savidov 2005). A recent survey among commercial aquaponics operators showed that the average size of farms has an acreage of 0.01 ha, with the median respondents investing 5000–9999 USD and reporting profits in nearly 40% of cases (Love et al. 2015).

7.3 System Components and Management

A standard aquaponics system (Figs. 7.4 and 7.7) is constituted by fish tank/s, a mechanical filter to remove settable solids (fish faeces, uneaten feed) and suspended solids, a biofilter (optional, depending on the type of plant grow system used) to convert the ammonia excreted by fishes into nitrate through nitrification, and plant trough/s. Aquaponics can also have a sump where water from the plant troughs converge and is added with buffer from a dedicated buffer addition tank.

Aquaponics follows the evolution of the systems occurred during almost forty year of applied research carried out by scientists worldwide. Most of the current designs are built following the original outline developed by Professor James E. Rakocy at the University of the Virgin Islands, who started working on the integration of plants and fish since the late seventies. Through the years Professor Rakocy and his team designed a commercial scale system with appropriate component ratios based on optimized nutrient balance between fish and plants (Fig. 7.7).

7.3.1 Fish Tanks

Fish tanks follow the engineering of recirculating systems. The design should allow the solids to be quickly removed to maintain good water quality. The ideal shape is circular as water rotates with no turbulence towards the centre of the tank where the drain is. In terms of design the ideal radius-to-height ratio in circular tanks is 3:1–4:1, this allows centripetal forces to bring the solids towards the central drain. To further improve the self-cleaning capacity a slope towards the centre is suggested to help the solids to move and settle towards the centre of the tank. Other designs are possible, with raceways leading the options for their good space use and cleaning efficiency. Squared tanks are used, but the presence of corners creates death spots where wastes accumulate, especially in flat bottomed tanks. Therefore the highest solid removal efficiency must be always guaranteed.

The most common material used for fish tanks is fiberglass (Fig. 7.8) for its versatility to be used for wide recipients. However, it is the most expensive option and may be used for systems with a very long lifespan. Other material used, mainly for small volumes of water is HDPE/LDPE, which is moulded for 0.1–5 m^3 tanks. Alternatively HDPE or EPDM liners can be used for either backyard or commercial scale systems. Thick liners of 0.75–1 mm can be welded to make bigger containers and are fairly resistant to mechanical stress. Liners can be a very cheap option to cover metal or wooden/bamboo walls or iron-meshed frames (Fig. 7.9). Alternatively liners can be used for ponds, providing that good water circulation and drainage is guaranteed to efficiently remove the solids. A suitable solution for backyard systems is the intermediate bulk container (IBC) (Fig. 7.21), which is used to transport liquids, is fairly resistant and of adequate volume to host up to 15–20 kg of fish serving 2–4 m^2 plant area.

The tanks vary in stocking densities depending on the species reared and on the aeration-oxygenation technology. In general the density is chosen according to the

Fig. 7.8 Fiberglass tanks

Fig. 7.9 Fish tanks made of liners and bamboo stakes

oxygen concentration in the water, which is replenished by either new incoming oxygenated water or by aeration/oxygenation inside the tank.

The more sensitive to oxygen is the fish species the more water exchange or aeration-oxygenation is needed to support the animals' needs. Alternatively the fish stocking density must be reduced to maintain adequate dissolved oxygen concentrations. In general for tilapia 80–90% oxygen saturation can be sustained by standard aerators up to a fish density at harvest of 60–70 kg m^{-3} under a hydraulic retention time of 1–1.5 h. Higher stocking densities or higher saturation are possible providing that pure oxygen is supplied.

Water exchange in tanks is also important to wash solid out and to dilute the ammonia excreted by fish, which is successively oxidized by the biofilter into less harmful nitrate. Ammonia is harmful to fish at concentrations as low as 1 mg L^{-1}, but its toxicity depends on the pH, since acid conditions in water bring ammonia into its ionized and less toxic form, ammonium. On the contrary levels of pH above 7 increase the concentration of the unionized form thus resulting dangerous for the fish.

In commercial aquaponic systems tanks are mainly managed with a staggered production of same-size fish stocked in each tank. Staggered management allows to:

- Harvest one tank of same-size fish at one time
- Have a continuous supply of fish to the market depending on the number of tanks available
- Maintain a fairly constant fish biomass into the whole system, which eventually keeps constant the levels of nutrients for plants
- Avoid peaks in fish biomass into the system, which eventually result in excessive oxygen consumption and high ammonia production.

7.3.2 *Mechanical Filtration*

Solid removal is a fundamental part of any recirculating system. Solids are formed from fish excreta and uneaten feed and must be removed efficiently and quickly, to prevent them from releasing ammonia into the water or to create anaerobic spots that bring to the production of hydrogen sulphide, a very toxic gas for fish. In any recirculating system the presence of solid reduces the efficiency of the aeration/oxygenation, due to the increased oxygen consumption from mineralizing bacteria, which use dissolved oxygen to digest the wastes (biological oxygen demand, BOD).

Nevertheless, in aquaponics the efficiency in removing solids can be lower than RAS, because the presence of small quantities of fine suspended solids would add more nutrients to the plants trough mineralization. On the contrary RAS systems need to remove as much fine solids as possible to prevent any risk of clogging of the biofilter, which would deteriorate the water quality dramatically.

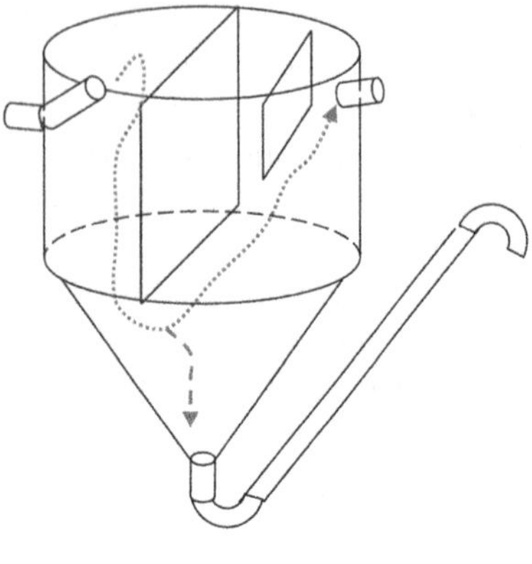

Fig. 7.10 Clarifier with central baffle

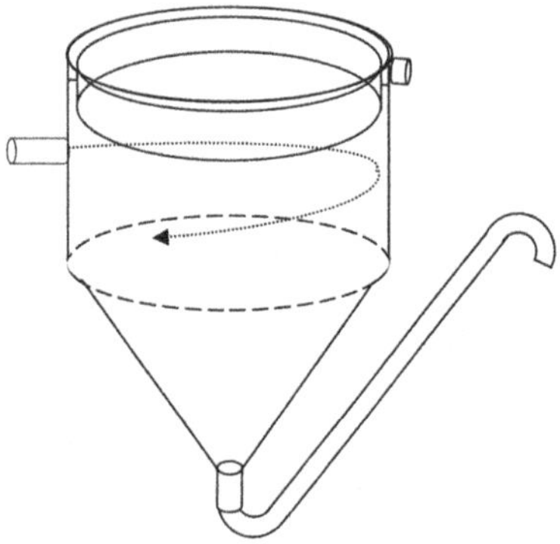

Fig. 7.11 Swirl separator

The most common solid removal devices are: clarifiers, swirl separators, radial flow clarifiers and drum filters. Clarifiers (Fig. 7.10) are conical-bottomed tanks where solid settle by gravity under a water retention time of 20 min (Rakocy 2007). In general the removal performance is 59%. On the other hand swirl separators (Fig. 7.11) use centrifugal force to settle solids but they seem to have lower removal efficiency (37.1%) than radial flow clarifiers (77.9%) (Fig. 7.12) (Davidson and Summerfelt 2005). Both systems have been already used in semi-commercial aquaponic systems with positive results.

Drum filters are the most used devices in RAS. Water is filtered through a micro screen of variable size (50–100 μm) in which solids are firstly trapped and then removed by a backflush of water that pushes the dirt out, through a dedicated outlet. Although very efficient, drum filters are not universally adopted in aquaponics due to economic reasons and because of the excessive removal of solids that sensitively reduce the pool of nutrients obtainable from fine wastes. Another technology use geotextiles to get rid of flocculated wastes. In this case concentrated sludge from an offline tank is mixed with an organic polymer that binds solid particles and precipitate them. The precipitate is then squeezed from the water into a permeable bag and successively removed once the bag is full.

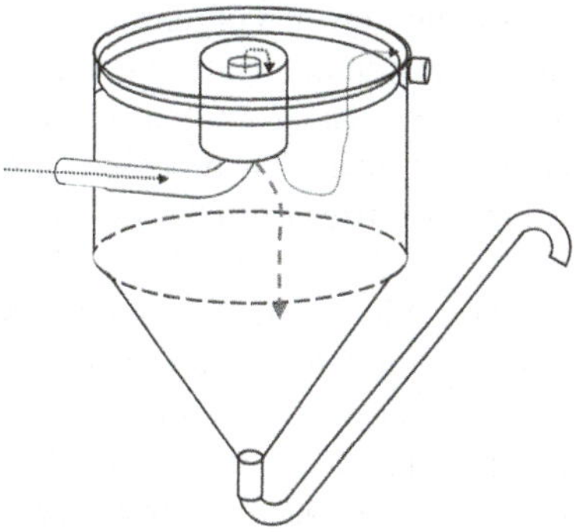

Fig. 7.12 Radial flow clarifier

Fig. 7.13 Clarifiers with 60° conical bottoms

Regardless the system solids need to be efficiently removed to prevent that they clog plant roots and increase the biological oxygen demand of the system. The overload of the system with solids is in fact negative because of the risks of anaerobic spots and hydrogen sulphide production. On the other hand moderate quantities of fine solids allow more nutrients to be released into the water through mineralization. This balance between fine solid removal and mineralization brought to different management strategies among aquaponic systems around the world.

In the UVI system (Fig. 7.7) fine solids escaping the clarifier (Fig. 7.13) are trapped in a filter tank made with orchard net (Fig. 7.14). At this stage most of the suspended solids settle on the mesh and become a growth substrate for the aerobic bacteria that produce biofilms. The increasing volume of the adhered solids and biofilm is periodically controlled by washing the nets to remove all the organic matter. The frequency of the washing of the nets is eventually used to create a controlled anaerobic environment (Rakocy 2008, personal communication). The increase of organic matter and aerobic bacteria on the nets brings in fact to an exploitation of the concentration of oxygen in the water that reach values next to zero, which favours organisms that thrive in oxygen-free conditions (anaerobiosis). In such condition denitrifing bacteria, which are anaerobic organisms, consume the nitrate nitrogen in the water and release nitrogen gas into the atmosphere. This controlled anaerobic process is important in the overall nutrient management of aquaponics because it keeps nitrogen to constant levels into the system, or even decrease it to adjust the N:K ratio to more favourable numbers for fruity plants. As mentioned in the previous section N:K ratios of 1 are optimal for fruits' settings and ripening, thus the overall reduction of the nitrogen in the water helps potassium to become relatively predominant. Being the anaerobic stage a critical condition due to the production of hydrogen sulphide gas, it is very important to intensively de-gassing the outgoing water before circulating back into the system.

In Canada the solid management follows a more intensive approach, in which the complete mineralization of the wastes is pursued. The rationale is that wastes contain good sources of nutrients that need to be released in order to supply optimal fertilization to plants. The systems make use of oxygen gas to supersaturate the water thus allowing both dissolved oxygen and oxidizing bacteria degrade the organic matter into simpler components that are used by plants.

A simpler approach makes use of media beds of adequate granulometry to mineralize the fine suspended particles. The media is contained in tanks that are constantly flooded and drained with the aquaponics water. The most common beds are made of inorganic substrates such as volcanic tuff, gravel, pumice, expanded clay. The media increases the surface available for oxidizing microorganisms to thrive and to degrade the organic matter into its simpler elements. The constant and regular flooding and draining of the media allows water and air to reciprocally penetrate the interstices and pores of the substrate and to supply with oxygen, water and nutrients the rich micro fauna.

7.3.3 Biofiltration

Biofiltration is the fundamental component of any recirculating system. As mentioned in the previous sections fishes release ammonia from their metabolism, and the concentration of released ammonia increases with the percentage of proteins contained in the feed. Ammonia concentrations would raise quickly due to the high densities of fish and the abundant feeding, thus resulting in the risk of toxicity and death of fish. Concentration as low as 1 mg L^{-1} are harmful especially if the pH in the water is basic, since ammonia would be in its unionized and more toxic form. To maintain good water quality it is necessary to oxidize this by-product into the less harmful nitrate. Two main bacteria species help to run this process: *Nitrosomonas* and *Nitrobacter*, the former oxidizes ammonia into nitrite, the latter nitrite into nitrate.

Fig. 7.14 Filter tanks to capture fine solids by means of orchard type nets

1. *Nitrosomonas:* Ammonia (NH_3) → Nitrite (NO_2^-)
2. *Nitrobacter*: Nitrite (NO_2^-) → Nitrate (NO_3^-)

These beneficial bacteria establish their colonies on every surface of the systems. They need both good water circulation to allow new ammonia molecules to come into contact with the colonies and good oxygenation to allow the prompt oxidation of ammonia.

Biofilter is any media that has a large surface per unit of volume (specific surface area—SSA) to let bacteria adhere over a large area. The biofilter presence in aquaponics system is facultative. The UVI system does not use any specific biofilter unit, but leaves to the system's surface in contact with water (the tanks, the submerged part of floating rafts, the pipes, the plant roots) the task to host nitrifying bacteria. The presence of biofilter is however suggested, if not recommended, to help the system to be more resilient against ammonia peaks or sudden changes from the optimal environmental conditions for nitrifying bacteria (temperature, salinity, oxygen). Common media used in both RAS and aquaponics are: bioballs, spherical plastic media with voids in the inside and a SSA of 600 m^2 m^{-3}; plastic beads that can reach SSA up to 1400 m^2 m^{-3}. Biofiltration in aquaponics can be also provided by media beds, whose substrate used to support the plants can also host nitrifying bacteria. Common media used is gravel (150–200 m^2 m^{-3}), volcanic tuff (300 m^2 m^{-3}), and expanded clay (200–250 m^2 m^{-3}). The correct sizing of the biofilter depends on the maximum feed intake of the fish stocked and the resulting ammonia produced. The optimal sizing also take into account of the climatic conditions, dissolved oxygen, salinity, pH and the type of biofilter media used with its specific SSA.

The bacteria in the biofilter work within optimal ranges of temperature (17–34 °C), good dissolved oxygen (>5 mg L^{-1}) low salinity, low dissolved solids. Such environmental conditions should adjust to the optimal of the plants and fish being cultured. Plants for example prefer pH ranges of 5.5–6.5, a level in which all the micronutrients for plants are in their maximum soluble form. On the other hand the higher temperatures required by bacteria do not perfectly match the optimal temperatures of certain vegetables. It is then necessary to find some compromises, being aware that the biofilter should be then oversized to compensate for the suboptimal working conditions of nitrifying bacteria.

7.3.4 Mineralization

Mineralization is the second most important microbiological process in the aquaponics system. It implies the progressive degradation of organic wastes into smaller components, the process releases nutrients otherwise not available to plants.

Many organisms are involved in mineralization: worms, nematodes, protozoa, fungi, bacteria. Each decomposer is involved in one step, to degrade wastes from bigger into smaller particles. In an aquaponics system it is common to spot the complex fauna and micro fauna degrading the organic matter, especially in substrate beds, in which the fine suspended solids accumulate and are degraded. The substrate beds, with their flood and drain cycles allow the decomposers to access air and at the same time capture new organic matter from the circulating water.

The process of mineralization requires oxygen to provide molecular oxidation and to let decomposers breath. This means that additional aeration should be provided to the system to maintain good dissolved oxygen.

During the mineralization proteins are degraded in amino acids and successively digested by bacteria that release nitrogen. However, plants can directly take up amino acids as well as any inorganic nitrogen form, with the exception of nitrogen gas. The mineralization of phosphorus is important because this element is not mobile and easily available. Mineralization converts organic phosphorus into phosphate (PO_4^-), which is assimilated by plants.

The mineralization is possible within the carrying capacity of the system. Too much waste is dangerous if not supported by the decomposers and by an adequate supply of oxygen, as it would build up into the system with the risk of anaerobic spots and the production of hydrogen sulphide.

7.3.5 Plant Beds

Plant beds in aquaponics mainly follow the same designs of hydroponics with the only difference that nutrients are not distributed by computers controlling the release of fertilizers from tanks into the circulating water, but simply by fish wastewater.

Fig. 7.15 DWC with tomato plants

7.3.5.1 Deep Water Culture

The most used plant bed is the floating system or deep water culture (DWC). In DWC water flows in long tanks of variable width (Figs. 7.6 and 7.15). The height of the water in the tank is 25–40 cm. In DWC plants float on polystyrene rafts with bare roots (Fig. 7.16), and access the nutritive solution through holes made on the floating sheets. The presence of fixed holes make the raft not flexible to adjust to different densities. A proper fish-to-plant ratio is maintained. In the case of the UVI system the surface ratio between plant beds and fish tanks is 7.3:1, while the volume ratio is 1:3.4 with an average fish-biomass-to-cultivable-area equivalent to 4 kg m^{-2} (Rakocy 2007). In DWC an intense aeration actively enhances the plants uptake by increasing the nutrient flow at root levels and by providing oxygen to nitrifying bacteria that convert ammonia into nitrate. Given the big volume of water the system requires more energy for pumping than any other types of aquaponic systems. On the other hand the big volume of water makes the system more resilient against ammonia peaks (dilution effect), while nutrients can accumulate into the water and serve the plants over a long period of time, even if fish biomass is consistently reduced or not present. Another advantage of DWC is in the thermal inertia of the system due to the big volume of water, which keeps the water under constant temperatures and prevents fish stress.

Fig. 7.16 Particular of bare roots in DWC

Advantages	Disadvantages
• Easy set up using plastic liners • Suitable for outdoor as well as indoor • Buffering capacity of ammonia through dilution in a large volume of water • Stable water temperatures • High quantities of nutrients diluted in the water • Systems can produce for a limited period of time with few or no fish • Highest productivity compared to other growth beds • Biofiltration surface provided by floating rafts • Simple management of rafts that can be pushed to one end of the tank for easy harvest • DWC can host additional species of aquatic animals to improve productivity • Plants do not wilt in case of black out • Easy maintenance for cleaning	• Need of food grade liners • Heavy system, not suitable for roof-top agriculture • The liners are prone to punctures if not thick enough • The systems need longer periods of time to reach adequate concentrations of nutrients • Construction costs may be higher if materials are not easily available (polystyrene rafts) • Polystyrene rafts degenerate quickly under UV if not protected with paint • Large amount of water to be pumped, higher cost of energy than other beds • More expensive water sterilization to comply with water safety regulations due to the water volume • Not suitable for some fruit crops and root crops • Tanks can breed mosquito larvae, control is needed • Tilapia damage crops if fry colonize the plant beds

7.3.5.2 Dynamic Root Floating Technique (DRFT)

The dynamic root floating technique is a variant of DWC with a shallower water column. DRFT is also called Taiwanese system and is quite widespread in South East Asia. DRFT is built on tables of variable lengths (Fig. 7.17) and has a water depth of just few centimetres (4–8) (Fig. 7.18). The rafts float like in DWC, but the presence of ridges from the bottom make it possible to decrease the water level and create an air chamber when the rafts settle on the top of the ridges. Air chambers increase air circulation at root level, prevent risks of rotting in plants and help the system to cool down the water during hot seasons, which is ideal in hot climates.

Fig. 7.17 DRFT in outdoor

The DRFT has considerable advantages against both traditional DWC and NFT, firstly because of its lighter weight than DWC that allows its use on rooftops, and secondly due to the presence of the polystyrene rafts and the air chamber underneath, which prevent any diurnal overheating of the circulating solution that constitutes one of the biggest issues in NFT instead (plant bolt and fish get stressed for extreme variations of day/night temperatures). Although not very commonly adopted, the system has been used by the author for three years with yields similar to traditional aquaponics DWC.

Fig. 7.18 The shallow water level in DRFT

Advantages	Disadvantages
• Easy set up using plastic liners on tables • Good for either small scale or commercial scale • Suitable for outdoor • Suitable for roof-top agriculture • Moderate buffering capacity of ammonia through dilution in the water • Insulating effect from polystyrene rafts • Additional cooling due to the air chamber • Lower root disease risks due to the presence of air • Passive aeration by the presence of air chamber • Higher concentrations of nutrients than DWC • System is resilient to black outs • Easy maintenance for cleaning • Water sterilization possible for food safety rules • Suitable for all types of leafy vegetables	• Need of food grade material • Higher cost of setting than DWC due to the presence of supporting structures • The liners need to be thick to avoid punctures • Proper care should be put in action to prevent polystyrene rafts from being damaged by UV light • Rafts do not provide surface for nitrifying bacteria when suspended on the water • Lesser buffering capacity of ammonia through dilution than DWC due smaller volume of water • Not suitable for certain fruit crops and root crops • Tanks can breed mosquito larvae, control is needed • Tilapia damage crops if fry colonize the plant beds • Need additional biofiltration due to the reduced surface and smaller volume of water than DWC

7.3.5.3 Nutrient Film Technique—NFT

The nutrient film technique—NFT is the most common system in hydroponics and the most used in aquaponics together with DWC. Like hydroponics the plants are placed in holes drilled on plastic pipes (Fig. 7.19 and 7.20) in which water flows in a shallow film to wet the roots. Given the small volume of the circulating water NFT is in general associated with a mechanical filter and a biofilter to efficiently remove wastes and convert the ammonia released by fish into nitrate. The outlet of the NFT pipes end into a sump where water is then poured back to the fish tanks after proper pH adjustments. This type of system offers the advantage that pipes can be moved and adjusted to increase/reduce the planting density according to the growth stage of the crop. In addition the lightweight is compatible for NFT to be developed on rooftops.

Although the system is very simple, it shows some drawbacks, which are found in the excessive daytime heating of the water flowing into pipes during the hot and in the vulnerability against black outs, as any lack of electricity immediately deplete the water into the pipes and stress/wilt, the plants. One solution to address the thermal excursions above mentioned, which also stress the animals, is to decouple the fish from the plant system in such a way that the two sub-systems could only communicate for limited period and for the strict time necessary to replenish nutrients into the plant sub-unit.

Fig. 7.19 NFT with round pipes positioned on A frames

Fig. 7.20 NFT with flat pipes

Advantages	Disadvantages
• Small flow of water, cost savings in pumping • Lightweight, suitable for roof-top agriculture • Suitable for both small or industrial scale farms • The concentration of nutrients can be adjusted real time (in hydroponics) to meet plant demand • High productivity if properly managed • Suitable for leaf vegetables, especially lettuces • Pipes can be moved to increase planting density • Suitable for water sterilization • Easy management • No media to handle	• Set up costs is high due to the number of pipes and supporting structures needed • The concentration of nutrients cannot be adjusted real time in aquaponics • Risks of black outs and loss of the whole production • Requires a biofilter and mechanical filtration • Extreme temperatures in water between night and day bring stress to the fish and make them sick • Higher risk of plant diseases with high water temperatures • Risk of lettuce bolt with high water temperatures

7.3.5.4 Media Beds

Media beds are very common in small backyard systems (Fig. 7.21). They are very versatile and can be used for both leaf, fruit and root vegetables. A whole range of inorganic media is used: expanded clay, tuff, pea gravel, and perlite.

Plants in such systems receive nutrients through surface irrigation by means of drippers or through flood-and-drain cycles. In flood-and-drain media is cyclically wetted by raising levels of water and then aerated when the water flows out from the tanks. Flood and drain can be operated by either the cyclical flushing of siphons that suck water out when water reaches a fixed height, or by the intermittent functioning of an inlet water pump given a constant, but smaller, outflow that allows the bed to be flooded. Media beds are very easy to manage, providing that fish wastewater is adequately clean from bulk solids to avoid organic matter build-ups and consequent anaerobic spots. This type of system is recommended for beginners who are neither experts in nitrification nor are constantly monitoring their systems. The presence of media helps the practitioners to have an adequate biofiltration and mineralization at the same time.

Fig. 7.21 Media bed obtained from IBC tanks: fish tank (*a*) with outlet serving the plant beds (*b*). Plant beds discharge water to the sump (*c*) where a submerged pump returns the water back to the fish

Advantages	Disadvantages
• Suitable for all types of plants (leaf, root, fruit) • Substrates used according to local availability • Carbonate-based substrates can buffer the water that tend to become acid with nitrification • Many water delivery options • Nitrification provided by media • Mineralization provided by media • Resilient to black outs	• Costs for transport of media • Not suitable for large scale farms • Heavy systems if using standard media (gravel) • Needs liners resistant to punctures

(continued)

(continued)

Advantages	Disadvantages
• Can be used for roof tops with lightweight media	• Media can clog with abundant fish solids • Plants can leave crop residues in the media • More labour during transplants • Media may damage stems in windy conditions • Water delivery may be not uniform, impair growth

7.3.5.5 Dutch Buckets

Dutch buckets are media-contained pots where plant grow by receiving the nutritive solution by means of drippers of by regular flood-drain cycles. This type of system is suitable for large fruiting plants or potted plants sold to the markets or to retailers. Being the plants growing in pots they can be easily moved to adjust the density according to their growth stage.

In aquaponics this type of systems has been developed with manifolds serving plants with adjustable flow of water, or by means of flood-drain cycles delivering water to plants positioned on a waterproof bed. The outflow from the pots directly converge to the fish tanks or to a sump. In general the pots with their media already provide biofiltration to the system water, but some degree of mechanical filtration is needed upstream, between the fish tanks and the pots, to avoid the accumulation of fish wastes and the clogging of the media. One positive aspect of this system is that irrigation by flood and drain does not require big investments. However, if the aquaponic water is delivered by micro-irrigation the delivery system may eventually clog due to the presence of organic matter and bacteria in the water that colonize the micro pipes. Should micro-irrigation be chosen a deep filtration with sand filter should then be guaranteed to get rid of all solids and secure a good quality water without any risk of clogging.

Advantages	Disadvantages
• Suitable for fruit plants and potted plants • Many water delivery options • Suitable for large-scale productions • Plant density adjustable • Nitrification/Mineralization provided by media • Resilient to temporary black outs	• Costs for transport of media • More labour for management • Needs liners resistant to puncture if pots are positioned above • Some water delivery options can clog (drip irrigation)

7.4 Aquaponics Versus Aquaculture Systems

At sustainability level aquaponics has the same advantages of any other aquaculture farming system. The production of fish brings in fact consistent advantages on water uses and ecological footprint, if compared to terrestrial animal husbandry. Being fishes cold blooded animals they do not consume most of the energy for heating their bodies, as it happens for mammals. This different physiology improves sensitively the fish efficiency in converting feed into body mass (feed conversion ratio—FCR). In fish FCR can be as low as 1 (1 kg of feed required to increase the animal body weight by 1 kg). On the contrary FCR values in chicken and monogastric warm-blooded animals are 3–4, while in ruminants they raise up to 6–7 (Verdegem et al. 2006). Low FCRs result in lower uses of land and water to produce the necessary feed for the livestock.

7.4.1 Traditional Systems

The evolution of traditional aquaculture to closed systems brought considerable advantages both in terms of pollution control and water use efficiency (Piedrahita 2003; Verdegem et al. 1999, 2006).

In many parts of the world aquaculture has not been managed with the necessary attention on environmental issues. Fish productivity strictly depends on the degree of farming intensification, which is a trade-off between land access, resource use (feed, energy, water) and waste production. In general traditional freshwater aquaculture is predominantly represented by pond systems.

The type of management intensification however affects the productivity of the aquaculture system, which varies accordingly to the use of inputs: *extensive* has no feed use but only fertilization is applied to increase the natural food production of ponds (phyto and zooplankton that is eaten by fish), *semi-intensive* with partial use of feed to integrate the natural pond productivity, *intensive* systems where feed fully covers the nutritional needs of the aquatic animals. Intensive systems make also use of energy to support the water oxygenation and water exchange to get rid of wastes and ammonia, which is toxic to fish (Barnabé 1990; Diana et al. 1997) (Table 7.6). However, new incoming water increases the risks of parasites and pathogens outbreaks, and put aquatic animals at risk of chemical contamination from outer waters, which makes it necessary to develop appropriate control strategies.

Table 7.6 Differences between aquaculture managements and productivity in tilapia growing in ponds (Diana et al., 1997)

	Extensive	Semi-intensive	Intensive
Fertilization	No fertilization or fertilization with manure or chemical fertilizers	Fertilization may be used	Fertilization may be used
Feed	No use of feed, fishes rely on pond natural food production	Partial use of feed to integrate natural pond food production	Complete use of feed to cover the nutritional needs of fish
Aeration	No aeration	No aeration	Aeration occurring
Water management	No water exchange	No water exchange	Water exchange in very intensive systems
Productivity per year	≤0.5 MT ha^{-1} with no fertilization; 1–3 MT ha^{-1} with fertilization	3–6 MT ha^{-1}	6–10 MT ha^{-1}; 10–20 MT ha^{-1} with water exchange

7.4.2 Advantages of Recirculating Systems

Closed recirculating systems have the considerable advantage that all the products from fish metabolism are processed within the system, although a certain amount of water is daily discarded to dilute the build-up of nutrients and to eliminate the bulk fraction of solids. However, this water exchange is minimal if compared against traditional aquaculture systems. In addition the small volumes of the incoming water is easily manageable through filtration and sterilization to control outer pathogens and parasites.

This closed management brings the overall water consumption per kilogram of fish produced to lower values (0.5–1.4 m^3) than traditional aquaculture, in which the overall water use depends on the management intensification: 4.7–7.8 m^3 for aerated ponds, up to 11.5 m^3 for extensive ponds, up to 30 m^3 for aerated pond with water exchange (Verdegem et al. 2006) (Table 7.8). In practical terms the very small volumes of incoming water required by closed systems can be easily and completely controlled and sterilized, bringing eventually to zero any risk to transmit diseases and pollutants into the systems.

RAS and aquaponic systems are a valuable method to grow fish in water scarcity conditions. The recirculation of water also limits the heat losses, which is beneficial in cold/hot climates as it saves major heating or cooling costs and maximize fish growth, since animals grow in optimal temperature ranges and optimal water quality.

7.4.3 Production and Quality in Recirculating Systems

In terms of performances the shift from traditional cage or pond culture to recirculating systems does not affect growth. In the case of tilapia FCR values from aquaponics range from 1.0–1.3 (Pantanella et al. 2012b) up to 1.7 (Rakocy, personal communication 2008) whilst recirculating tank systems show values of 1.4–1.8 (DeLong et al. 2009) and cage culture/earthen ponds can range from 0.82–0.98 (Ying and Lin 2001) up to 1.2–1.5 (El Sayed 2006). Likewise the fish growth rate, measured as specific growth rate (SGR: % of daily body weight increase), is 0.91–5.1% in aquaponics (Seawright et al. 1998; Al-Hafedh et al. 2008, Pantanella et al. 2012b) versus 1.43–3.22% in earthen ponds, but under higher feeding regimes (Pruginin et al. 1988).

In the case of other warmwater species, such as young African catfish SGR from aquaponics (1.36–2.13%) (Endut et al. 2010; Pantanella et al. 2012b) is similar to recirculating systems (1.24–1.94%) (Pantazis and Neofitou 2003; Ahmad 2008). Likewise FCR in aquaponics (0.97–1.39) (Endut et al. 2010, Pantanella et al. 2012b) is similar to either earthen ponds (0.98–1.54) (De Graaf and Janssen 1996) and recirculating systems (0.94–1.29) (Degani et al. 1988). For some other species aquaponics has shown interesting growth rates and FCR, as shown in Table 7.7.

On a qualitative point of view RAS and aquaponics proved that the fish containment helps to prevent any risks of parasites or chemical/biological pollution from external water sources. On the other hand the rearing of fish in closed systems proved no risks of heavy metal build-ups in the flesh, if compared to the levels found in animals reared with traditional systems (Martins et al. 2011).

There is currently a wide debate about the genetic contamination of wild stocks with farmed fish, the difficulties in preventing the mutual transmission of parasites and pathogens between farmed and wild fish, and the raise in tolerance of parasites against common drugs. These issues are now bringing the industry and policy makers to consider different ways to produce. The farming of fish in closed systems could be undoubtedly a valid solution to address the environmental problems that are affecting the industry. However, given the higher investment costs the returns must be guaranteed by farming high-value fish.

In terms of costs both RAS and aquaponics require higher investments, but aquaponics gets some advantages for the slightly lower technology used and the conversion into profits of the water treatment costs normally occurring in RAS. In aquaponics the plant production part is eventually the main source of income for farmers, who may differentiate their output by combining animal with vegetable crops. However the combination of the two components may limit the management choices of either fish or plants, since the optimal environmental conditions of one can differ from those of the other crop.

Aquaponics, as well as recirculation, may not be convenient for farming fish that can be produced extensively or semi-intensively in ponds, as their selling prices barely cover the feed and energy costs occurring in closed systems. The higher investments and operating costs from intensive systems may not be covered unless

Table 7.7 Performances of fish species in aquaponics

Fish species	Stocking density (kg m^{-3})	Fish weight at stocking (g)	Feeding regime (% CP)	Diet (% BW)	FCR	SGR	Author
Bester sturgeon (*Acipenser ruthenus* × *Huso huso*)	7.56	95	46	2	1.01–1.25	1.38–1.66	Dediu et al. (2012)
Hybrid sex reversed tilapia (*O. mossambicus* × *O. niloticus*)	8.7	434	32	1.8–0.6	na	na	Mc Murty et al. (1997)
Sex reversed Nile tilapia (*O. niloticus*)	1.3–16	3.8	41.6	5.0–4.4	1.0–1.1	4.4–5.1	Seawright et al. (1998)
Sexed male Tilapia (*S. aurea*)	1.12	62	32	4.0–1.2	1.59	1.2	Watten and Buschs (1984)
Nile tilapia (*O. niloticus*)	4.86	32.4	41	0.6	na	0.8–1.1	Tyson et al. (2008)
Mixed sex blu tilapia (*S. aureus*)	1.68	na	na	na	1.32	na	Mc Murty et al. (1990)
Mixed sex Nile tilapia (*O. niloticus*)	6.8–39.7	42.5–248	34	3 to 2	1.0–1.7	0.7–1.8	Al-Hafedh et al. (2008)
Nile tilapia (*O. niloticus*) (1) Red tilapia (*Oreochromis* sp.) (2)	5.4 (1) 10.8 (2)	70	32	ad libitum	1.79	na	Rakocy et al. (2004a)
Nile tilapia (*O. niloticus*) (1) Red tilapia (*Oreochromis* sp.) (2)	6.1 (1) 9.1 (2)	79.2 (1) 58.8 (2)	32	ad libitum	1.7 (1) 1.8 (2)	1.39 (1) 1.29 (2)	Rakocy et al. (2004b)
Mixed sex Nile tilapia (*O. niloticus*)	8 (1) 20 (2)	24 (1) 90 (2)	43 (1) 40 (2)	2 (1) 1.7 (2)	1.0 (1) 1.3 (2)	2.7 (1) 1.4 (2)	Pantanella et al. (2010)
Nile tilapia (O. niloticus)	10.7	100	na	2.5–1.25	na	na	Savidov (2005)
Red tilapia (*Oreochromis* sp.)	7.1–8.8	38.9	32	6–1.2	1.76	na	Rakocy et al. (1997)
GM Nile tilapia (O. niloticus)	15	10	40	2.5–1.1	1.57–3.9	4.2–0.08	Pantanella et al. (2011c)
Tilapia (*T. mossambica*) (T) & common carp (*Cyprinus carpio*) (C)	40	na	na	5		3.14 (T) 2.56 (C)	Naegel (1977)

(continued)

Table 7.7 (continued)

Fish species	Stocking density (kg m^{-3})	Fish weight at stocking (g)	Feeding regime (% CP)	Diet (% BW)	FCR	SGR	Author
African catfish (*Clarias gariepinus*)	9 (1) 21 (2)	81 (1) 183–193 (2)	31–40	2 (1) 1.5 (2)	0.97–1.1 (1) 1.25–1.3 (2)	2.0–2.1 (1) 1.36 (2)	Pantanella et al. (2011a, b, c,d)
African catfish (*Clarias gariepinus*)	na	30–40	32	2–4	1.23–1.39	1.68–1.83	Endut et al. (2010)
Murray cod (*Maccullochella peelii peelii*)	10	120–220	43	1.0–1.5	0.8–1.1	0.9–1.1	Lennard and Leonard (2006)
Mullet (*Mugil cephalus*)	7.4–8.1	83.2 (1) 84.9 (2) 100.4 (3)	54	0.5–0.6	3.5–4.5 (1) 2–2.2 (2) 3.1–3.2 (3)	0.1–0.2 (1) 0.4–0.5 (2) 0.25 (3)	Pantanella et al. (2011b)
Largemouth bass (*Micropterus salmoides*)	11	33.6	44	0.9	1.5	0.77	Pantanella, unpublished data

%CP percentage of crude protein, *%BW* percentage of body weight, *FCR* feed conversion ratio, *SGR* specific growth rate

Table 7.8 Comparative advantages and disadvantages, water use and productivity of aquaponics and RAS against traditional aquaculture (Verdegem et al. 2006—modified)

	Semi intensive ponds	Intensive aerated pond	Intensive aerated pond and water exchange	Aquaponics, RAS
Advantages	– Easy management – Low investments – Natural feed and pellet – Low production costs of fish – Integrated aquaculture	– Relatively low management – Pellet as main feed source, but natural feed still plays a role – Control on aeration	– Higher yielding – Control on aeration – Water monitoring – Ammonia monitoring – Use of pelleted feed	– Small land footprint – High environmental and water control (indoor) – High productivity – Safety (indoor) – Low need of water – Farming everywhere
Disadvantages	– Land demanding – Consistently rely on water sources – Contaminant risk – Disease risk (outdoor)	– Rely on feed – Rely on energy – Sludge accumulation – Limited choices to raise density – Contaminant risk – Disease risk (outdoor)	– Rely mainly on feed – Energy intensive – Rely on water – Sludge accumulation – Pollution of water (water discharge) – Contaminant risk – Disease risk (outdoor)	– Rely only on feed – Energy intensive – Management intensive – Multiple skills – Higher risk of failure for any breakdown – High investments – High running costs
Productivity (MT ha^{-1} y^{-1})	2–8	4–20	15–35	>100
Water use (m^3 kg $fish^{-1}$)	11.5	4.7–7.8	30	0.15–3.2

valued fish are cultured or commercial species are sold in markets where premium prices are applied for quality fish sold with no residues and pollution issues.

Aquaponics can be a valuable option in fish nursery productions, as the high turnover of the fry, the high stocking densities achievable and the higher degree of biosecurity guarantee for good incomes and reduced losses than traditional pond management.

A comparison of advantages and disadvantages of aquaponics/RAS against traditional systems outlines the small footprint and higher productivity of such advanced systems (Table 7.8), however the final decision for their use depends on the degree of risk that the fish farmers are willing to take.

7.5 Aquaponics Versus Agriculture Systems

7.5.1 Advantages of Soilless Systems Against Traditional Agriculture

Aquaponics and hydroponics overcome some of the problems commonly occurring in soil-based agriculture, which have been increased by the adoption of monoculture practices in greenhouses (Table 7.9). Typically with soil-based agriculture farmers have to carry out a series of tasks to prepare and manage their crops, which imply ploughing, removal of weeds and fertilization, irrigation, weed control. In addition agriculture farmers do not have control on the release of nutrients from soil, which is affected by the soil texture, its chemical characteristics in binding nutrients (cation exchange capacity), and environmental conditions (temperature). At the same time farmers have limited strategies to cope with salinity.

In terms of productivity soilless cultivation increases the crops' water use efficiency up to ten times, while the crop productivity can be more than doubled than conventional agriculture (Resh 2004).

Soilless cultivation can be developed in urban and suburban areas, which sensibly reduces transport costs. Furthermore aquaponics can be developed in areas not suitable for traditional agriculture due to exhaust soil conditions or bad water quality. Aquaponics fits particularly well the needs to produce food wherever there is no fertile land or access to land, but in terms of economic competitiveness the adoption of aquaponics in fertile areas has to be carefully assessed due to the higher investment and production costs than traditional agriculture.

To summarize soilless systems show some advantages over conventional soil-based systems:

- Increased yields due to cultivation in protected environments, in which it is easier to control the climatic parameters optimal for plants
- Lack of competition for nutrients from weeds
- Control of nutrients according to the growth stage and nutrient requirements of the crops

- Increase in quality of productions, due to optimal plant nutrient balances
- Reduction in residues and pesticide due to integrated/biologic management
- Better organoleptic and nutraceutical characteristics due to optimal nutrient management
- Lack of crop rotation to avoid "soil tiredness" that reduce crop yields over the years
- Better control of soil-borne diseases through physical avoidance of micro-organisms and biological management.

7.5.2 *Differences Between Aquaponics and Hydroponics*

Aquaponics differs from hydroponics. Although similar management can be applied in both systems, aquaponics is a more complex agroecosystem in which the presence of both fish and micro-organisms play a key role in plant Growth (Table 7.9). In aquaponics the levels of nutrients are lower than hydroponics, and most of the times the level of nitrogen used is 20–40% of the concentrations normally in use in hydroponics. The reason of such productivity in lieu of low concentrations of nutrients stands in a constant, but continuous, supply of micro and macro elements by fish. Nevertheless the more complex dynamics in aquaponics allow plants to uptake also free amino acids from water and fulfil equally their nutritional purposes (Ghosh and Burris 1950).

The presence of microorganisms in the water lead to a different management of aquaponics systems. Contrarily to hydroponics, which is mainly kept sterile to avoid pathogens' contamination, in aquaponics the complex habitat created by beneficial bacteria and fungi makes the system less prone to diseases, due to the high competitive environment the pathogens have to face.

The system complexity however raises the need to have higher levels of knowledge and expertise from operators, who should be aware of the different needs of both plants, fish and bacteria/fungi. The integration of these three living elements raises the need to get some compromises in either water (pH, temperatures, nutrient levels) and ambient/climate management.

One drawback of aquaponics, however, is found in the need to combine two different management at one time, which results in suboptimal conditions for either fish or plants. If the presence of the aquatic animals from one side testimonies for the safety of the products, on the other hand it severely limits the choices for disease and pest management. Many of the remedies in use in organic agriculture could not be applied to plants due to toxicity for fish, and have to be refrained unless they are used under strict control in limited cases. The aquaponic ecosystem is manageable providing that a bunch of preventive remedies are put into action to avoid any spread of pests or diseases into the system. Therefore preventive management requires high expertise in people who should know the dynamics of fish and plants, as well as be aware of epidemiologic factors.

Table 7.9 Comparative advantages and disadvantages of aquaponics against traditional hydroponics and soil cultivation

	Aquaponics	Hydroponics	Soil
Flexibility to new areas/climates	Possible. But needs to consider fish requirements	Possible. Agriculture even in extreme conditions (desert)	Not possible if soil is not fertile
System control	More complex, need also to consider animal optimum	Adapted to the crop (greenhouse)	Adapted to the crop (greenhouse)
Ambient conditions	Can be affected by higher humidity due to fish tanks	Standards of protected controlled agriculture (greenhouse)	Standards of protected controlled agriculture (greenhouse)
Water use	Low, as it compensates only for evapotranspiration and evaporation	Very low, as it compensates only for plant evapotranspiration	Medium, high. Depends on the delivery system and soil characteristics
Water physical characteristics	Should compromise plants and fish needs for either pH, temperatures	Optimal for plants	No water control
Nutrient concentrations	Can be low, due to constant fish supply	High	Medium to high. Release of nutrients affected by soil and weather/climate, leaching
Nutrient use efficiency	High, similar or better nutrient use efficiency than hydroponics due to lower concentrations of nutrients in the solution	High, minimal amount of fertilizers used, uniform distribution and real time adjustable flow of nutrients. No leaching-dispersal	Low, affected by soil, leaching and weather/climate
Use of space	Optimal, at least twice as productive as soil	Optimal, at least twice as productive as soil	Medium, depends on soil fertility
Soil-borne disease and pests	Not affected	Not affected	Affected
Epidemiology	Minimal risk of soil borne diseases, due to system resiliency. Higher risk of leaf diseases (fungi) due to higher humidity from fish tanks	Risk of disease spread higher due to use of same aqueous media to all plants (closed system)	Risk of disease spread higher due to runoff of contaminated water

(continued)

Table 7.9 (continued)

	Aquaponics	Hydroponics	Soil
Sanitation	Risk of contamination due to use of water with coliforms from fish. Sterilization of water is possible. No *E. coli* risks because fish are cold blooded animals	No risk of contamination for human health	Risk of contamination if using water from canals
System management	Organic-like but not certifiable organic (in EU)	Chemical or integrated	Chemical, integrated, organic
Product quality	Similar or higher to hydroponics. Higher presence of metabolic compounds may benefit	Full control with appropriate nutrient delivery and adjustable nutrients and salinity at fruiting stage	Variable, depends on the soil and management. Fruits may be better tasting
Investment costs	Higher than hydroponics due to the need to set up the fish system as well	High, due to setting up of protected environment, nutrient delivery and monitoring	Low, but needs investments on machineries
Management	Highly expert knowledge needed, due to fish/plant presence	Expert level is needed to cope with instruments and settings	Medium level
Risk	High risk of failure if electricity is down with no backup system	Medium, depends on automation	Medium, more affected by environmental conditions

In terms of productivity the scientific literature has already proven that aquaponics is as efficient as hydroponics both in terms of yields and quality, providing that certain nutrient ratios are maintained. In terms of market aquaponics has a better outlook than hydroponics for its organic-like management. In USA aquaponics can benefit from organic certification, which opens up produce to premium price markets. On the contrary EU regulations limit organic productions to soil-based agriculture despite the biological outlook of aquaponics and its full cycling of nutrients. Nevertheless the expansion of aquaponics is still limited due to higher investment costs than hydroponics that prevent farmers from considering this technique a much profitable alternative.

Aquaponics could experience good growth if it is developed as decoupled system, in which one farmer specializes in fish production while the another gets the aquaculture wastewater for plant production. The separation of the two subunits increases the opportunities to improve both pest and disease management, since there is no cross-contamination between the subsystems.

7.6 Production Systems in Use

7.6.1 Commercial Productions

A recent survey on commercial scale producers (Love et al. 2015) traced the identikit of the average aquaponic farmer. The majority of commercial farms are based in the USA with the owner having a leading role in the venture for at least 49% of the cases. Most of the commercial farms produce on DWC (77%) and media beds (76%), the data also testimony that a combination of different systems is the norm. The average farms are not big: size of 100 m^2, investment of 5000–10,000 USD, no cold storage room at least for half of them and no food safety plan for 38% of them. Aquaponic farms appear more vegetable-oriented rather than fish-oriented due to the length of the fish crops, though 69% of farms reared tilapia, a fish that can be harvested up to commercial size in only six months. The prevalence of leafy greens and herbs witnesses the orientation of farmers to high-return crops.

There is a big interest worldwide in adopting aquaponics for commercial horticulture. On this point research and pilot scale projects (Fig. 7.22) at different latitudes are focusing at demonstrating the economic feasibility of integrated systems and at optimizing their sub-components. There are ongoing collaborations between research institutions and the industry. A European Cooperation in Science and Technology—COST program FA1305 started in 2014 gathering research institutes and private companies with the scope to organize a comprehensive aquaponics platform for research and commercial development. Likewise in North and South America universities are partnering and fostering research and development for the support of the aquaponics sector.

Fig. 7.22 A pilot scale system in Europe

Big farms are increasing in number especially in North America. In Canada many firms are extensively producing aquaponics lettuces with DWC often using a degree of mechanization in harvest. In USA the leading design is the UVI system, with many farms replicating the ratios and the components. The success in North America is also driven by the possibility to certify organic the produce, which brings high revenues and good returns on investments. In Europe there is an expansion of aquaponic farms especially in the northern countries. The success stands in the green outlook of the technology and the awareness of consumers for safe products. Aquaponics has however good potential for expansion in water scarce countries. In the Middle East many countries are strategically planning production systems with water saving technologies as a way to guarantee their food security and to make the countries as much self-sufficient as possible. In UAE at the Zayed Agricultural Centre, an UVI type system of nearly 2000 m^2 of cultivable area, was built to show the potential for integration of fish and plants. Currently across the whole region private entrepreneurs and trusts are planning to build aquaponic systems for commercial operations following their respective country directives for food security, water security and self-sufficiency.

7.6.2 Small Scale for Backyard Consumption, Market and Food Security

Most of the small scale systems are meant for home consumption. Although not directly involved in commercial scale operations these systems proved to be supportive for family needs either for the supplement of chemical-free vegetables or to reduce the family retail expenditures at the grocery. According to a survey carried out in 2013 (Love et al. 2014) the average size of backyard farms are 15 m^2 with vegetables playing the main role. Interestingly most of the farmers do aquaponics more as a hobby and the main drivers in the production are sustainability and the production of own (safe) food. Aquaponics is growing mostly in urban and peri-urban areas due to the fact that micro scale agriculture is either considered a

Fig. 7.23 Micro scale system

leisure and because of the characteristic of soilless production: intensive outputs within small acreages wherever fertile land or other inputs are scarce (Fig. 7.23).

Aquaponics is also considered a strategy for food security not only because of the production per se, but because of the cash derived from selling small amounts of vegetables in local markets. In recent years FAO, the Food and Agriculture Organization of the United Nations, implemented microscale aquaponics projects for food security in the Middle East and Africa and showed that household food production is improved, can empower women and be more sustainable and resilient whenever production is organized at community level throughout the whole production chain (from seed to market) and under the credit support of revolving funds. Recent FAO workshops also witnessed the growing interest in promoting aquaponics as a water-saving food production technology to be used on islands or in conditions of scarce water resources, which is particularly important in many climate-change affected areas.

Advantages	Weaknesses
• High productivity with limited spaces • Landless food production • Seasonal-free productions in protected environments • Valorisation of household work • Women empowerment • Improved access to the markets • Improved value of products • Improvement of household food security • Household cash for health and education	• Higher initial investment costs than other traditional but low yielding systems • Higher degree of skills needed for the management of both plants and fishes • Grow out systems, need to rely on constant supply of inputs and fingerlings • Electricity not reliable in some areas • Need to produce high quality and high value crops and fish to be highly profitable • Market access to be developed

7.7 Alternative System Designs

Aquaponics combines fish and plant species within the same environment. This integration includes also beneficial bacteria involved in the nitrification, which work in optimal ranges of pH, temperature. Most species are adaptable to variable conditions, but not always it is possible to let each crop to grow under optimal environmental parameters. One of the main aspects that additionally limit the choices of the management in aquaponics is the risk of cross-toxicity in using any biological remedies for pest control, which results in some cases in crop failures or fish losses. In recent years there has been a constant and growing interest in separate the fish and plant units to ease the management through decoupled or hybrid systems.

7.7.1 Decoupled Aquaponics Systems

The rationale of these systems is that both fish and plants live in separate environments that are temporarily connected just to bring nutrients from fish to plants or reclaimed water back to fish. Basically a decoupled system is the combination of a RAS and a hydroponic unit using fish wastewater as source of nutrients. In terms of management such solution ease the farming of fish that are always reared in optimal conditions of temperatures and water parameters. On the other hand the standalone plant growing areas are set with optimal pH, humidity and temperatures set for the vegetable crops. The decoupled aquaponics systems can be managed with periodical recirculation of water between the fish and plant subsystem (two-way, or hybrid system) (Fig. 7.24) or can be run in unidirectional way with water going

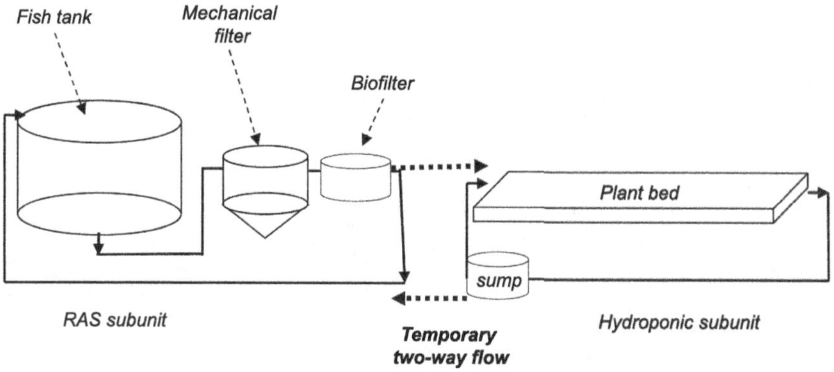

Fig. 7.24 A two-way decoupled system (hybrid RAS). The RAS and hydroponic subunits work as standalone units but are temporary connected to allow nutrient-rich water to go to the plant bed and reclaimed water from the plant to the fish subsystem. The plant bed still works as a biofilter and supports the nitrification needs of the fish subunit

only from the fish to the plant subsystem (one-way system) (Fig. 7.25). In terms of practical management this second option is seen more favourably by plant growers, who do not have to be worried about any risk of toxicity to fish. Such type of open system well suits the needs of outdoor agriculture with ferti-irrigation lines serving rows of plants.

One of the advantages of decoupled systems is that they do not necessarily need experts in both fish and plants, as one farmer just specializes in recirculating aquaculture, while another one specializes in soilless cultivation by using fertilized water from the fish producing neighbour.

In terms of investment this solution would ease the adoption of aquaponics, as farmers would not necessarily need to double the investment on both fish and plants, but can outsource one of the two while concentrating in their main and single core business.

7.7.2 Low-Tech Designs

At backyard level there is interest in developing low-tech systems that are simple and of immediate understanding by farmers. The Indonesian *Yumina-Bumina*, for example, re-thinks at aquaponics in a very simple and comprehensible way for whoever is used to pond culture. Surrounding pots all-around the banks or walls of the tank help the water quality to be maintained at optimal levels while delivering fertilized water to the plants (Figs. 7.26 and 7.27). The media contained in the pots procure at the same time solid entrapment, biofiltration and mineralization of the fish water. Yumina-Bumina uses higher fish stocking densities than traditional aquaponics, also because the size of the harvested fish is rather small and targets the single person portion sizes. The Yumina-Bumina stocks fingerlings of catfish at 300–500 fish m^{-3}, 50 fish m^{-3} for Nile tilapia.

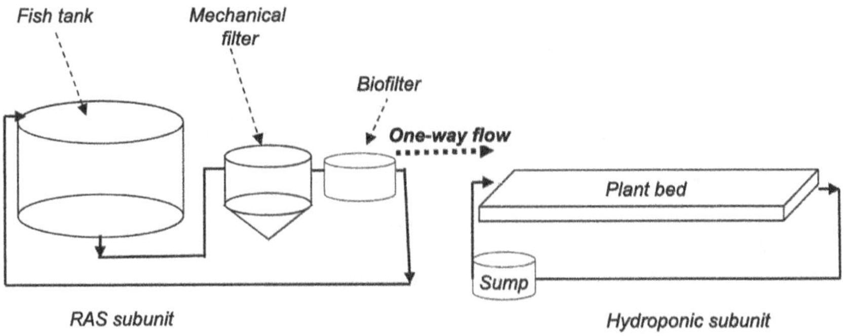

Fig. 7.25 A one-way decoupled system. The RAS and hydroponic subunits work independently. The plant bed receives the fertilized water from the fish subunit. This system allow farmers to have more freedom in their integrated pest/disease management

Fig. 7.26 A yumina-bumina system with pots

Fig. 7.27 A floating yumina system

In general such systems use a fish pond:plant ratio of 0.25 m^3 m^{-2}, which means that every cubic meter of pond water corresponds to 0.25 m^2 of plant growing area. The production cycle is quite fast as the harvest in catfish occurs in only 2–2.5 months, while gourami and pangasius in 6–12 months. For tilapia the harvest of mix-sex fish occurs before they reach sexual maturity at the age of 4–5 months, which completely bypasses the need to carry out sex reversal in these fish.

The focus on low-tech systems is important in emerging countries where the limited access to money for investment and the lack of knowledge of the dynamics of aquaponics prevent many from adopting backyard systems. Nevertheless systems that approach the traditional way agriculture is managed and that require low maintenance are ideal to meet the limited skills of local households. In Myanmar a demonstration facility with a tank serving a gravel bed proved that a 30 m^2 system that includes a bamboo nethouse and a solar system for standalone energy supply would cost as low as 25 USD m^{-2} and bring a net profit of 1.6–2.2 USD a day from vegetables and secures a fish consumption of 400 grams per day (Pantanella et al. 2014). Similarly Dr. Wilson Lennard from his researches could produce aquaponic kits for backyard farming in developing countries serving a few square meters of plant beds for as less as 100 USD.

The aquaponics concept could be applied to staples or even to traditional agriculture. The key factor stands in the recirculation of the water that allows for the build-up of nutrients to levels that guarantee for commercial productions of crops without any addition of chemical fertilization. Some experiments carried out with tilapia and rice growing on sand beds proved higher rice yields than chemically fertilized paddies (respectively 8.1 and 5.1 MT ha^{-1}) in 120-day crops with a fish:plant ratio of 1–1.5 kg. Interestingly, considering the short duration of fast growing varieties of rice, it would be possible to perpetually support the daily rice needs of a family of 5 members with only 100 m^2 of growing area (Pantanella et al. 2011c).

7.8 Saline Aquaponics

Saline water provides new opportunities to farm fish and plants in a more sustainable way. Despite its wide diffusion cage farming has never obtained a full acknowledgement due to pollution issues, the risk of genetic contamination of wild stocks due to escapees and disease outbreaks, the competition with other water uses for recreational purposes. In the last decades the integrated multi-trophic aquaculture (IMTA) provided some solutions to the control of the pollution from fish cages, but it has obtained limited impact due to the high water dilution of nutrients in open bodies.

Aquaponics with its build-up in nutrients provides opportunities for marine and brackishwater aquaculture to control the potential source of pollution by preventing organic wastes to be released into the environment and to obtain at the same time additional incomes from plant production. Turning fishes out from cages into recirculating systems not only does maintain the optimal growth parameters of fish, but would also achieve higher levels of biosecurity against pollutant and pathogens, which eventually guarantee for higher yields and safer aquaculture productions.

Saline aquaponics does not differ much from the freshwater aquaponics, with the only exceptions that biofiltration has to be scaled up to compensate for the lower nitrification efficiency of bacteria under higher salinity, and the need to increase the concentrations of nutrients to compensate for the reduced plant uptake due to higher osmotic pressure in the water.

The salinity level in water definitely affects the type of system in use and the fish and plants choice. Some marine species such as European seabass (*Dicentrarchus labrax*) and seabream (*Sparus aurata*) respectively grow well at salinity of 5–7 g L^{-1} and 10 g L^{-1}, while mullet (*Mugil* spp.) and Asian seabass (*Lates calcarifer*) can reach nearly freshwater conditions.

The decrease in salinity, within certain physiological limits, rather than being a depressing factor can improve the growth performances of fish, which do no spend energy for balancing the osmotic pressure of highly saline water.

Fig. 7.28 Basil growing at 3 g L^{-1} salinity can achieve similar yields per m^2 of plants growing on freshwater hydroponics by simply improving density, climatic control and use of anti stress factors

Although salt is not the best element for plant growth due to the toxicity of sodium and the reduced capacity of plants to uptake water against a negative osmotic pressure, it is possible to crop some traditional horticultural plants that show a degree or resistance at salinity levels of 0.5–7 g L^{-1}. In addition, tailored agronomic strategies such as reduction of water stress conditions, improved fertilization, the grafting of commercial varieties on salt-resistant rootstocks, the use of anti-stress factors can greatly help to increase productivity up to the yields achievable in freshwater hydroponics (Fig. 7.28).

Besides, salt-tolerant plants (halophytes) can tolerate concentrations up to marine strength and show interesting commercial opportunities for leaf productions. The most known are *Salsola* spp. (Fig. 7.29), *Atriplex* spp, *Kochia scoparia*, sea fennel, *Salicornia* spp, seabeet (Fig. 7.30). There are also at least fifty different species of grain crops that can be simply cultivated with irrigation lines in outdoor conditions. There is also growing interest in seaweed for their nutritional and nutraceutical characteristics. They can be cultivated in closed systems and can

Fig. 7.29 Salsola optimally grows within 10–20 g L^{-1} salinity

Fig. 7.30 Seabeet growing at 10 g L^{-1} salinity on DRFT

greatly benefit from nutrients released by fish. Closed conditions also secure controlled production standards and compliance to food safety regulations.

The type of farmed aquatic animal strictly affects the choice of plants or seaweeds that can be cultivated based on the salinity ranges. In the case of saline aquaponics it is thus important to develop preliminary market studies to assess the demand of crops, the margin of profitability and the risk factors to secure economical sustainability of the ventures. Besides, the development of systems with cost-effective designs and technologies are the safest strategy to guarantee quick returns on investments.

7.9 Future Research

Most of the research on aquaponics has been carried out on the nutrient balances between different species of fish and plants. Contrarily to the past, when the focus was more on the engineering aspect of the system there is nowadays raising interest to determine the quality of the productions and to develop effective growth strategies. There is indeed a great deal of research topics that need to be explored, most of them pertaining the optimal nutrition of plants and the ways to modulate the concentrations of nutrients according to the growth stage of the plants. Secondly, research is also targeting new designs that can best meet the crop needs and be energy-saving.

Since 2014 the European funded EU COST FA1305 action has gathered the academic, research and development sectors with the SMEs from many European and no EU countries to evaluate the state of the art of aquaponics and to join the efforts in innovation and education. The action of research and development is mainly oriented towards the water quality management, the alternative sources of feed for aquatic animals, the best combinations of fish and plants in different latitudes also in a perspective of food security, in the assessment of the economic feasibility of aquaponics against other alternatives, and in the review of indicators for ecological, social and economic sustainability.

The assessment of the economic feasibility and the research of alternative system designs that best suit the need of the industry and small farmers is at the top of the agenda of many researchers and stakeholders. Recent aquaponic workshops carried out by FAO have arisen the need to develop clear assessment of the costs and benefits of aquaponics to let farmers and entrepreneurs be informed of the advantages/disadvantages of adopting this integrated system under their respective climatic and environmental conditions.

The engineering research is currently looking at alternative ways of running aquaponics from recirculating systems. On this large interest is now put in decoupled aquaponics, where the fish and plants subunits are managed separately, but temporarily communicate for the delivery of nutrients or for the return of reclaimed water back to the fish. Therefore there is the need to optimize the fish sub-units to let them reach good balances of nutrients for plants to be grown in similar conditions of traditional hydroponics.

The idea to make the aquaponics systems as much self-reliant as possible is another key research topic. In Canada the research team lead by Dr. Nick Savidov is working on the 5th generation aquaponics with the aims to produce zero-waste through complete mineralization of fish solids. On the other hand a team of researchers in Europe are focusing on the internal production of supplementary food to reduce the costs from feeds.

In developing countries the focus is on building systems suitable for the spending capacity of the locals and are of adequate simplicity to be used by low-educated farmers with hassle-free management and with low energy demand. Besides, there is the need to widen the potential of aquaponics to grow staple crops, which can guarantee for food security in areas where traditional agriculture cannot be done for either natural or anthropic causes. All these research solutions however need to be assessed against the costs and the economic advantages they can bring to the production system to make them really sustainable and adoptable.

7.10 Concluding Remarks

Aquaponics is a valid production system that meets the need to produce more with less inputs. The research in the past years has proven that systems are robust to handle both fish and plants and the productive traits of the crops are competitive against soil-based agriculture and hydroponics, even with lower levels of nutrients.

Aquaponics well suits the need of the fish industry for more sustainable productions, as it consumes less water and reduces down to zero the impact of wastes on the environment by re-using them in substitution of chemical fertilizers. The use of water in aquaponics is the lowest in aquaculture and comparable to advanced RAS without the hassle of more sophisticated and expensive technologies for water treatments. Aquaponics also complies with the need for high quality food, as its closed recirculating system reduces any risks of contamination with outer pollutants and prevents the contact with pathogen and parasites from unprocessed water, which eventually offsets any use of drugs.

There is however a number of areas of research that still need to be addressed to fully improve aquaponics to be adopted at industrial scale. This include the choice of plants to best meet each environmental condition, the climatic control of aquaponics, the decoupled technology for easier management of plant-only or fish-only systems, and the food safety issues for the retail sector. Also one of the key requirements for the expansion of aquaponics is the validation of its cost effectiveness.

Integrated systems have a great potential to improve agroecosystems efficiency. Nevertheless performances and economic sustainability are always factors influenced by environmental conditions, sub-system design and management. Increased productivity and sustainability of agroecosystems should further consider the optimal management of input, output and by-product as an important factor to improve overall system efficiency.

The system integration is the key factor for low input productions. However, the complexity of agroecosystems due to fish and plant integration requires increased management and environmental needs of plants, animals and their surrounding habitat. Aquaponics is as efficient as hydroponics in producing high quality food. However the full expansion of every integrated system would be only possible when products have lower production costs than traditional agriculture, or when aquaponics brings higher and faster returns on investments than hydroponics or traditional aquaculture. The key for the long term success of aquaponics would eventually be the perfect trade-off among environmental, social and economic sustainability.

References

Adams REW (2005) Prehistoric mesoamerica. University of Oklahoma press, Norman, OK, USA, p 544

Adler PR, Summerfelt ST, Glenn DM, Takeda F (2003) Mechanistic approach to phytoremediation of water. Ecol Eng 20:251–264

Ahmad MH (2008) Response of African catfish, *Clarias gariepinus*, to different dietary protein and lipid levels in practical diets. J World Aquac Soc 39:541–548

Al-Hafedh YS, Alam A, Beltagi MS (2008) Food production and water conservation in a recirculating aquaponic system in Saudi Arabia at different ratios of fish feed to plants. J World Aquac Soc 39:510–520

Barnabé G (1990) Aquaculture, (II volume), 2nd edn. Ellis Horwood Limited, Chichester, England, p 1104

Brown L, Kane H (1994) Full House: reassessing the earth's population carrying capacity. Worldwatch Environmental Alert Series, W.W. Norton, New York, NY

Costa-Pierce BA (1996) Environmental impact of nutrients from aquaculture: towards the evolution of sustainable aquaculture systems. In: Baird JD, Beveridge MCM, Kelly LA, Muir JF (eds) Aquaculture and water resource management. Blackwell, UK, pp 81–109

Davidson J, Summerfelt ST (2005) Solids removal from a coldwater recirculating system—comparison of a swirl separator and a radial-flow settler. Aquacult Eng 33:47–61

Dediu L, Cristea V, Xiaoshuan Z (2012) Waste production and valorisation in an integrated aquaponic system with bester and lettuce. Afr J Biotech 11:2349–2358

Degani G, Ben-Zvi Y, Levanon D (1988) The effect of different dietary protein sources and temperatures on growth and feed utilization of African catfish *Clarias gariepinus* (Burchell). Isr J Aquac Bamidgeh 40:113–117

De Graaf G, Janssen H (1996) Artificial reproduction and pond rearing of the African catfish *Clarias gariepinus* in Sub-Saharan Africa—a handbook. FAO Fisheries Technical Paper, 362. Food and Agriculture Organization of the United Nations

DeLong DP, Losordo TM, Rakocy JE (2009) Tank culture of tilapia. SRAC Publication 282. South Regional Aquaculture Center

Diana JS, Szyper JP, Batterson TR, Boyd CE, Pedrahita RH (1997) Water quality in ponds. In: Egna HS, Boyd CE (eds) Dynamics of pond aquaculture. CRC Press, Boca Raton, New York, p 437

Downing JA, Baker JL, Diaz RJ, Prato T, Rabalais NN, Zimmerman RJ (1999) Gulf of Mexico hypoxia: land and sea interactions task force report No. 134. Council for Agricultural Science and Technology, Ames, IA

El Sayed AFM (2006) Intensive systems. In: El Sayed AFM (ed) Tilapia culture. CABI Publishing, Wallingford, UK, pp 70–94

Endut A, Jusoh A, Ali N, Wan Nik WB, Hassan A (2010) A study on the optimal hydraulic loading rate and plant ratios in recirculation aquaponic system. Biores Technol 101:1511–1517

FAO (2014) Small-scale aquaponic food production. Integrated fish and plant farming. FAO Fisheries and Aquaculture Technical Paper No. 589. Rome, FAO, p 262

Ghosh BP, Burris RH (1950) Utilization of nitrogenous compounds by plants. Soil Sci 70:187–203

Gloger KC, Rakocy JE, Cotner JB, Bailey DS, Cole WM, Shultz KA (1995) Waste treatment capacity of raft hydroponics in a closed recirculating fish culture system. World Aquaculture Society, Book of Abstracts, pp 126–127

Graber A, Junge R (2009) Aquaponic systems: nutrient recycling from fish wastewater by vegetable production. Desalination 246:147–156

Jensen MN (1997) Hydroponics. HortSci 32(6):1018–1021

Jensen MN (1981) New developments in hydroponics systems: descriptions, oprating characteristics, evaluation. In: Proceedings: Hydroponics: where is tt growing? Hydroponics society of america, Brentwood, CA, pp 1–25

Jones JB (1997) Hydroponics, a practical guide for the soilles grower. St Lucie Press, Boca Raton, FL USA, pp 9–141

Lennard WA (2004) Aquaponics the theory behind integration. Gippsland Aquaculture Industry Network—GAIN, Traralgon Victoria. Australia http://www.growfish.com.au/content.asp?contentid=1060. Accessed 18 Feb 2007

Lennard WA, Leonard BV (2004) A comparison of reciprocating flow versus constant flow in an integrated, gravel bed, aquaponic test system. Aquacult Int 12:539–553

Lennard WA, Leonard BV (2006) A Comparison of three different hydroponic sub-systems (gravel bed, floating and nutrient film technique) in an aquaponic test system. Aquacult Int 14:539–550

Lennard WA (2013) Aquaponic system design parameters. Fish to plant ratios. Aquaponics fact sheet series. http://www.aquaponic.com.au/Fish%20to%20plant%20ratios.pdf. Accessed on 28 Jan 2016

Leoni S (2003) Colture senza suolo in ambiente mediterraneo. le nuove tecniche per l'orticoltura e la floricoltura da serra. Edagricole, Bologna, Italy, p 278

Losordo TM, Masser MP, Rakocy JE (1998) Recirculating aquaculture tank production systems an overview of critical considerations. SRAC Publication No. 451, September 1998—Revised

Love DC, Fry JP, Li X, Hill ES, Genello L, Semmens K, Thompson RE (2015) Commercial aquaponics production and profitability: findings from an international survey. Aquaculture 435:67–74

Love DC, Fry JP, Genello L, Hill ES, Frederick JA, Li X, Semmens K (2014) An international survey of aquaponics practitioners. http://journals.plos.org/plosone/article?id=10.1371/journal.pone.0102662. Accessed on 28 Jan 2016

Massantini F (1968) Le Culture Idroponiche. L'Italia Agricola, Numero speciale Culture Protette 11–12:1137–1172

Masser MP, Rakocy JE, Losordo TM (1999) Recirculating aquaculture tank production systems: management of recirculating systems. SRAC Publication No. 452 (revision March 1999) Department of Agriculture, USA

Martins CIM, Eding EH, Verreth JAJ (2011) The effect of recirculating aquaculture systems on the concentrations of heavy metals in culture water and tissues of Nile tilapia *Oreochromis niloticus*. Food Chem 126:1001–1005

Mc Murty MR, Nelson PV, Sanders DC, Hodges L (1990) Sand culture of vegetables using recirculated aquacultural effluents. Appl Agric Res 5:280–284

Mc Murty MR, Sanders DC, Cure JD (1997) Efficiency of water use of an integrated fish vegetable co-culture system. J World Aquac Soc 28:420–428

Naegel LCA (1977) Combined production of fish and plants in recirculating water. Aquaculture 10:17–24

Pantanella E, Cardarelli M, Colla G (2010) Aquaponics vs hydroponics: potentials for sustainable aquaculture. Paper presented at Aquaculture Europe 2010, Oporto, Portugal, 6–9 October 2010

Pantanella E, Fabrizi F, Cardarelli M, Colla G (2011a) Catfish and sweet basil aquaponics: a comparison of fish and plant growth under two different protein diets. Paper presented at World Aquaculture 2011, Natal, Brazil, 6–10 June 2011

Pantanella E, Bergonzoli S, Fabrizi F, Cardarelli M, Colla G (2011b) Saline aquaponics, new challenges for marine aquaculture. Paper presented at Aquaculture Europe 2011, Rhodes, Greece, 18–21 Oct 2011

Pantanella E, Fabrizi F, Cardarelli M, Colla G (2011c) Closed rice-fish recirculating system (CRIFIRS): an innovative and low-tech production method for environmental friendly productions and food security. Paper presented at World Aquaculture 2011, Natal, Brazil, 6–10 June 2011

Pantanella E, Cardarelli M, Colla G, Danieli PP, MacNiven A (2011d) Integrated aquaculture—floating agriculture: is it a valid strategy to raise livelihood? Acta Hortic 921:79–86

Pantanella E, Cardarelli M, Colla G, Rea E, Marcucci A (2012a) Aquaponics vs hydroponics: production and quality of lettuce crop. Acta Hortic 927:887–893

Pantanella E, Colla G, Appelbaum S (2012b) Sustainable aquaculture and farming systems: water use efficiency in aquaponics. Paper presented at the Drylands, Deserts and Desertification conference. Ben-Gurion University of the Negev, Sede Boqer Campus, Israel. 12–15 Nov, 2012

Pantanella E, Cardarelli M, Colla G (2012c) Yields and nutrient uptake from three aquaponic sub-systems (floating, NFT and substrate) under two different protein diets. Paper presented at the European Aquaculture Society AQUA2012 Global Aquaculture Securing our Future. Prague, Czech Republic. 1–5 Sept 2012

Pantanella E, Colla G, Bhujel RC Myint W (2014) Aquaponics for livelihoods? Paper presented at the World Aquaculture Society conference Aquaculture America 2014, Seattle, USA. 9–12 Feb 2014

Pantazis PA, Neofitou CN (2003) Feeding frequency and feed intake in the African catfish *Clarias gariepinus* (Burchell 1822). Isr J Aquac Bamidgeh 55:160–168

Piedrahita RH (2003) Reducing the potential environmental impact of tank aquaculture effluents through intensification and recirculation. Aquaculture 226:35–44

Pruginin Y, Fishelson L, Koren A (1988) Intensive tilapia farming in brackishwater from an Israeli desert aquifer. In: Pullin RSV, Bhukaswan T, Tonguthai K, Maclean JL (eds) The second international symposium on tilapia in aquaculture. ICLARM conference proceedings vol 15, pp 75–81

Rakocy JE (1989) Tank culture of tilapia. SRAC Publication No. 282 Department of Agriculture, USA

Rakocy JE, Losordo TM, Masser MP (1992) Recirculating aquaculture tank production systems. integrating fish and plant culture. SRAC Publication No. 454, Department of Agriculture, USA

Rakocy JE, Hargreaves JA (1993) Integration of vegetable hydroponics with fish culture: a review. In: Techniques for modern aquaculture. Proceedings of an aquacultural engineering conference, ASAE. Spokane, Washington, USA, 21–23 June 1993, pp 112–136

Rakocy JE, Bailey DS, Shultz KA, Cole WM (1997) Evaluation of a commercial-scale aquaponic unit for the production of tilapia and lettuce. In: Fitzsimmons K (ed) Tilapia aquaculture: proceedings of the fourth international symposium on tilapia in aquaculture at Orlando, Florida. Northeast Regional Agricultural Engineering Service, Ithaca, New York, USA, pp 357–372

Rakocy JE (1999a) The state of aquaponics. Part 1. Aquac Mag 25(3):83–88

Rakocy JE (1999b) The state of aquaponics. Part 2. Aquac Mag 25(4):64–70

Rakocy JE, Shultz RC, Bailey DS, Thoman ES (2004a) Aquaponic production of tilapia and basil: comparing a batch and staggered cropping system. Acta Hortic 648:63–69

Rakocy JE, Shultz RC, Bailey DS, Thoman ES (2004b) update on tilapia and vegetable production in the UVI aquaponic system. new dimensions on farmed tilapia. Proceedings from the 6th international symposium on tilapia in aquaculture, Manila, Philippines September 12–16, 2004. http://ag.arizona.edu/azaqua/ista/ista6/ista6web/pdf/676.pdf. Accessed on 15 Jan 2016

Rakocy JE, Masser MP, Losordo TM (2006) Recirculating aquaculture tank production systems: aquaponics-integrating fish and plant culture. SRAC Publication No. 454 (revision November 2006) Department of Agriculture, USA

Rakocy JE (2007) Aquaponics, integrating fish and plant culture. In: Simmons TB, Ebeling JM (eds) Recirculating aquaculture. Cayuga Aqua Ventures, Ithaca, NY, USA, pp 767–826

Resh HM (2004) Hydroponic food production. A definitive guidebook for the advanced home gardener and the commercial hydroponic grower, 6th edn. New Concept press, Mahwah, NJ, USA, pp 567

Savidov N (2005) Evaluation and development of aquaponics production and product market capabilities in Alberta. Phase II. Final Report—Project #2004-67905621. http://www.ecogrow.ca/pdf/CDC_Report_Phase_II.pdf. Accessed on 10 Jan 2016

Seawright DE, Stickney RR, Walker RB (1998) Nutrient dynamics in integrated aquaculture–hydroponics systems. Aquaculture 160:215–237

Schram E, Roques J, Abbink W, Yokohama Y, Spanings T, de Vries P, Bierman S, Van de Vis H, Flik G (2012) Exposure of fish to high nitrate in recirculating aquaculture systems affects growth, feed intake and physiology. Paper presented at aqua 2012 global aquaculture, securing our future. Prague, Czech Republic, 1–5 Sept 2012

Sonneveld C, Straver N (1989) Nutrient solutions for vegetables and flowers grown in water or substrates. Voedingsoplossingen Glasstuinbouw Bull. N. 8. Naaldwijk, The Netherlands

Sutton MQ, Anderson EN (2004) Introduction to cultural ecology. Altamira Press, Lanham, MD, USA, p 352

Tesi R (2002) Colture fuori suolo in orticoltura e frutticultura. Italy, Edagricole, Bologna, p 112

Tillman D, Cassman KG, Matson PA, Naylor R, Polasky S (2002) Agricultural sustainability and intensive production practices. Nature 418:671–677

Tyson RV, Simonne EH, White JM, Lamb EM (2004) Reconciling water quality parameters impacting nitrification in aquaponics: the pH levels. Proc Fla State Hortic Soc 117:79–83

Tyson RV, Simonne EH, Treadwell DD (2008) Reconciling pH for ammonia biofiltration and cucumber yield in a recirculating aquaponic system with perlite biofilters. HortScience 43 (3):719–724

UN-Water (2012) Managing water under uncertainty and risk. The United Nations World Water Development Report 4. UNESCO, Paris

Van Anrooy R (2002) Marketing opportunities for aquaculture products in the Lesser Antilles. In: Lovatelli A, Walters R, van Anrooy R (eds) Report of the subregional workshop to promote sustainable aquaculture development in the small Island developing states of the lesser antilles FAO fisheries report no. 704 SLAC/FIRI/FIPP/R704. FAO, Rome, Italy

Van Rijn J (1996) The potential for integrated biological treatment systems in recirculating fish culture—a review. Aquaculture 139:181–201

Van Rijn J (2010) Chapter 9: Denitrification. In: Timmons MB, Ebeling JM (eds) Recirculating aquaculture, 2nd edn. Cayuga Aqua Ventures, New York, pp 387–424

Verdegem MCJ, Eding EH, van Rooij JM, Verreth JAJ (1999) Comparison of effluents from pond and recirculating production systems using formulated diets. World Aquac 30:28–36

Verdegem MCJ, Bosma RH, Verreth JAJ (2006) Reducing water use for animal production through aquaculture. Water Resour Dev 22:101–113

Waggoner PE (1994) How much land can ten billion people spare for nature? Council for Agricultural Science and Technology. Task Force Report No. 121

Watten BJ, Busch RL (1984) Tropical production of tilapia (*Sarotherodon aurea*) and tomatoes (*Lycopersicon esculentum*) in a small scale recirculating water system. Aquaculture 41:271–283

Ying Y, Lin CK (2001) Effects of biomass of caged Nile tilapia (*Oreochromis niloticus*) and aeration on the growth and yields in an integrated cage-cum-pond system. Aquaculture 195:253–267

Chapter 8
Estimating Carbon Footprint Under an Intensive Aquaculture Regime

Sara Gonzalez-Garcia, Pedro Villanueva-Rey, Gumersindo Feijoo and Maria Teresa Moreira

Abstract This chapter presents a method to assess the carbon footprint and edible protein content Energy Return On Investment ratio (ep-EROI) of one of the most important and representative fish species, namely, turbot (*Scophthalmus maximus*) cultivated in Spain under an intensive aquaculture regime. The analysis was performed considering a cradle-to-farm gate life cycle assessment (LCA). To do so, representative hatcheries, nurseries and turbot farms were inventoried in detail. The relative contribution to overall greenhouse gases (GHG) emission and cumulative energy demand (CED) were evaluated. The results indicated that intensive turbot farming has significant GHG emission and energy requirements mainly due to electricity and feed requirements. The subsystem related with hatching and nursing reports the highest contribution to the impacts under study. Regarding the ep-EROI, an average level of 0.38% was obtained for Spanish turbot, which is considerably low in comparison with other aquaculture species. Results from this study can be used to optimise and promote more sustainable turbot production chains.

Keywords Edible protein energy · Environmental *hotspots* · Life cycle assessment · Spanish aquaculture · Turbot culture system

8.1 Introduction to the Aquaculture Sector and Life Cycle Assessment

Environmental concerns specifically focused on effects derived from greenhouse gases (GHG) emission and fossil fuels depletion have developed a society increasingly aware of environmental preservation (Ribeiro et al. 2013). Special attention is being paid on fishery sector and seafood supply chains (Thrane 2004; Henriksson et al. 2012; Parker 2012). Historical increments on fishery capture

---

S. Gonzalez-Garcia · P. Villanueva-Rey · G. Feijoo · M. T. Moreira (✉)
Department of Chemical Engineering, Institute of Technology, University of Santiago de Compostela, 15782 Santiago de Compostela, Spain
e-mail: maite.moreira@usc.es

F. I. Hai et al. (eds.), *Sustainable Aquaculture*, Applied Environmental Science and Engineering for a Sustainable Future,
https://doi.org/10.1007/978-3-319-73257-2_8

yields have been achieved by means of increasing fishing efforts as well as by new fishing grounds exploration (Henriksson et al. 2012). The main direct impact on the environment from the fishing sector is related with the decrease of the stock sizes (Schau 2012), deriving on a non-sustainability exploitation of fishery resources (Hilborn et al. 2015). Despite multiple scientific discussions on the concept and meanings of sustainability as well as its application on fisheries, there is not consistency regarding sustainable seafood definition (Hilborn et al. 2015). In this context, a fast growth in aquacultural systems has been evidenced (FAO 2006; Henriksson et al. 2012; Iribarren et al. 2012; Parker 2012). Different species of shellfish (mussels, oysters), fish (salmon, trout, tilapia) and aquatic plants (microalgae) are farmed in a variety of culture environments and production systems (Ayer and Tyedmers 2009). However, the development of marine fish farming is confronted with many environmental limitations (Aubin et al. 2006) mostly related with nutrient emissions into aquatic ecosystems and accessibility to coastal areas. As a result, land-based fish farming is receiving special development, mainly due to lower requirements of water and efficient wastewater management within the farm boundaries (Aubin et al. 2006; Liu et al. 2016). Moreover, environmental impacts considerably depend and vary according to the cultivated species and the cultivation system (Ayer and Tyedmers 2009; Martins et al. 2010). Fish meal as well as oil as feed ingredients and electricity requirements (Martins et al. 2010; Iribarren et al. 2012) have been identified as critical factors in aquaculture production systems and for that reason, efforts must be performed in order to evolve these systems into more environmental friendly systems (Roque d'Orbcastel et al. 2009).

Measuring, understanding and improving GHG emission of fishery and aquaculture based products is an important part of the seafood industry's labors to improve environmental profiles, labels and market products to consumers, meet government regulations, and promote the environmental and economic sustainability. In the process of better understanding the environmental impacts of aquaculture, Life Cycle Assessment (LCA) has become more repeatedly used to identify not only the best practices but also to assess the environmental performance and critical activities (Pelletier and Tyedmers 2008). LCA is an international standardized method (ISO 2006) managed to evaluate the impact that a production system has on the environment. Since it involves the term "life cycle", the assessment takes into account all the stages carried out throughout the production system that is, including raw materials and energy production, manufacturing, distribution, use and final disposal. It is based on a compilation of inventory data (energy and material flows) corresponding to each stage and it quantifies the contributions of these flows to a set of resources use and emissions related with environmental impact categories (ISO 2006).

In recent decades, LCA has been applied to environmentally assess a range of aquaculture based products (Pelletier and Tyedmers 2010) from a life cycle operational perspective under extensive or intensive regimes. Fish species such as rainbow trout (Seppala et al. 2001, Papatryphon et al. 2004, 2005; Aubin et al. 2009; Roque d'Orbcast et al. 2009), turbot (Aubin et al. 2006, 2009; Iribarren et al. 2012), tilapia (Pelletier and Tyedmers 2010), salmon (Ellingsen and Aanondsen

2006; Ayer and Tyedmers 2009; Pelletier et al. 2009; Liu et al. 2016), sea-bass (Aubin et al. 2009) or the Arctic char (Summerfelt et al. 2006) are examples of species analysed from a LCA perspective. Regarding shellfish, shrimp (Mungkung et al. 2006; Cao et al. 2011) and mussels (Iribarren et al. 2010a) have been also analysed following the LCA principles.

As previously reported, species cultivated as well as cultivation practices affect the environmental results. Decreasing energy dependence on farming activities as well as the nutrients loading must receive special attention if more environmentally friendly systems have to be developed. Novel production systems are evolving towards more efficient water filtering technologies in order to reduce water requirements, allowing decreasing nutrients and solid release into the environment (Blancheton 2000; Ebeling et al. 2006; Aubin et al. 2009). In addition, recirculating systems are being promoted as an interesting opportunity to also reduce water use and control water quality. The environmental consequences derived from these systems as well as their comparison with conventional flow through system have been determined in the literature and environmental improvements have been demonstrated (Summerfelt et al. 2006; Ayer and Tyedmers 2009; Roque d'Orbcast et al. 2009).

Therefore, LCA methodology could become a valuable management and forward planning tool for aquaculture systems since it could contribute to quantify and prioritise strategies of improvement. The combination of this approach with economical and social analyses could allow defining the concept of aquaculture sustainability.

8.2 Turbot Aquaculture Sector

European aquaculture products destined to human consumption represented around 90.4 Mt in 2012, a high and outstanding value when compared with fishery captures (68.5 Mt). Moreover, and according with statistics, world aquaculture is increasing year by year (Acuicultura en España 2014). The rapid expansion of aquaculture sector seems to be related with many sustainability worries such as GHG emissions, introduction of non-indigenous species, dependence on fishery captures and socio-economic features (Henriksson et al. 2012).

Within the marine species produced in Europe, turbot occupies the sixth position in terms of production tonnes. Spanish turbot aquaculture provided in 2013 around 6800 tonnes of turbot (88.3% of European production), which supposed a reduction of 14.5% regarding the production in 2012 (Acuicultura en España 2014). Turbot (*Scophthalmus maximus*) is one of the most important fish species cultivated in Spain under an aquaculture regime (Acuicultura en España 2014), together with red sea bream (*Pagellus bogaraveo*), gilt-head bream (*Sparus aurata*) and sea-bass

(*Dicentrarchus labrax*). Spanish aquaculture production was around 0.25 Mt in 2013, of which 94% corresponded to marine species. The remaining 6% corresponded to continental species such as rainbow trout.

Spanish turbot farms are mainly located in Galicia (NW Spain). Galician turbot aquaculture supplies 99% of total Spanish production (Acuicultura en España 2014) and this species represents around 20% of the Spanish finfish production from marine aquaculture (Iribarren et al. 2012). Regarding market values, the average price for Spanish turbot in 2013 was around 8.42 € kg^{-1}, which turned in an economic turnover of 57 M€ (Acuicultura en España 2014).

Although Spanish turbot sector holds a top position in Europe, only one study can be found in the literature regarding the estimation of the environmental profile derived from the culture practices (Iribarren et al. 2012). Nevertheless, few studies can be found in the bibliography regarding turbot aquaculture in France (Aubin et al. 2006, 2009) despite of being an important finfish product. In addition, turbot fishing fleets have been environmentally analysed (Schau et al. 2009). According to the turbot fishery study, this finfish species is one of the most intensive fossil fuel, requiring more than 2 kg fuel per kg of caught fish (Schau et al. 2009). High rates of energy requirements involve serious consequences from both environmental and economical point of views.

Aquaculture systems are often described as sustainable and environmentally friendly alternatives to conventional fishery practices (Henriksson et al. 2012). However, they frequently require outsized system boundaries including fisheries and agricultural practices (Henriksson et al. 2012) and for that reason, the development of low-intensive farming systems must be considered. According to the reported studies on turbot aquaculture, carbon footprint is a mandatory impact category to be considered for evaluation. Production of feed and energy requirements can be considered as the environmental critical factors but not only on turbot systems but also any type of fish species analysed (e.g. rainbow trout and sea-bass) (Aubin et al. 2009; Iribarren et al. 2012). The feed conversion ratio, the feed ingredients ratios as well as the culture regime (e.g. inland water re-circulating system vs offshore marine fish-cage farming system) considerably affect the environmental profiles and therefore, they should be assessed in detail in order to promote best practices (Aubin et al. 2009; Iribarren et al. 2012). Costs are also important factors to be evaluated, considering that the construction costs of a land-based system are significantly higher than the floating cages the costs of maintaining the fish in good quality water is also higher in land-based systems due to the pumping costs. However, land-based farming systems are easier controlled than offshore farming systems, allowing a better environmental control (Person-Le Ruyet 2001; Seafish.org 2012).

8.3 Case Study: Turbot (*Scophthalmus Maximus*) Culture System

As aforementioned, aquaculture systems require an environmental evaluation since aquaculture sector has become the highest growing animal production sector all over the world as an alternative to conventional fishery captures (Cao et al. 2011; Henriksson et al. 2012). In the process of better understanding the environmental consequences from aquaculture implementation, LCA has been converted into more regularly used tool to identify the best practices and to assess the overall macro-level environmental performance. Therefore, evaluating macro-level environmental impacts requires a full evaluation of activities or processes that comprise the whole production chain. This study employs LCA to quantify the carbon footprint (CF) as well as the edible protein content energy return on investment (ep-EROI) associated with turbot (*Scophthalmus maximus*) production in Galicia under intensive culture regime in a land-based farming system from a cradle-to-farm gate perspective. The main objectives of this study are to (1) identify key stages and hotspots with the highest contribution to CF and, (2) to obtain a dimensionless ratio between energy requirements all over the life cycle and the energy provided by the turbot meat that is, ep-EROI estimation.

8.3.1 *Functional Unit*

The functional unit is the reference basis regarding which the impacts are quantitative described in a LCA (ISO 2006). In this study, the unit chosen was 1 kg of adult turbot ready for consumption at households.

8.3.2 *System Boundaries Description*

The aquaculture system under study was defined considering all the typical activities carried out in the productive process of a turbot farm. Thus, the production chain (*foreground system*) was divided in three main subsystems: Hatching and nursing (SS1), Growing (SS2) and Ongrowing and final operations (SS3). Information regarding the activities performed in the different subsystems was based on information supplied by representative Galician turbot farming plants corresponding to the practices carried out in the reference period 2010–2011. Figure 8.1 depicts the system boundaries of the case study.

Production of different inputs to the foreground system such as fossil fuels, fish feed, chemicals and electricity were also considered within the system boundaries and included in the background system. In addition, the treatment of waste produced in the different activities was also considered as displayed in Fig. 8.1.

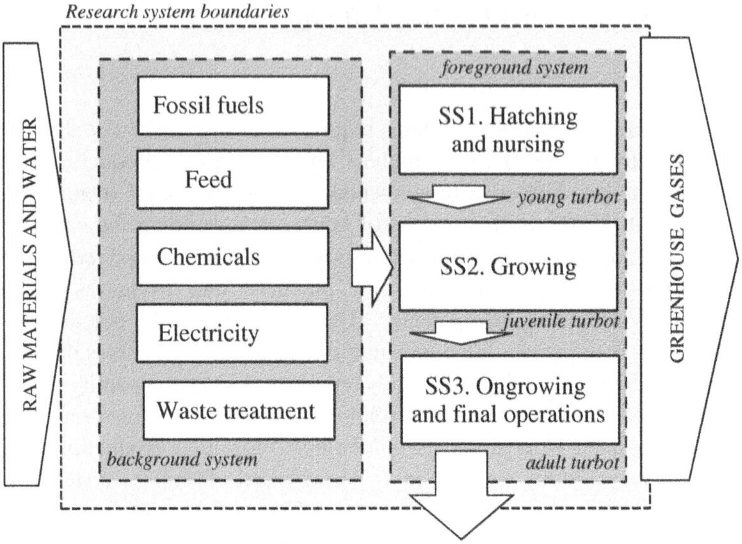

Fig. 8.1 System boundaries for a cradle-to-farm gate LCA of turbot production in Spain

8.3.3 *Life Cycle Inventory Quality*

The quality and representation of inventory data managed is an important issue to be addressed in a LCA study, since they directly affect to the quality of environmental results obtained. As previously reported, information regarding the *foreground system* was directly supplied by Galician turbot farms for the period 2010–2011. Two hatcheries located in the Ria de Vigo (NW Spain), two growing farms located in O Grove and Xove (NW Spain) and one full farm including hatchery and growing unit placed in Valdoviño (NW Spain) were the main sources of inventory data. These data were contrasted with data reported in a previous study (Iribarren et al. 2012) corresponding to the reference years 2007 and 2008. A summary of the most representative inventory data corresponding to the turbot farm is depicted in Table 8.1.

However, secondary data were also managed but corresponding to the *background system*. Thus, the ecoinvent database (Frischknecht et al. 2007) was the main source of inventory data for the following related processes: chemicals production (Althaus et al. 2007), fossil fuels production, electricity production and waste management (Dones et al. 2007). Regarding fish feed production, inventory data was taken from Iribarren et al. (2012), where a detailed environmental assessment of aquafeed production was performed.

Table 8.1 Summary of aggregated LCI for turbot farming in Spain

Inputs from technosphere		
Materials		
Liquid oxygen	3.48	kg
Fish feed	1.55	kg
Energy		
Diesel	1.02	L
Electricity	20.04	kWh
Inputs from environment		
Freshwater	20.40	kg
Seawater	17.16	kg
Outputs to technosphere		
Products		
Adult turbot	1.00	kg
Waste to treatment		
Paper and cardboard	1.72	g
Wood	4.02	g
Plastic	16.75	g
PP filters	0.135	g
Mineral oil	0.134	g
Water-hydrocarbons mixture	0.411	g
Plastics	0.349	g
Oil filters	0.009	g
Metal containers	0.007	g
Lab waste	0.066	g
Batteries	0.022	g
Sanitary waste	0.018	mL
Fluorescent lamps	0.034	g
Medicated feed	0.561	g
Outputs to environment		
Emissions into air		
SO_2	3.95	g
CO	0.77	g
CO_2	5.99	g
NO_x	5.33	g

8.3.4 Carbon Footprint Methodology

The Life Cycle Impact Assessment characterises environmental impacts based on LCI data. In this study, the environmental assessment has been reported in terms of GHG emission that is, the carbon footprint of the turbot farming system. Carbon Footprint can be defined as the potential impact of gaseous emissions on heat-radiation absorption in the atmosphere (Aubin et al. 2009). The environmental results are reported in terms of kg CO_2 eq per functional unit (1 kg adult turbot at farm gate ready for consumption). Characterisation factors reported by IPCC with a time frame of 100 years (IPCC 2007) were managed.

8.3.5 Edible Protein Content Energy Return on Investment Ratio (ep-EROI) Methodology

Energy Return On Investment (EROI) is a term that appeared in the early 1970s and gained relevance in the 70s and 80s due to the fuel crisis (Hall 1972; Gupta and Hall 2011). Although the first applications of EROI estimations were focused on energy sector, it has recently applied to other activities such as food sector (Pelletier and Tyedmers 2011; Vázquez-Rowe et al. 2014). EROI estimates the energy that is returned from an energy-collecting process as compared to the energy that is required to provide this energy (Gupta and Hall 2011).

Tyedmers (2000) and Hall (2011) introduced the application of the EROI concept into the food sector by calculating a dimensionless ratio of the edible protein energy content of an animal relative to the energy used in its production that is defined as the edible protein energy return on investment (ep-EROI) ratio (Vázquez-Rowe et al. 2014). Specifically, the estimation of ep-EROI is relevant in seafood products, where farming activities are in many cases highly energy intensive (Vázquez-Rowe et al. 2014).

In this study, ep-EROI estimation was accomplished by means of the calculation of the ratio between the edible protein energy output of the turbot meat and the energy inputs linked to the turbot farming activities following the formula reported by Vázquez-Rowe et al. (2014):

$$\text{ep-EROI} = \text{Energy inputs}/\text{Energy outputs}$$

Energy inputs were estimated considering the Cumulative Energy Demand (CED) that is, the renewable and non-renewable energy invested throughout the life cycle of the production chain under study displayed in Fig. 8.1.

8.4 Carbon Footprint Analysis

SimaPro 7 was the software chosen for computing the CF and CED results by means of the implementation of inventory data (Goedkoop et al. 2010). The principles established by ISO standards (ISO 2006) and ILCD handbook (European Commission 2010) from an attributional perspective were followed.

8.4.1 Carbon Footprint of Turbot Aquaculture

According to the results obtained for the assessment regarding to the production of 1 kg of adult turbot at farm gate under an intensive aquaculture system performed on land, the carbon footprint corresponds to 19.9 kg CO_2 eq. One of the main aims of this study is not only determining the CF but also identifying the environmental hotspots. Figure 8.2 depicts the distribution of equivalent CO_2 emissions produced throughout the production chain between the three subsystems considered in the *foreground system*. This figure indicates that SS1, which involves all the activities performed at the hatchery and nursing unit where the spawns are incubated, weaned and grown on to a suitable size for transferring to the turbot farm (young turbot), involves the highest contributing ratio to the GHG emission (48% of total CF). This large contributing ratio could be expected since it is the subsystem with the highest electricity requirements (74%) as well as around 35% of total aquafeed requirements are consumed in this stage.

SS3—ongrowing and final operations is the second most important subsystems in terms of CF, being responsible for 40% of total contributions. In this step, the juvenile turbot from the growing farm (SS2) is grown until turbot gets the market weight. Commonly, this activity is performed in onshore tanks as difference to other marine finfish species. This stage involves the highest ratio of aquafeed requirements (57%) and around 10% of electricity consumption.

SS2—growing, is the third responsible of contributions to CF with a contributing ratio of 12% of total GHG emission. In this stage the young turbots are grown in closed recirculation systems in order to control the growing environment until they present an optimum size (juvenile turbot), moment in which they are

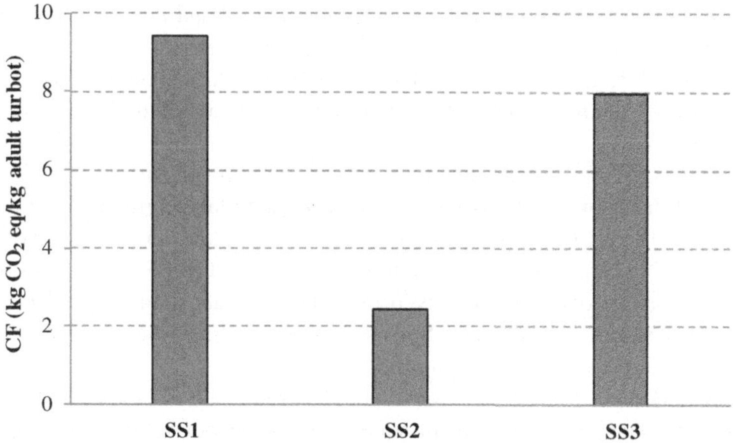

Fig. 8.2 Distribution of CF between subsystems involved throughout the production chain under study

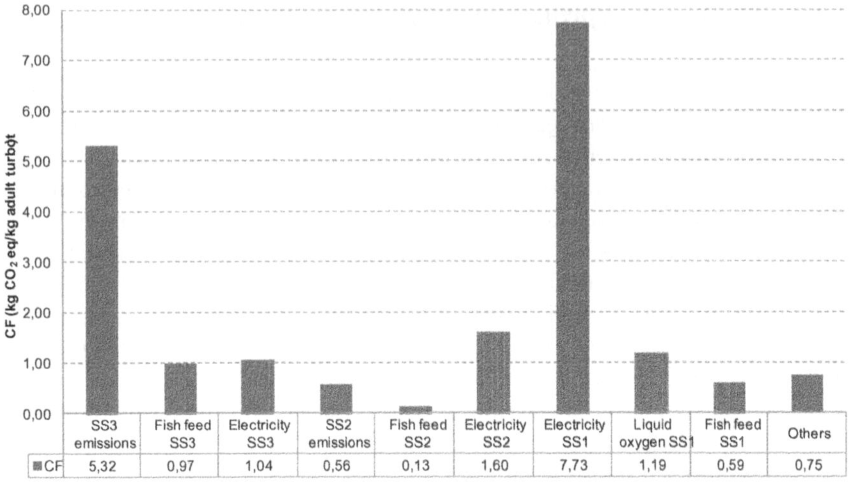

Fig. 8.3 Distribution of CF per processes involved throughout the turbot production chain under study

transferred to outdoor tanks. This stage requires 16% of the total electricity consumption.

Figure 8.3 displays the distribution of CF between the different processes involved in the cradle-to-farm gate life cycle of turbot aquaculture. By means of this figure, the environmental hotspots that is, the processes with the highest responsibility on GHG emission can be identified. Thus, improvement alternatives or proposals focused on the minimisation of CF should consider these processes.

In Fig. 8.3, the contributing factors involves the production of the corresponding inputs (fish feed, electricity in the Spanish national grid and liquid oxygen) as well as on-site emissions produced within the activities carried out in the different subsystems (SS1, SS2 and SS3). The factor reported as "others" includes the remaining contributing processes with a contribution ratio to the global profile lower than 3% (e.g. waste treatment related activities).

According to this figure, the high consumption of electricity in SS1 corresponding to the spawns incubation at the hatchery and larvae grown at the nursery, represents 39% of total GHG emissions, followed by the on-site emissions produced in SS3 (27%). Within the different GHG emitted in SS3, it is important to highlight the emissions of CO_2 derived from diesel combustion. The production of the fish feed required in SS1 and SS3 is also outstanding from an environmental perspective with a total contributing ratio of 8%.

The remarkable contributions from aquafeed production does not results unusual since it is general trend not only in aquaculture systems (Aubin et al. 2009; Cao et al. 2011) but also in other types of animal rearing (Baumgartner et al. 2008; Castanheira et al. 2010; González-García et al. 2015).

8.4.2 ep-EROI of Turbot Aquaculture

As aforementioned, ep-EROI results an interesting indicator to take into account when environmental studies are performed on seafood products since this item allows a deeper understanding of the energy efficiency not only on the aquaculture systems but also on the fishing sector (Vázquez-Rowe et al. 2014). For the estimation of this indicator, the knowledge of two items is mandatory: the CED that is the energy consumed throughout the whole production chain (see Fig. 8.1) as well as the edible protein energy content in the product under study.

Table 8.2 reports the energy consumption (considering renewable and non-renewable sources) in the cradle-to-gate system under study (CED) as well as the edible protein energy, both in terms of MJ per functional unit (323 and 1.21 MJ, respectively). Thus, the ep-EROI corresponding for the Spanish turbot is 0.38%, which means that 0.38% of the invested energy is returned in protein content.

8.4.3 Discussion of CF and ep-EROI Results

Regarding the CF level, activities with an excessive contribution to the GHG emission during the production chain were identified and could be used to develop mitigation strategies to promote more environmental sustainable turbot production. Production of feed and electricity requirements emerged as environmental hotspots, which have been also identified in other aquaculture species such as shrimp (Cao et al. 2011) or even in other related studies of turbot production (Aubin et al. 2009). The CF value obtained in our study (19.9 kg CO_2 eq) is considerably higher than the one obtained by Aubin et al. (2009): 5.62 kg CO_2 eq. Differences on the energy profiles could considerably be responsible for this difference. Research into renewable energy sources such as solar or wind power could help to reduce the impact. Differences on the aquafeed composition could also affect the CF value. The large ratio of fish meal in the composition and the corresponding impacts derived from its production is an important issue that must be considered. In fact, Cao et al. (2011) reported that fish derived ingredients are more impactful than these derived from crops. Therefore, the substitution (if possible) of fish based ingredients by crop based ingredients could be considered as an interesting alternative to bear in mind.

Table 8.2 Summary of energy values and ep-EROI corresponding to the Spanish turbot under aquaculture production

Cumulative energy demand	323 MJ
Edible protein energy	1.21 MJ
ep-EROI	0.38%

The ep-EROI obtained for the Spanish turbot aquaculture is 0.38%. However, and in order to correctly understand the relevance of this value, it is mandatory to have background information regarding the ep-EROI obtained for other species. There are multiple studies focused on the ep-EROI estimation for livestock species such as chicken, swine, beef cattle and lamb (Pimentel and Pimentel 2003; Váquez-Rowe et al. 2014), for fishing species such as mackerel, tuna, shrimp or swordfish (Ramos et al. 2011; Tyedmers 2001; Parker and Tyedmers 2012) and even for aquaculture species such as mussel and shrimp (Tyedmers 2001; Troell et al. 2004; Iribarren et al. 2010b).

Chicken presents a really high ep-EROI of around 25% mainly due to the high production yields (Pimentel and Pimentel 2003). On the contrary, beef cattle derive on a low value, ≈2.5% (Pimentel and Pimentel 2003). Literature data suggest that livestock products, such as lamb (1.8%) or beef (2.5%), and intensive aquaculture products (shrimp, 1.4%), have similar ep-EROI levels to fish species from offshore trawling and long lining fisheries such as tuna, swordfish or shrimp (5.9, 3.4 and 4.1% respectively), which demonstrate the vulnerability of these fleets (Vázquez-Rowe et al. 2014). Nevertheless, extensive aquaculture products (mussels) report ep-EROI levels comparable to pelagic species captured with purse seiners such as mackerel and tuna (68.9 and 14%, respectively). In the case of the Spanish turbot, its ep-EROI value is really low in comparison with other marine species which can provide useful information for the future management of aquaculture practices performed on its culture.

8.5 Concluding Remarks

The increasing interest on seafood involves higher production yields and alternative culture strategies required to satisfy the worldwide fish consumption demands. However, special attention must be provided to fish farming activities specifically from an environmental and energy point of view. Farming activities demand large amount of inputs such as aquafeed and energy, production of which can involve negative environmental consequences. In this sense, LCA is a valuable tool since it allows identifying the critical processes, which should be considered for the proposal of improvement alternatives focused on the minimisation of environmental profile and energy demand.

Acknowledgements This study was carried out in the framework of the project FUNDACIÓN OESA-ACUICULTURA Project reference: 220/01078. Dr. S. González-García would like to express her gratitude to the Spanish Ministry of Economy and Competitivity for financial support (Grant reference RYC-2014-14984). Dr. P. Villanueva-Rey wishes to thank the Galician Government for financial support (postdoctoral student grants programme). The authors belong to the Galician Competitive Research Group GRC2013-032 and to the CRETUS Strategic Partnership (AGRUP2015/02). All these programmes are co-funded by FEDER (UE).

References

Acuicultura en España (2014) http://www.acuicultura-descubrela.es/images/acuicultura_en_esp2014.pdf. Accessed Feb 2016

Althaus HJ, Chudacoff M, Hischier R, Jungbluth N, Osses M, Primas A (2007) Life cycle inventories of chemicals. Ecoinvent report No. 8. Swiss Centre for Life Cycle Inventories, Dübendorf

Aubin J, Papatryphon E, van der Werf HMG, Petit J, Morvan YM (2006) Characterisation of the environmental impact of a turbot (*Scophthalmus maximus*) re-circulating production system using life cycle assessment. Aquaculture 261:1259–1268

Aubin J, Papatryphon E, van der Werf H, Chatzifotis S (2009) Assessment of the environmental impact of carnivorous finfish systems using life cycle assessment. J Clean Prod 17:354–361

Ayer NW, Tyedmers PH (2009) Assessing alternative aquaculture technologies: life cycle assessment of salmonid culture systems in Canada. J Clean Prod 17:362–373

Baumgartner DU, de Baan L, Nemeck T (2008) European grain legumes—Environment-friendly animal feed: life cycle assessment of pork, chicken meat, egg and milk production. Grain legumes integrated project report. Agroscope Reckenholz-Tänikon Research Station ART, Zürich

Blancheton JP (2000) Developments in recirculation systems for Mediterranean fish species. Aquacult Eng 22(1–2):17–31

Cao L, Diana JS, Keolian GA, Lai Q (2011) Life cycle assessment of Chinese shrimp farming systems targeted for export and domestic sales. Environ Sci Technol 45:6531–6538

Castanheira EG, Diaz EC, Arroja L, Amaro R (2010) The environmental performance of milk production on a typical Portuguese dairy farm. Agric Syst 103:498–507

Dones R, Bauer C, Bolliger R, Burger B, FaistEmmenegger M, Frischknecht R, Heck T, Jungbluth J, Röder A, Tuchschmid M (2007) Life cycle inventories of energy systems: results for current systems in Switzerland and other UCTE Countries. Ecoinvent report No. 5. Swiss Centre for Life Cycle Inventories, Dübendorf

Ebeling JM, Welsh CF, Rishel KL (2006) Performance evaluation of an inclined belt filter using coagulation/flocculation aids for the removal of suspended solids and phosphorus from microscreen backwash effluent. Aquacult Eng 35(1):61–77

Ellingsen H, Aanondsen SA (2006) Environmental impacts of wild caught cod and farmed salmon: a comparison with chicken. Int J Life Cycle Assess 11(1):60–65

European Commission (2010) International reference life cycle data system (ILCD) handbook. General guide for life cycle assessment. detailed guidance. 1st edn. March 2010. EUR 24708 EN. Publications Office of the European Union, Luxembourg

FAO—Food and Agriculture Organization (2006) State of world aquaculture: 2006. Fisheries technical paper 500. FAO, Rome

Frischknecht R, Jungbluth N, Althaus HJ, Doka G, Heck T, Hellweg S, Hischier R, Nemecek T, Rebitzer G, Spielmann M, Wernet G (2007) Overview and methodology. Ecoinvent report No. 1. Swiss Centre for Life Cycle Inventories, Dübendorf

Goedkoop M, de Schryver A, Oele M, Durksz S, de Roest D (2010) Introduction to LCA with SimaPro 7. Pré Consultants, The Netherlands

González-García S, Belo S, Dias AC, Várzea Rodrigues J, da Costa RR, Ferreira A, de Andrade LP, Arroja L (2015) Life cycle assessment of pigmeat production: Portuguese case study and proposal of improvement options. J Clean Prod 100:126–139

Gupta AK, Hall CAS (2011) A review of the past and current state of EROI data. Sustainability 3:1796–1809

Hall CAS (1972) Migration and metabolism in a temperate stream ecosystem. Ecology 53:585–604

Hall CAS (2011) Introduction to special issue on new studies in EROI (energy return on investment). Sustainability 3:1773–1777

Henriksson PJG, Guinée JB, Kleijn R, de Snoo GR (2012) Life cycle assessment of aquaculture systems—a review of methodologies. Int J Life Cycle Assess 17:304–313

Hilborn R, Fulton EA, Green BS, Hartmann K, Tracey SR, Watson RA (2015) When is a fishery sustainable. Can J Fish Aquat Sci 72:1433–1441

Iribarren D, Moreira MT, Feijoo G (2010a) Revisiting the life cycle assessment of mussels from a sectorial perspective. J Clean Prod 18:101–111

Iribarren D, Hospido A, Moreira MT, Feijoo G (2010b) Carbon footprint of canned mussels from a business-to-consumer approach. A starting point for mussel processors and policy makers. Environ Sci Pol 13:509–521

Iribarren D, Moreira MT, Feijoo G (2012) Life cycle assessment of aquaculture feed and application to the turbot sector. Int J Environ Res 6:837–848

ISO (2006) ISO 14040:2006. Environmental management—life cycle assessment —principles and framework, 2nd edn. Geneva, Switzerland

Liu Y, Rosten TW, Henriksen K, Skontorp Hognes E, Summerfelt S, Vinci B (2016) Comparative economic performance and carbon footprint of two farming models for producing Atlantic salmon (*Salmo salar*): land-based closed containment system in freshwater and open net pen in seawater. Aquacult Eng 71:1–12

Martins CIM, Eding EH, Verdegem MC, Heinsbroek LT, Schneider O, Blancheton JP, Verreth JAJ (2010) New developments in recirculating aquaculture systems in Europe: a perspective on environmental sustainability. Aquacult Eng 43:83–93

Mungkung R, Udo de Haes H, Clift R (2006) Potentials and limitations of life cycle assessment in setting ecolabeling criteria: a case study of Thai shrimp aquaculture product. Int J Life Cycle Assess 11(1):55–59

Papatryphon E, Petit J, Van der Werf HMG (2004) The development of life cycle assessment for the evaluation of rainbow trout farming in France. In: Proceedings of the 4th international conference on: life cycle assessment in the agri-feed sector, Horsens, Denmark, pp 73–80, 6–8 Oct 2003

Papatryphon E, Petit J, Hayo V, Kaushik SJ, Claver K (2005) Nutrient balance modelling as a tool for environmental management in aquaculture: the case of trout farming in France. J Environ Manage 35:161–174

Parker R (2012) Review of life cycle assessment research on products derived from fisheries and aquaculture: a report for Seafish as part of the collective action to address greenhouse gas emissions in seafood. Final report. Sea Fish Industry Authority

Parker R, Tyedmers P (2012) Fuel consumption and greenhouse gas emissions from global tuna fisheries: a preliminary assessment. ISSF technical report 2012-03. International Seafood Sustainability Foundation, McLean, Virginia, USA

Pelletier NL, Tyedmers PH (2008) Life cycle considerations for improving sustainability assessments in seafood awareness campaigns. Environ Manage 42:918–941

Pelletier NH, Tyedmers P (2010) Life cycle assessment of Frozen Tilapia fillets from Indonesian lake-based and pond-based intensive aquaculture systems. J Ind Ecol 14(3):467–481

Pelletier NH, Tyedmers P (2011) An ecological economic critique of the use of market information in life cycle assessment research. J Ind Ecol 15:342–354

Pelletier N, Tyedmers P, Sonesson U, Scholz A, Ziegler F, Flysjo A, Kruse S, Cancino B, Silverman H (2009) Not all salmon are created equal: life cycle assessment (LCA) of global salmon farming systems. Environ Sci Technol 43(23): 8730–8736

Person-Le Ruyet J (2001) Water quality requirement for seawater fish. In: Arzul G (coord) (ed) Aquaculture, environment and marine phytoplankton. Ifremer, Actes Colloq., Brest, vol 34. pp 71–75, 21–23 May 2001

Pimentel D, Pimentel M (2003) Sustainability of meat-based and plant-based diets and the environment. Am J Clin Nutr 78:660–663

PRé Consultants (2014) http://www.pre.nl

Ramos S, Vázquez-Rowe I, Artetxe I, Moreira MT, Feijoo G, Zufia J (2011) Environmental assessment of the Atlantic mackerel (Scomber scombrus) season in the Basque Country. Increasing the timeline delimitation in fishery LCA studies. Int J Life Cycle Assess 16:599–610

Ribeiro I, Peças P, Henriques E (2013) A life cycle framework to support materials selection for Ecodesign: a case study on biodegradable polymers. Mater Design 51:300–308
Roque d'Orbcastel E, Blancheton JP, Aubin J (2009) Towards environmentally sustainable aquaculture: comparison between two trout farming systems using life cycle assessment. Aquac Eng 40:113–119
Schau EM, Ellingsen H, Endal A, Aanondsen SA (2009) Energy consumption in the Norwegian fisheries. J Cleaner Prod 17(3):325–334
Schau EAM (2012) Environmental life cycle assessments of fish food products with emphasis on the fish catch process. Thesis for the degree of philosophiae doctor, Department of Industrial Economics and Technology Management, NTNU, Norwegian University of Science and Technology, Trondheim, Norway
Seafish.org (2012) http://www.seafish.org/media/Publications/TURBOT_HYPERBOOK_SHOW_print_comp.pdf. Accessed Feb 2016
Seppala J, Silvenius F, Gronroos J, Makinen T, Silvo K, Storhammar E (2001) Rainbow trout production and the environment. Finnish Environmental Institute, Helsinki, pp 164 (in Finnish; abstract, tables and figures in English)
Summerfelt ST, Wilton G, Roberts D, Rimmer T, Fonkalsrud K (2006) Developments in recirculating systems for Arctic char culture in North America. Aquacult Eng 30:31–71
Thrane M (2004) Energy consumption in the Danish fishery. J Ind Ecol 8(1–2):223–239
Troell M, Tyedmers P, Kautsky N, Ronnback P (2004) Aquaculture and energy use. In: Cleveland C (ed) Encyclopedia of energy, vol 1. Elsevier, St. Louis, MO, pp 97–108
Tyedmers P (2000) Salmon and sustainability: the biophysical cost of producing salmon through the commercial salmon fishery and the intensive salmon culture industry. Ph.D. Dissertation, University of British Columbia, Vancouver
Tyedmers P (2001) Energy consumed by North Atlantic fisheries. In: Zeller D, Watson R, Pauly D (eds) Fisheries impacts on North Atlantic ecosystems: catch, effort, and national/regional datasets, vol 9. pp 12–34 (Fisheries Centre Research Reports)
Vázquez-Rowe I, Villanueva-Rey P, Moreira MT, Feijoo G (2014) Edible protein energy return on investment ratio (ep-EROI) for Spanish seafood products. Ambio 43:381–394

Chapter 9
Impact of Pharmaceutically Active Compounds in Marine Environment on Aquaculture

Muhammad B. Asif, Faisal I. Hai, William E. Price and Long D. Nghiem

Abstract Occurrence of pharmaceutically active compounds (PhACs) in marine ecosystems has been confirmed in recent studies. These PhACs include human pharmaceuticals, veterinary medicines and those used in aquaculture. Levels of PhACs in the marine environment are suspected to rise in future due to the increase in anthropogenic activities. This chapter critically discusses the occurrence, sources and fate of PhACs in marine environment. A particular focus has been given to the adverse impacts of PhACs on marine biota and the potential exposure to human. Data related to the distribution of PhACs in seawater, sea sediments and marine biota is presented to elucidate their bioconcentration and bioaccumulation potential. Impacts of PhACs at cellular, molecular and species level are summarised here to understand their mode of action. Lastly, potential biomarkers for effective biomonitoring of marine environment are highlighted to anticipate the risks related to marine ecosystem and human health.

Keywords Adverse impacts · Aquaculture · Bioconcentration and bioaccumulation · Biomarkers · Ecotoxicology · Mode of action · Pharmaceutically active compounds (PhACs)

M. B. Asif · F. I. Hai (✉) · L. D. Nghiem
Strategic Water Infrastructure Laboratory, School of Civil, Mining and Environmental Engineering, University of Wollongong, Wollongong, NSW 2522, Australia
e-mail: faisal@uow.edu.au

M. B. Asif
e-mail: mba409@uowmail.edu.au

L. D. Nghiem
e-mail: longn@uow.edu.au

W. E. Price
Strategic Water Infrastructure Laboratory, School of Chemistry, University of Wollongong, Wollongong, NSW 2522, Australia
e-mail: wprice@uow.edu.au

F. I. Hai et al. (eds.), *Sustainable Aquaculture*, Applied Environmental Science and Engineering for a Sustainable Future,
https://doi.org/10.1007/978-3-319-73257-2_9

9.1 Introduction

Human life expectancy across the globe has increased considerably over the last few decades due to the invention of more than 4000 different pharmaceuticals. Similarly, different medicines are being used to effectively control the spread of epidemics in livestock and aquaculture so that food demand of ever growing global population can be met (Petrovic et al. 2013; Richardson et al. 2005). According to an estimate, up to 3000 new compounds/molecules are being introduced in global market every year to treat human and animal related diseases (Verlicchi et al. 2012). Pharmaceutically active compounds (PhACs) comprising illicit drugs, veterinary medicines and human pharmaceuticals are discharged into the sewerage system upon consumption. Ineffective biodegradation of PhACs in wastewater treatment plants (WWTP) often results in poor aqueous phase removal. Treated wastewater is then discharged into the aquatic ecosystem without any effort to further remove PhACs, making it the main source of contamination for aquatic environment (Daughton and Ruhoy 2009; Lahti et al. 2012; Glassmeyer et al. 2008; Arpin-Pont et al. 2016). Depending upon the availability of PhACs in solid waste, manure and wastewater treatment sludge, PhACs may percolate down to contaminate groundwater (Daughton and Ternes 1999; Díaz-Cruz et al. 2003). Furthermore, some PhACs are directly discharged in surface water bodies such as lakes and rivers during swimming and bathing (Balmer et al. 2005). Various recently published reviews have reported the occurrence of PhACs in groundwater and freshwater bodies throughout the world (Brausch and Rand 2011; Arpin-Pont et al. 2016; Lapworth et al. 2012; Pal et al. 2010; Luo et al. 2014).

Occurrence and impacts of PhACs in freshwater bodies have been studied extensively in the last decade. However, the potential impacts of PhACs released into the coastal and marine ecosystems require more attention and understanding because all the PhACs present in freshwater bodies would ultimately enter marine/coastal ecosystem (Gaw et al. 2014). Coastal and marine ecosystems are one of the most unique, diverse and productive habitat in the world, providing vital services such as protecting against extreme events (such as floods and storms) and removing different type of pollutants (Barbier et al. 2011; Costanza et al. 1998). Moreover, the economic value of marine ecosystems cannot be ignored as over 500 million people are directly or indirectly related to the fishing industry (FAO 2014). Due to poor management of coastal ecosystems, we have already lost approximately 29, 30, 35 and 50% of sea grass, corals reefs, mangroves and marshes, respectively (Barbier et al. 2011). Degradation of marine ecosystems results mostly from anthropogenic sources such as shoreline construction projects/coastal engineering, elevated fishing trends, coastal population increase and inadequately treated sewage discharge (Small and Nicholls 2003; Crain et al. 2009; Li 2003). Acidification, rise in see levels and algal blooms can further deteriorate some marine ecosystems (Martínez et al. 2007).

While evaluating different anthropogenic stressors to coastal/marine ecosystems based on a species level assessment, Crain et al. (2009) reported pollution as one of the most significant and widespread threat to the marine ecosystem. Presence of PhACs in wastewater discharged into the marine ecosystem is of particular significance due to their ability of inducing different biological effects in marine biota (such as fish and shellfish) even at a very low concentration (Franzellitti et al. 2014, 2015). The change in pH and temperature of the marine environment due to climate change can influence the physicochemical state of PhACs and accordingly their impact. Therefore, the mechanism of deterioration of marine ecosystem by PhACs can be different globally (Banni et al. 2015; Nichols et al. 2015). It is expected that the impact of PhACs on coastal environment would increase over the years due to population increase in coastal areas and increasing aquaculture activities (Neumann et al. 2015; Burridge et al. 2010). PhACs have the ability to accumulate in fish, posing serious health concerns for human due to the consumption of seafood/fish. Exposure to PhACs may also induce antibiotic resistance in human and animals (Love et al. 2011; Le et al. 2005).

This chapter aims to summarize the data available on the occurrence, fate and sources of PhACs in coastal/marine environment with their potential impacts on human health and marine biota. Moreover, potential biomarkers responses to the PhACs in marine fauna are also highlighted.

9.2 Sources and Occurrence of Pharmaceutically Active Compounds in Coastal and Marine Environment

9.2.1 Human Pharmaceuticals

Wastewater is the major source of human related PhACs in marine ecosystem. PhACs and their metabolites can find their way into the wastewater through different routes such as hospital discharges, community consumption, and manufacturing waste disposal (Daughton and Ruhoy 2009; Trautwein et al. 2014; Emke et al. 2014). Similarly, sources of illicit drugs in wastewater include waste disposal, excretion after consumption and sometimes the dumped contrabands (Rosi-Marshall et al. 2015; Pal et al. 2013).

Removal of PhACs can vary widely (<10 to >99%) depending on their physicochemical properties, and the type of treatment process (Hai et al. 2014a, 2016). In general, PhAC removal is poor as conventional wastewater treatment plants (WWTP) were designed for efficient removal of pathogens, solids and nutrients (Kümmerer 2009; Evgenidou et al. 2015) but not PhACs. Such ineffectively treated wastewater is discharged directly into the ocean through marine outfalls and/or indirectly through the rivers (Lara-Martín et al. 2014; Benotti and Brownawell 2007). Membrane bioreactor (MBR), an alternative to conventional biological treatment process, has been thoroughly investigated in the last decade for water reuse applications (Hai et al. 2014c). However, water reuse may not be possible due to the presence of PhACs in the final product. Therefore, it is important to understand the fate and removal mechanism of PhACs during biological treatment. Hai et al. (2014a) has reviewed the potential of current and emerging treatment technologies for PhAC removal. In general, removal of PhACs during different treatment processes depends on their physicochemical properties (such as pH, hydrophobicity, functional groups and halogen contents) and the operational parameters (such as solid retention time and temperature) of the biological treatment process (Hai et al. 2014b, 2011). For example, hydrophobic PhACs are generally well removed (>80%) by an MBR (Wijekoon et al. 2013). Notably, unexpected events such as power loss, pH change and maintenance problems can influence the performance of a biological treatment process (Phan et al. 2015). Another approach can be the introduction of new microorganisms such as white rot fungi having the potential of enhancing the removal of non-hydrophobic PhACs. For example, Yang et al. (2013a) achieved >50 and 80–90% removal of bisphenol A and diclofenac in a non-sterile fungal MBR. Performance of white rot fungi for the removal of PhACs has already been reviewed, providing an interesting insight on controlling factors (Yang et al. 2013b; Asif et al. 2017a, b).

Approximately one fifth of the world population lives in coastal areas. Moreover, 21 out of 33 megacities including New York, Mumbai and Guangzhou are situated in coastal areas (Li 2003). According to a rapid response assessment study, up to 90% of the untreated wastewater from coastal areas is directly discharged into marine environment (Corcoran 2010). Hong Kong and Los Angeles are the examples of major coastal cities, discharging their treated wastewater directly into the marine ecosystem through marine outfalls (Maruya et al. 2012; Xu et al. 2011). Substantial amount of PhACs are being discharged into the coastal ecosystem on daily basis. For instance, a study showed Victoria Harbour in Hong Kong receives approximately 14 kg/day of PhACs (Minh et al. 2009). Large rivers also contribute as a source of PhACs in marine ecosystem. For example, 150 tonnes of PhACs are discharged into the marine ecosystem through Yangtze River in China (Qi et al. 2014).

Marine ecosystem may also receive wastewater directly from ships, boats and cruise liners. Cruise liners can carry as many passengers as equivalent to the population of a small town, and can discharge treated wastewater into the marine ecosystem within a distance of 4 nautical miles from coastal waters (Organisation 2003). Moreover, small boats may release wastewater into the marine ecosystem without any treatment. Kookana et al. (2014) noted that many coastal cities in Asia treat wastewater through septic tanks, resulting in the deterioration of groundwater quality. The effluent from septic tanks is then discharged into the coastal water without further treatment. Another source of PhACs in coastal waters is the leachate originating from landfills and seafills. Rodríguez-Navas et al. (2013) reported that 27 μg/L of total PhACs was measured in the leachate of a landfill located at Mallorca Island, Spain. In some regions of the world, pharmaceutical manufacturing waste and solid waste were historically discarded at sea (Lee and Arnold 1983; Son et al. 2011).

9.2.2 Veterinary Medicines

Veterinary medicines are employed for variety of purposes such as protection of animal health, enhancement of animal growth rate and control of parasites in crops (Sarmah et al. 2006). These medicines are essential in livestock industry to eliminate the breakout of an epidemic that may result in the loss of animal lives (Kemper 2008). There can be several ways to administer these PhACs such as implant, injection, drench, and/or adding them in animal feed or drinking water. Use of veterinary medicines for growth enhancement has been banned in Europe. However, growth enhancement additives are still used in some countries (Kools et al. 2008; Du and Liu 2012).

Animals excrete 30–90% of the administered PhACs depending upon the type of the PhAC (Sarmah et al. 2006). These excreted PhACs are either released directly into the environment or used as manure in agricultural fields. PhACs present in animal manure can leach due to rainfall and/or irrigation water, contaminating groundwater and/or surface water (Kools et al. 2008). The use of PhAC for the control of bacterial diseases in crops is legal in some countries such as New Zealand (Vanneste 2013; Kümmerer 2009). In short, contamination of groundwater and surface water during agricultural and horticultural activities ultimately results in the release of these compounds in marine ecosystem. Moreover, surface runoff from agricultural lands in coastal areas can also serve as a direct source of PhACs in coastal waters.

9.2.3 Pharmaceutically Active Compounds for Aquaculture

Aquaculture is now one of the biggest industry of the world as it is the major source of protein for more than 500 million people around the globe (FAO 2014). More than 600 species of fish are matured in freshwater and marine environment, making it the largest source of protein in the world (Larsen and Roney 2013). Although aquaculture is spread all over the world, >90% of all aquaculture is practiced in Asia (FAO 2014). Aquaculture of marine fish is carried out in land ponds or in holding pens situated in the sea (Rico and Van den Brink 2014).

A wide range of PhACs including antibiotics (e.g., tetracycline, flumequin and Oxolinic acid), disinfectants (e.g., hydrogen peroxide and organophosphates) and anthelmintic drugs (e.g., pyrethroids and avermectins) are used in aquaculture activities for disease control (Rawn et al. 2009). PhACs can be added directly to feed or water. Quantity of PhACs used during aquaculture is dependent on the type of fish, fish density and water-exchange rate (Rico and Van den Brink 2014).

Antibiotics in aquaculture production are selected based on their ability to enhance bacteriocidal activity. Antibiotics such as metronidazole and penicillin induce bacteriocidal effects by attacking the cell wall or cell components in bacteria. Bacteriostatic effects are induced by antibiotics such as chloramphenicol and tetracycline which result in bacterial growth inhibition via DNA damage, limited protein production and limited metabolic activity. Antibiotics, due to their selective behaviour, imparts a very little effect on the multicellular or higher organisms (Guardabassi and Courvalin 2006; Todar 2002; Nikaido 2009).

Parasiticides comprising of disinfectants and anthelmintic drugs are used to avoid the epidemic break out of infectious viral and parasitic diseases. Sea-lice are one of the eco-parasites in fish such as salmon. *Lepeophtheirus salmonis*, *Caligus teres* and *Caligus elongatusin* are some of the salmon species attacked by sea-lice. Parasite infestation in aquaculture results in significant loss of fish due to subepidermal haemorrhage. Hydrogen peroxide, pyrethroids, organophosphates and avermectins are most commonly used parasiticides to mitigate the problem of eco-parasites in aquaculture (Burridge et al. 2010; Roth et al. 1993).

According to one estimate, up to 75% of the PhACs administered during aquaculture find their way into aquatic environment through the dispersion of food pellets and excretion of PhACs. These food pellets dispersed in marine environment can be consumed by other marine species, spreading PhACs further in the marine ecosystem (Grigorakis and Rigos 2011). Concentration of PhACs in marine ecosystem due to aquaculture is several folds higher than the concentration of PhACs present in treated wastewater. For instance, relatively high concentration of PhACs (2.5 mg/L) has been observed in Vietnam mangroves (Le and Munekage 2004).

Antibiotics and parasiticides applied in aquaculture are discharged into aquatic and/or marine environment, exposing marine biota to a wide range of PhACs. Due to therapeutic effects and lack of specificity, parasiticides may pose a serious threat to the non-target indigenous species in marine ecosystem. Therefore, aquaculture medicines has been recognised as a major environmental problem (Nash 2003).

9.3 Fate of PhACs in Marine Ecosystem

9.3.1 Seawater and Marine Sediments

Occurrence, distribution and fate of PhACs have been studied in marine and coastal ecosystem around the globe with >70% of these studies published in last five years (Langford and Thomas 2011; Emnet et al. 2015; Gaw et al. 2014). However, there is a need to focus more on South America and Africa due to the lack of data availability. It is interesting to note that no data is available for India, the second largest population of the world (Ahmad et al. 2015). Most of the studies published since 2000 has focused on the marine ecosystems of Europe (20 studies) and Asia (21 studies). The scope of these studies vary significantly i.e. some studies focus on the occurrence of PhACs based on therapeutic classes (Nödler et al. 2014), some focus on specific type of PhAC (Jia et al. 2011) and a few worked on detection methods and their validation (Pintado-Herrera et al. 2013). Around 110 PhACs and

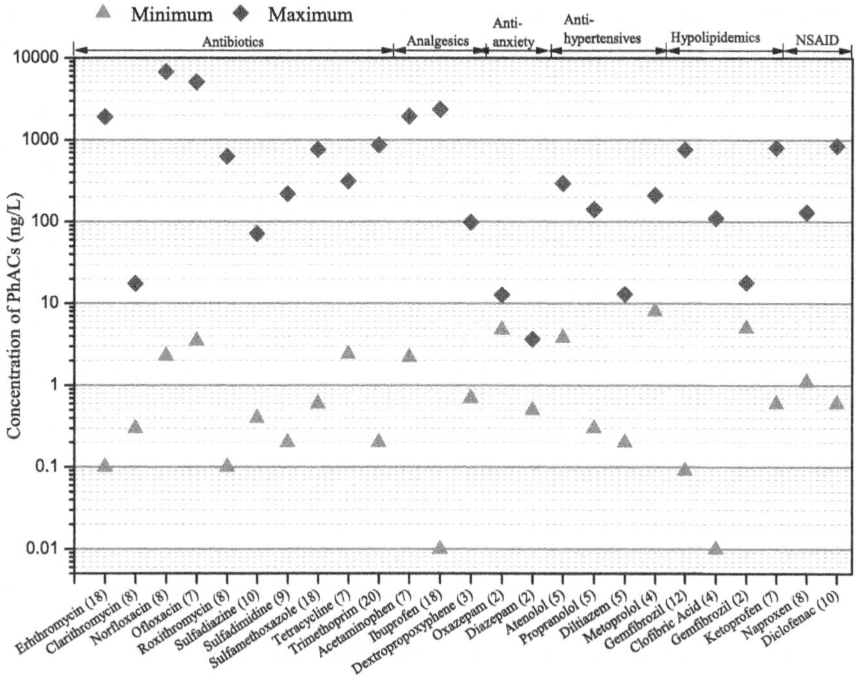

Fig. 9.1 Minimum and maximum concentration of common PhACs detected in Seawater. PhACs (no. of studies) are arranged based on their therapeutic classes. NSAID: nonsteroidal anti-inflammatory drug. Data extracted from the reviews published by Gaw et al. (2014), and Fabbri and Franzellitti (2015)

their metabolites have been identified in seawater with concentration as low as 0.01 ng/L (Ibuprofen and Clofibric Acid) and as high as 6800 ng/L. (Norfloxacin). A wide variations in the detected concentration of PhACs can be attributed to different detection/analytical method as explained in the in the following paragraph. Notably, the concentration of identified PhACs is often higher than the predicted threshold value or "no effect concentration" (0.01 μg/L) of PhACs in surface water (Hughes et al. 2012). Occurrence of PhACs in marine environment has been reviewed recently (Gaw et al. 2014, Fabbri and Franzellitti 2016). Minimum and maximum values of the most frequently detected PhACs classified based on their therapeutic classes are presented in Fig. 9.1. It can be observed from Fig. 9.1 that antibiotics are most frequently detected in seawater.

During the review of the contemporary literature, it is noted that a number of different analytical methods for the detection of PhACs has been employed, making it difficult to compare the results and findings of these studies. The threshold limit of detection, reliability and repeatability of an analytical method depend on different factors namely, (a) type of extraction process; (b) internal standards; and (c) choice of analytical method. Solid phase extraction (SPE) has been the most commonly used extraction method for liquid samples. Moreover, HLB sorbent is employed to effectively retain different polar and non-polar organic compounds (Pichon 2000). However, effective recovery of some highly polar compounds such as ranitidine may require adjustment in pH of the water sample (Gómez et al. 2006). Ratio of using gas chromatography spectrometry (GC-MS) to liquid chromatography-mass spectrometry (LC-MS) for the detection of PhACs in marine environment is approximately 1:1 (Petrović et al. 2005). However, matrix effects during LC-MS and/or GC-MS analysis may interfere with the final output (Caban et al. 2012). In this regard, an internal standard can help to overcome matrix effect in addition to injection and extraction errors (Vanderford and Snyder 2006; Boden and Reiner 2004).

Concentration of PhACs higher than the predicted threshold value as described above nullify the assumption that dilution due to the discharge of freshwater in sea would reduce the concentration of PhACs in marine ecosystem. For instance, the concentration of gemfibrozil and ketoprofen in the coastal waters of Costa Rica was ranged from 77–758 and 185–805 ng/L, respectively (Spongberg et al. 2011). The impact of hydrodynamics of water on the dispersion route of PhACs in seawater has been demonstrated recently. For instance, Bayen et al. (2013) studied the dispersion of PhACs using Delft 3D hydrodynamic model in marine system of Singapore and observed that the dispersion of PhACs was greatly influenced by coastal mixing patterns. While investigating the predicted concentration of oxcarbazepine and carbamazepine in submarine outfall using the MARS 3D model, Fenet et al. (2014) observed the influence of stratification on the dispersion of oxcarbazepine and carbamazepine along the depth of the marine system. While studying the dispersion of PhACs in coastal waters, hydrodynamic model of water masses should be taken into account.

The fate and the partition of PhACs are influenced by the pH of seawater and acid dissociation constant (pK_a). For instance, it has been observed that the octanol-water partition coefficient (K_{ow}) and solid water portioning coefficient (K_d) of fluoxetine and propranolol increase linearly with pH (Owen et al. 2009; Brooks et al. 2003). It has been predicted that the lipophilicity of ionisable PhACs would enhance at the typical pH (8.0) of the marine system, resulting in sorption and bioaccumulation of PhACs onto solids and marine organisms, respectively. For example, trimethoprim ($pK_a = 6.6$) cannot dissociate completely in marine environment (McEneff et al. 2014). Log D values can also be used to predict the bioaccumulation of both neutral and ionisable PhACs at given pH. The compounds with high log D value would more likely to adsorb and/or accumulate in marine organisms and/or suspended solids (Fu et al. 2009). For example, PhACs such as estrogens, paroxetine, vastatin and bisphenol A are adsorbed to suspended solids due to their high log D values (Bayen et al. 2013).

As evident from above discussion, sorption and bioaccumulation of many PhAC may increase in marine environment (alkaline pH and saline conditions). Therefore, PhACs with high log D value may sink on marine sediments (Gilroy et al. 2012). Although the partitioning of PhACs onto the marine sediments has not been extensively examined, occurrence of over 50 PhACs in marine sediments has been reported in literature. PhACs commonly detected in marine sediments are presented in Table 9.1. Among all PhACs, ethinyl estradiol (up to 130 ng/g) is detected most frequently in marine sediments worldwide.

Table 9.1 Occurrence of different PhACs classified according to their therapeutic class in marine sediments and marine biota

Therapeutic class[a]	PhACs	Marine sediments (ng/g)	Marine biota (ng/g)	References
Estrogens	Diethylstilbestrol	11–63	2.9–11.4	Zhang et al. (2011), Pojana et al. (2007)
	Ethinyl estradiol	0.15–130	3–38	Pojana et al. (2007), Wang et al. (2012), Robinson et al. (2009), Bertin et al. (2011), Emnet et al. (2015)
Antibiotics	Flumequine	0.2–0.6	–	Lalumera et al. (2004)
	Oxytetracycline	0.2–0.8	9.5	Lalumera et al. (2004), Na et al. (2013)
	Metronidazole	36–54	–	Pintado-Herrera et al. (2013)
	Trimethoprim	0.1–734,000	0.6	Klosterhaus et al. (2013), Le and Munekage (2004)
	Tetracycline	0.6–7.1	1.9–9.5	Na et al. (2013), Zheng et al. (2011), Li et al. (2012)
	Sulfamethazine	3.67	3.9	Na et al. (2013)
	Sulfamerazine	1.76–3.24	16.2	Na et al. (2013)
	Sulfameter	56.65	43	Na et al. (2013)
	Sulfamethiazole	1.89	2.1	Na et al. (2013)
	Sulfamethoxazole	0.4–820,000	2.3–20	Li et al. (2012), He et al. (2012), Klosterhaus et al. (2013), Zheng et al. (2011)
NSAID[b]	Diclofenac	<0.1–10	1.3–5.3	Pintado-Herrera et al. (2013), Maranho et al. (2015), Alvarez-Muñoz et al. (2015)
	Mefenamic acid	6–23	–	Pintado-Herrera et al. (2013)
	Naproxen	0.6–15.8	–	Pintado-Herrera et al. (2013)
Analgesics	Fenoprofen	>0.1–26	–	Pintado-Herrera et al. (2013), Maranho et al. (2015)
	Ibuprofen	98–100	–	Pintado-Herrera et al. (2013), Long et al. (2013)
	Acetaminophen	<0.1–25.5	–	Pintado-Herrera et al. (2013), Maranho et al. (2015), Stewart et al. (2014)

(continued)

Table 9.1 (continued)

Therapeutic class[a]	PhACs	Marine sediments (ng/g)	Marine biota (ng/g)	References
Anti-hypertensive	Atenolol	<0.1–0.3	0.3–13	Maranho et al. (2015)
	Propranolol	0.1–0.9	–	Maranho et al. (2015)
	A-Hydroxy metoprolol	1–3	–	Langford and Thomas (2011)
Antidepressant	Amitriptyline	<0.1–0.4	–	Maranho et al. (2015)
	Fluoxetine	<0.1–0.7	–	Maranho et al. (2015)
Hypolipidemic drug	Clofibric acid	<0.1–0.1		Maranho et al. (2015)
	Gemfibrozil	<0.1–0.9	–	Maranho et al. (2015)
Antilipemic	Fenofibrate	0.18–0.2		Maranho et al. (2015)
Nervous stimulant	Caffeine	1.9–12.2	–	Maranho et al. (2015)
Lipid-lowering agent	Simvastatin	2–4	–	Langford and Thomas (2011)
Illicit drug	Amphetamine	3.3	4.2–20	Klosterhaus et al. (2013), Long et al. (2013), Dodder et al. (2014)
Diuretics	Triamterene	0.3–10.8	–	Klosterhaus et al. (2013), Long et al. (2013)
Anticonvulsant	Carbamazepine	<0.1–88.8	–	Pintado-Herrera et al. (2013), Maranho et al. (2015), Stewart et al. (2014)

[a]Therapeutic classes. *Source* www. http://www.drugbank.ca/
[b]NSAID: Nonsteroidal anti-inflammatory drug

9.3.2 *Marine Biota*

Unionized portion of PhACs can bioaccumulate in marine fish, shellfish and mollusk species due to their high affinity towards lipophilic matter. Some marine species such as krills, marine benthics, baleen whales and fish (including some sharks) are termed as filter feeders because of their unique filtering structure, allowing them to strain food and other particulate matter from water. However, the straining mechanism in filter feeders cannot prevent the bioaccumulation of dissolved PhACs (Gomez et al. 2012; Bueno et al. 2013). Despite the unavailability of effective detection method for bioaccumulated PhACs in marine biota (Gaw et al. 2014), a number of studies have reported the occurrence of PhACs in marine biota as evident from Table 9.1. Antibiotics such as sulfamethoxazole and tetracycline were the most detected therapeutic class of PhACs both in seawater and marine biota as reported by Li et al. (2012) after analysing the samples from 9 coastal cities along the Chinese Bohai Sea. Occurrence and bioaccumulation of PhACs in edible marine species such as Fish, crabs and shrimps pose a serious threat to the human health particularly in China because 80% of world aquaculture occurs in China. Moreover, samples collected from South China also confirmed the occurrence of 32 different PhACs in edible marine species (Chen et al. 2015).

Although it has been established that PhACs can bioaccumulate in fish biota posing a serious threat to human health, bioaccumulation can vary in different species. For example, Chen et al. (2015) noted that PhACs bioaccumulate more in fish than mollusks. Several studies investigated the bioaccumulation factor (BAF) and the bioconcentration factor (BCF) to further evaluates the fate and impact of PhACs on marine biota. BAF is the ratio of the concentration of a compound in biota (flora or fauna) to the concentration of the same compound in the surrounding media (water or soil). Whereas bioconcentration is referred to the accumulation of dissolved compounds in aquatic fauna (Shenker et al. 2011). Na et al. (2013) investigated the bioaccumulation factor (BAF) for antibiotics and classified them into: (a) bioaccumulative (BAF > 5000 L/Kg) such as sulfamethazine, sulfamethiazole, sulfamonomethoxine; and (b) potentially bioaccumulative (2000 L/Kg > BAF < 5000 kg) such as doxycycline and nomethoxine. Klosterhaus et al. (2013) investigated the occurrence of 104 PhACs and personal care products in wild ribbed horse mussels. Quite interestingly, PhACs such as atenolol, sulfamethoxazole and gemfibrozil frequently detected in both seawater and marine sediments were not found in mussels, whereas the concentration of carbamazepine, nonylphenol and digoxigenin was high in the tissues of mussels. Therefore, BAF may also depend on the physicochemical properties (pH, stability and ionization) of PhACs. BAF of frequently detected PhACs along with their sediment-water distribution coefficient is presented in Fig. 9.2. Minimum and maximum BCF of commonly detected compounds and their occurrence in relevant compartment of biota in are given in Table 9.2.

Bioaccumulation of PhACs in marine biota depends on the hydrophobicity of compounds, meaning that hydrophobic PhACs (log K_{ow} > 3) are more likely to bioaccumulate in marine biota than hydrophilic PhACs (OECD 2008). Howard and Muir (2011) investigated the bioaccumulative potential of 275 PhACs and 92 of them were rated as potentially bioaccumulative. Therefore, concentration of highly hydrophobic PhACs such as ionophore antibiotics can be higher in marine sediments than water. Similarly, high concentration of diclofenac, ibuprofen and gemfibrozil are commonly found in sewage sludge as sorption is the main removal pathway for hydrophobic compounds in biological wastewater treatment processes (Kim and Carlson 2006; Yu and Wu 2012). Fate of PhACs in sewage sludge/biosolids and the impacts of operational conditions in conventional and sludge processing technologies on the removal of PhACs has been reviewed comprehensively (Semblante et al. 2015). In addition to pH and physicochemical properties of compounds, contact time with the sorption medium also govern the bioaccumulation rate of PhACs (Ingram et al. 2011).

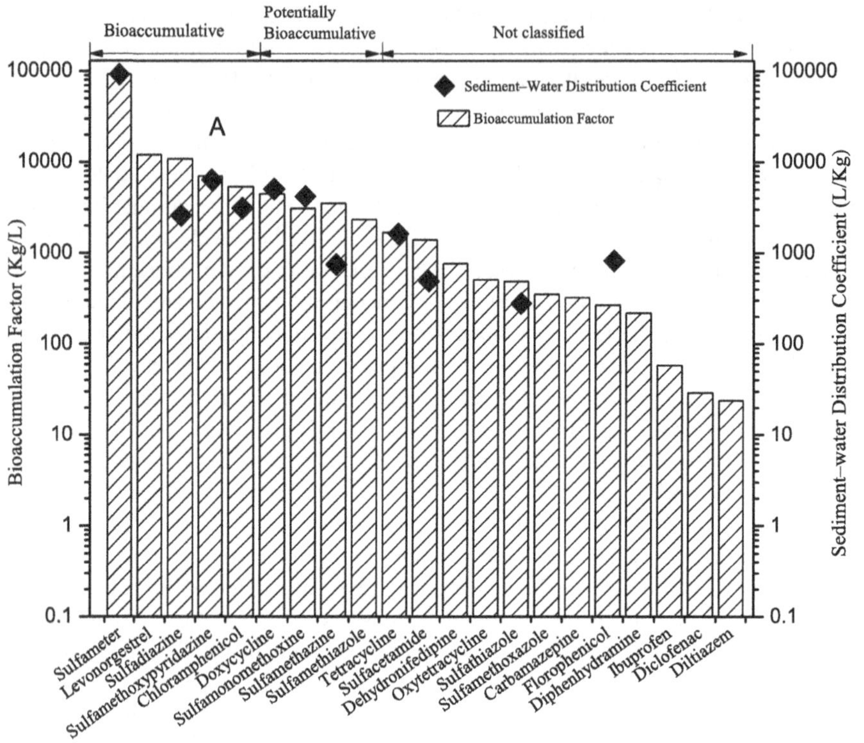

Fig. 9.2 Maximum value of bioaccumulation factor (BCF) for PhACs arranged in the order of their bioaccumulative potential in addition to their sediment-water distribution coefficients. *Sources* Na et al. (2013), Fick et al. (2010b), Garcia et al. (2012), and Zenker et al. (2014)

9.4 Impacts of Pharmaceutically Active Compounds (PhACs)

9.4.1 Marine Biota

Marine biota may get exposed to pharmaceuticals through food, grills or via contact with marine sediments. Extent of their exposure to PhACs broadly depends on four factors: (a) dilution rate; (b) physicochemical properties of PhACs (hydrophobicity, stability and ionization); (c) distance of coastal waters and/or seawater from potential source (aquaculture, agricultural land or sewage treatment plant); and (d) contact time (Nödler et al. 2014; Zheng et al. 2011; Comeau et al. 2008). Impact on marine biota resulting from the exposure of PhACs can be severe depending on the concentration of PhACs and favourable environmental condition for PhACs to persist (Emnet et al. 2015).

Table 9.2 Minimum–Maximum bioconcentration factor (BCF) for PhACs along with the compartment(s) of their occurrence in marine biota

PhACs	Compartment(s) of marine biota	BCF
Bezafibrate	Blood plasma	6.3–17
Carbamazepine	Blood plasma, body, muscle, liver	0.8–4.6
Cilazapril	Blood plasma	>1000
Diclofenac	Blood plasma, bile, liver, kidney, gill, muscle	0.3–2732
Diltiazem	Blood plasma	24–139
Fexofenadine	Blood plasma	5–13
Haloperidol	Blood plasma	3.2
Ibuprofen	Blood plasma, body	0.08–5.8
Ketoprofen	Blood plasma	3.5–48
Levonorgestrel	Blood plasma	>10,000
Meclizine	Blood plasma	200–1400
Memantine	Blood plasma	38–164
Mianserin	Blood plasma	<50
Naproxen	Blood plasma	22–26
Orphenadrine	Blood plasma	64–100
Oxazepam	Blood plasma	0.7–3.6
Fluoxetine	*Gammarus* sp., body	8.8–900
Norfluoxetine	Body	80–650
Sulfamethazine	Muscle	0.61–1.19

BCF were determined by exposing the test specie(s) to different PhAC(s) in a controlled laboratory environment

Sources Fick et al. (2010a), Lahti et al. (2011), Mehinto et al. (2010), Wang and Gardinali (2013), Garcia et al. (2012), and Zenker et al. (2014)

9.4.1.1 Human Pharmaceuticals

A peculiar characteristic of PhACs is to remain biologically active even at very low dose/concentration to achieve their therapeutic effect by interacting with target mediators. However, high concentration of PhACs may sometime interact with off-target mediators. Furthermore, exposure to PhACs would present these compounds with an opportunity to target specific mediators in marine biota. Thus, undesired and accidental exposure to PhACs can raise the potential of their eco-toxicological effects (Gunnarsson et al. 2008; Schmitt et al. 2010) Environmental implications of PhACs were addressed by developing conceptual models. These conceptual models assume that physiology of test species, therapeutic effect of PhACs and their toxicity effects along with all relevant information must be integrated before reaching the conclusion. Understanding of the functional and evolutionary conservation of PhACs in species and impacts on physiological pathways are of particular importance for the selection of experimental duration and exposure limits in addition to the prediction of mode of action (MOA) (Schmitt et al. 2010; Christen et al. 2010). The evolutionary conservation and possible interaction of human PhACs with marine biota has been confirmed recently (Franzellitti et al. 2011, 2013). Various MOA based studies have also verified the biological read-across hypothesis (BRAH) i.e. conservation of therapeutic targets would not always result in the conservation of their functions across different marine species (Rand-Weaver et al. 2013). Moreover, it was observed that PhACs will only affect non-targeted species with conserved molecular mediator(s) (enzymes, metabolic components or receptors) and with plasma concentration equivalent to human. This hypothesis is most suited for fish species and least applicable for invertebrates. Significance of this hypothesis would enhance exponentially if it can be applied to all therapeutic classes of PhACs because this would help to understand, predict and mitigate potential environmental impact of a compound at manufacturing stage (Rand-Weaver et al. 2013; Ford and Fong 2016).

Relationship between ecological and molecular endpoints can also be developed using MOA approach. For example, MOA based conceptual model was applied to understand the evolutionary effects of fluoxetine on aquatic organisms. The MOA of fluoxetine initiated with the targeting of a physiological controller, serotonin (5-HT), which was predicted to have severe impacts on the reproduction, locomotion, metabolism of marine organisms (Fabbri and Franzellitti 2016). Moreover, such impacts on marine species can occur even at very low concentrations i.e. concentration below environmental limits (Ford and Fong 2016).

The MOA of ibuprofen in *Ruditapes philippinarum* was studied by employing ligo-DNA microarray with >11,000 transcripts to ensure in depth analysis (Milan et al. 2013). Differentially transcribed genes analysis of ibuprofen confirmed its negative impacts on signalling mechanisms in clams. Ibuprofen was also lethal for arachidonic acid signalling mechanism due to altered expressions of several transcripts. Moreover, ibuprofen induced cyclo-oxygenase inhibition was due to the over-transcription of phospholipase A2 enzyme, a catalyst for cellular phospholipids hydrolysis (Knight et al. 1999).

Despite the lack of understanding about the pharmacological mechanisms of carbamazepine, its MOA has been clearly described. Carbamazepine inhibits channel currents of Na^+ and Ca^{2+}. However, the extent of inhibition is influenced by the voltage of current. The inhibition can be divided in two components: (a) inhibition via interaction with adenylyl cyclase system; and (b) lowering the generation of cyclic adenosine monophosphate in brain, reducing the activity of protein kinase (PKA). Carbamazepine induced inhibition affected all tissues of mussels due to reduced levels of cyclic adenosine monophosphate (cAMP) and limited activity of PKA (Montezinho et al. 2007; Martin-Diaz et al. 2009). Presence of β-adrenergic receptor (AR) mediated pathway in filter feeders particularly bivalves has been confirmed based on the analysis of AR encoded transcripts. Moreover, it was observed that the occupancy of AR helped in enhancing the concentration of cAMP in the tissues of different bivalve species. Hence, AR mediated transduction pathway can be conserved significantly in bivalves (Fabbri and Capuzzo 2010; Koutsogiannaki et al. 2006; Shpakov et al. 2005). However, the exposure of mussels to propranolol resulted in reduced PKA activity and lowered cAMP levels due to AR blockage, induced by propranolol, in digestive glands

(Franzellitti et al. 2011). In short, marine biota cannot cope with the change in environmental conditions because of the exposure to PhACs having the ability to conserve their therapeutic effects, resulting in the reduced levels and activity of cAMP and PKA, respectively. MOA for PhACs in non-targeted spices is mostly unknown due to their widespread occurrence around the globe. Therefore, it is essential to further assess the unknown effects of PhACs on non-targeted species.

In addition to impacts discussed above, other reported side effects of PhACs on the membrane stability, immune system and DNA are mostly based on lab scale studies. These studies are useful for selecting the biomarkers and bioassays to evaluate the impacts of PhACs in marine and coastal ecosystem. Moreover, these studies, mainly focusing invertebrates, provide additional evidence on the lethal impact of PhACs on marine biota. Biological effects of PhACs on marine biota and exposure conditions are briefly summarised in Table 9.3.

Carbamazepine is usually prescribed for the treatment of neural hyperexcitability and epilepsy. Its high dose and prolonged consumption may cause hepatotoxicity (Santos et al. 2008). Carbamazepine has been reported to interfere with the stability of marine biota. For example, exposure to carbamazepine, 0.1 and 10 μg/L for 7 days, reduced the stability of lysosome membrane stability (LMS) in marine mussels by 60 and 80%, respectively (Martin-Diaz et al. 2009). Similarly, LMS of mussel declined linearly with the increase in the concentration of propranolol (0.3–30 ng/L) after 7 days (Franzellitti et al. 2011). Furthermore, 40% reduction in LMS of mussels was observed at very low fluoxetine concentration (0.03 ng/L) after 7 days (Franzellitti et al. 2014). LMS reduction due to PhACs exposure is not limited to mussels. Ibuprofen exposure, in the range of 0.1–50 mg/L, to the clam (*R. Philippinarum*) significantly deteriorated LMS (Aguirre-Martínez et al. 2013b). LMS reduction was also observed in other marine species such as crab (*Carcinus maenas*) and colonian ascidian (*Botryllus schlosseri*) (Aguirre-Martinez et al. 2013a, b, c; Matozzo et al. 2014). As a sensitive biomarker, LMS is a good indicator of cellular health, growth and survival in mussel, providing early warning regarding the side effects of PhACs. Aguirre-Martínez et al. (2013b) observed an increase in the lysosomal to cytoplasm ratio in addition to LMS reduction, confirming the bioaccumulation of fluoxetine mussel in tissues.

The prolonged exposure to PhACs or high dose may generate reactive oxygen species (ROS) such as glutathione reductase (GR), catalase (CAT), and/or superoxide dismutase (SOD). For example, 75 ng/L of ibuprofen increased the activities of ROS in Mediterranean mussel after 3–7 days. Likewise, exposure to ibuprofen also resulted in the increase of DR and CAT after 3 days. Interestingly, ROS returned to their base level after certain amount of time, exhibiting bell shape behaviour. Cytoprotective response may be the reason for the return of ROS to their base level (Gonzalez-Rey and Bebianno 2012, 2014). Poor responses to antioxidant enzymes were observed in *Hediste diversicolor* after its exposure to fluoxetine, ibuprofen, propranolol and ethinylestradiol (Maranho et al. 2014). Depending on the physiological conditions, species can develop antioxidant responses. Hence, the capability of PhACs to produce oxidative stress in marine biota can be monitored based on their antioxidant activity.

Table 9.3 Impacts of commonly detected PhACs on different marine biota under different exposure conditions

Biomarker	Marine biota Scientific name	PhACs	Experimental conditions			Impacts	References
			Exposure levels	Exposure time	Exposure type		
Lysosomal membrane stability	Manila clam *R. philippinarum*	Ibuprofen Novobiocin Caffeine Carbamazepine	1–50 µg/L	35 days	In vivo	Reduced haemocytes	Aguirre-Martínez et al. (2013b)
	Mussel *M. galloprovincialis*	Carbamazepine Propranolol Fluoxetine	0.1–10 ng/L 0.3–1000 ng/L 0.03–30,000 ng/L	7 days	In vivo	Reduced haemocytes	Martin-Diaz et al. (2009), Franzellitti et al. (2011, 2014)
	Mussel *P. perna*	Triclosan Oxytetracycline	1 µg/L, 100 µg/L, 1.2–12,000 ng/L	4 and 3 days	In vivo	Reduced haemocytes at 1.2–12,000 ng/L Reduced digestive gland	Banni et al. (2015), Cortez et al. (2012)
	Colonian Ascidian *B. schlosseri*	Ibuprofen Novobiocin Caffeine Carbamazepine	100 and 1000 µg/L	60 min	In vivo	Reduced haemocytes	Matozzo et al. (2014)
	Crab *C. Maenas*	Ibuprofen Novobiocin Caffeine Carbamazepine	0.1–50 µg/L	28 days	In vitro	Reduced haemocytes at >15	Aguirre-Martinez et al. (2013a, b, c)
Peroxidation of lipid	Mussel *M. galloprovincialis*	Acetaminophen	25 and 403 µg/L	10 days	In vivo	Enhancement in digestive gland	Solé et al. (2010)
	Clam *R. decussata*	Carbamazepine	0.03–9 µg/L	96 h	Acute test	Reduced lipid levels	Almeida et al. (2014)
	Clam *R. philippinarum*	Carbamazepine	0.03–9 µg/L	96 h	Acute test	Increased lipid levels	Almeida et al. (2014)

(continued)

Table 9.3 (continued)

Biomarker	Marine biota Scientific name	PhACs	Experimental conditions			Impacts	References
			Exposure levels	Exposure time	Exposure type		
Phase I metabolism	Mussel *M. galloprovincialis*	Acetaminophen Propranolol	23 and 403 μg/L	10	In vivo	Activity of carboxylesterase increased in digestive glands	Solé et al. (2010)
	Crab *C. Maenas*	Carbamazepine	0.1–50 μg/L	28 days	In vitro	Increase in ethoxyresorufin O-deethylase and dibenzyl fluorescein dealkylase activities observed in muscles, grills and hepatopancreas	Aguirre-Martínez et al. (2013a)
DNA damage	Crab *C. Maenas*	Carbamazepine Novobiocin	0.1–50 μg/L	28 days	In vitro	Increase in muscle, hyhepatopancreas and grills	Aguirre-Martínez et al. (2013a)
	Colonian ascidian *B. schlosseri*	Ibuprofen	100 and 1000 μg/L	60 min	In vivo	Increased hemocytes	Matozzo et al. (2014)
	Marine polychaete *H. diversicolor*	Ibuprofen Propranolol	0.05–500 ng/g	17 days	In vivo sediment spiking	Increased hyhepatopancreas (ibuprofen) and decreased hyhepatopancreas (propranolol)	Maranho et al. (2014)
Antioxidant resistance	Mussel *M. galloprovincialis*	Carbamazepine	0.1 and 10 μg/L	10 days	In vivo	Enhanced glutathione S-Transferase activity in gonads and glands	Martin-Diaz et al. (2009)
		Acetaminophen	25 and 403 μg/L	10 days	In vivo	Enhanced glutathione S-Transferase activity in grills and glands	Solé et al. (2010)
		Fluoxetine	0.03–300 ng/L	7 days	In vivo	Catalyse activity showed bell shape behaviour and Enhanced glutathione S-Transferase activity in grills and glands	Franzellitti et al. (2014)
	Crab *C. Maenas*	Carbamazepine	0.1–50 μg/L	28 days	In vivo	Enhanced activities of glutathione S-Transferase and glutathione peroxide in grills, hepatopancreas and muscle	Aguirre-Martínez et al. (2013a)
		Fluoxetine	0.5–750 ng/L	7 days	In vivo	Enhanced glutathione S-Transferase activity in hepatopancreas	Mesquita et al. (2011)
	Clam *R. philippinarum*	Carbamazepine	0.03–9 μg/L	96 h	Acute test	Enhanced glutathione levels, reduced reduced/oxidized glutathione ratio and bell shape trend for superoxide dismutase activity	Almeida et al. (2014)
		Acetaminophen	0.05–5 mg/L	96 h	Acute test	Increased glutathione S-Transferase activity at mg/L and reduced glutathione S-Transferase activity at 0.05 mg/L	Antunes et al. (2013)

(continued)

Table 9.3 (continued)

Biomarker	Marine biota Scientific name	PhACs	Experimental conditions			Impacts	References
			Exposure levels	Exposure time	Exposure type		
Acetylcholinesterase activity	Mussel *M. galloprovincialis*	Acetaminophen	23 and 403 μg/L	10 days	In vivo	Enhancement in grills	Solé et al. (2010)
		Diclofenac	250 ng/L	0–15	time course in vivo	Enhancement in grills after 7 days	Gonzalez-Rey and Bebianno (2014)
		Fluoxetine	0.03–300 ng/L	7 days	In vivo	Grills reduction	Franzellitti et al. (2014)
	Marine polychaete *H. diversicolor*	Ibuprofen Carbamazepine	0.05–500 ng/g	17 days	In vivo sediment spiking	Increased acetylcholinesterase activity	Maranho et al. (2014)
		17-α-ethynylestradiol Fluoxetine	0.01–100 ng/g	17 days	In vivo sediment spiking	Acetylcholinesterase activity increased in case of Fluoxetine and followed bell-shaped trend for 17-α-ethynylestradiol	Maranho et al. (2014)

PhACs also interfere with the functionality of neuromuscular system by increasing the activity of acetylcholinesterase (AChE). For example, activity of AChE was enhanced in *V. philippinarum* after 3 days exposure to ibuprofen (1 mg/L) (Santos et al. 2010). AChE controls neuromuscular system in order to regulate the movement of muscles in all animals, making it another biomarker for the exposure of marine biota to the neurotoxic compounds (Viarengo et al. 2007). PhACs can affect the immune system of marine biota. For instance, 60 min exposure of ibuprofen at concentration 100–1000 mg/L deteriorated the immune system in *B. schlosseri* (Montezinho et al. 2007).

9.4.1.2 Aquaculture Medicines

As described earlier (Sect. 9.2.3), antibiotics and parasiticides applied for disease control in aquaculture may pose significant threat to aquaculture species and marine biota. Due to the lack of awareness, aquaculture medicines are sometimes routinely applied even without the presence/threat of a disease (Cabello 2006). For instance, shrimp farmers in Thailand were interviewed regarding the use of antibiotics and frequency of application. The interviews revealed that 52 out of 76 farmers used >10 types of antibiotics, mostly on prophylactic basis to prevent the break-out of a disease. Some farmers affirmed the use of antibiotics on daily basis (Holmström et al. 2003).

Data on the use of antibiotics in aquaculture is scarce. Available data shows variation in the amount of medicine used from one country to another. For example, Smith (2008) reported that the use of antibiotics ranges from 1 g/tonne (in Norway) to 700 g/tonne (in Vietnam). Aquaculture antibiotics may impart different side effects on host and indigenous biota, resulting in antimicrobial resistance in aquaculture pathogens.

Aquaculture pathogens develop resistance against antibiotics due to their frequent and unwarranted use. Exposure to sub-lethal dose of antibiotics may produce microbial resistant genes in pathogens. These pathogens also acquire some antibiotic inhibiting enzymes that inactivate different antibiotics through acetylation and phosphorylation. For instance, beta lactamases inactivates penicillin by destroying beta lactam (Romero et al. 2012; Defoirdt et al. 2011). While examining the bacterial strains isolated from patients in UK, Pakistan and India, rise of antibiotic resistance in bacteria was attributed to beta lactamase enzyme (Kumarasamy et al. 2010). Moreover, some pathogens develop resistance against multiple antibiotics due to: (a) the accumulation of multiple resistant genes in their cells; and (b) the increase in gene expressions having the codes for multidrug efflux (Romero et al. 2012). Therefore, the ability of antibiotics to induce their therapeutic effects would compromise in the presence of antibiotic resistant pathogens. Common aquaculture antibiotics with their significance to human pharmaceuticals and resistant pathogens isolated from aquaculture farms are presented in Table 9.4.

Microbes present in the intestinal tract of healthy fish species may be important for nutrition, digestion and disease control. For example, Bates et al. (2006)

Table 9.4 Use of different antibiotic classes in aquaculture production and their significance for human pharmaceuticals along with the examples of isolated antibiotic resistant

Antibiotic class	Resistant pathogen	Multiple resistance	Aquaculture settings	Significance for human medicine[a]	References
Tetracyclines	*A. hydrophila*	✓	Tilapia and mullet aquaculture farms, Egypt	High	Ishida et al. (2010)
	A. salmonicida	✓	Atlantic Salmon cultures, Canada	High	McIntosh et al. (2008)
Aminoglycosides	*E.* ictaluri	✓	Catfish farm, Vietnam	Critical	Dung et al. (2008)
Beta-lactams	*V. harveyi*	✓	Coastal waters and shrimp aquaculture ponds, Indonesia	Critical	Teo et al. (2000)
	Aeromonas sp.	✓	Different fish farms, Australia	Critical	Akinbowale et al. (2006)
Amphenicols	*Pseudomonas* sp. *Enterobacter* sp.	✓	Salmon aquaculture facilities, Chile	Moderate	Fernández-Alarcón et al. (2010)
Nitrofurans	*V. anguillarum*	✓	Infected sea breams, Greece	Critical	Smith and Christofilogiannis (2007)
Macrolides	*Salmonella* sp.	✓	China fish market	Critical	Broughton and Walker (2009)
Fluoroquinolones	*T. maritimum*	✓	Infected sole and turbot in Portugal and Spain	Critical	Avendaño-Herrera et al. (2008)
Sulphonamides	*Aeromonas* sp.	✓	Infected mrigel and punti, India	Moderate	Das et al. (2009)

[a]Heuer et al. (2009). Frequent non-human use of human medicine would develop antibiotic resistance in pathogens. Categorization based on significance shows the level of importance given to these antibiotics while developing risk management strategies

highlighted that gut microbes in zebrafish can be important due to their potential involvement in metabolism, immune system and proliferation. The impacts of macrobiotics on beneficial host microbes can be detrimental due to alterations in their functionalities (Navarrete et al. 2008). Thus, antibiotics with the potential of altering the functionalities of host microbes should not be applied. A few studies have focused on this aspect of antibiotics. For example, effect of oxytetracycline on intestinal microbial diversity present in juvenile salmon has been investigated using the sequencing of 16S rDNA amplicons and restricted fragment length polymorphism. The results revealed that treatment with oxytetracycline resulted in significant reduction of intestinal microbes (Navarrete et al. 2008). It has been suggested that microbial diversity including beneficial and innocuous microbes is essential for successful aquaculture activities because the loss in microbial diversity would facilitate the invasion/proliferation of opportunistic microbes such as phylotypes (Schulze et al. 2006; Romero et al. 2012). In another study, impact of different antibiotics such as sulfafurazole, penicillin, oxolinic acid, erythromycin and oxytetracycline on gut microbes was examined. It was observed that microbial diversity increased with sulfafurazole, oxytetracycline, and oxolinic acid treatments. In contrast, penicillin and erythromycin reduced microbial diversity in gut (Austin and Al-Zahrani 1988). Therefore, it can be concluded based on the presented evidence that each antibiotic imparts different impacts on microbial diversity, changing the composition of intestinal and gut microbes in fish species.

Parasiticides such as hydrogen peroxide, pyrethroids and emamectin benzoate are being used in aquaculture farms to treat infectious virus and parasites. Emamectin benzoate (EB) dose of 1–25 μg/Kg is used in salmon farming (Roy et al. 2000). Although EB doses as high as 10 folds of the recommended dose may not result in mortality of salmon and trout, it can cause several side effects such as dark coloration, lethargy, lack of appetite and toxicity. High dose of EB can also result in its interaction with non-target species (Waddy et al. 2002). For instance, premature molting of American lobsters was observed due to EB ingestion. However, the molting response was observed in limited number of species (Waddy et al. 2007). Similar to antibiotics, frequent use of parasiticides can result in the formation of anti-parasiticides genes, neutralizing its anti-parasitic effects (Consortium 2006; Burridge et al. 2010).

9.4.2 *Human Health*

Fish is the largest source of protein on this planet and also the source of income for more than 500 million people. Edible marine life forms are increasingly used in coastal areas, potentially resulting in the exposure of human to PhACs (Small and Nicholls 2003). Regular monitoring of imported seafood is being carried out to check the presence of permitted veterinary medicines. However, these monitoring programs are not effective in addressing those PhACs that may be present in the coastal waters and/or seawater due to the discharge of treated wastewater or the direct surface runoff from agricultural lands. Moreover, health risk assessment due to the consumption of seafood cannot be possible without the guidelines to determine the daily uptake limit of all PhACs detected in seafood. European Union (EU) took the initiative and set the maximum residue limits (MRL) for 23 antibiotics present in seafood (Commission 2010). Concentration of fluoroquinolone and trimethoprim in the samples of fish and caged mussels was below the MRL (Zheng et al. 2012; McEneff et al. 2014). Similarly, fish samples taken from Czech supermarkets were tested for the presence of antibiotics. The concentration of antibiotics was found well below EU MRL (Fedorova et al. 2014). In contrast, different studies reported that the concentration of erythromycin, sarafloxacin in shrimps and mollusc, respectively exceeded the EU MRL for these antibiotics (Li et al. 2012; Chen et al. 2015). Since the chronic dietary effects of PhACs are not completely understood, existing MRL for antibiotics may not be effective for sensitive population (Love et al. 2011).

Occurrence of antibiotics is usually monitored in raw seafood rather than on as-consumed basis. McEneff et al. (2014) observed that the concentration of antibiotics was enhanced up to 20 folds after cooking of fish and shellfish. The increase in concentration may be attributed to the cleaving of different metabolites such as acyl glucuronide, resulting in the release of respective parent antibiotic. Therefore, analytical methods should incorporate the steps to determine the concentration of reversible metabolite in seafood. Moreover, the fate of antibiotics and their metabolites after cooking and digestion should be assessed during the human risk assessment studies.

World Health Organisation (WHO) has recognised that the development of antibiotic resistance in bacteria is an emerging global health emergency. Antibiotic resistance genes, developed from the exposure of bacteria to the non-lethal dose of antibiotics, may transfer from non-pathogenic to pathogenic bacteria due to horizontal gene transfer (Taylor et al. 2011). Marine ecosystem receives antibiotics from multiple sources such as aquaculture activities and wastewater discharge, making it a reservoir for antibiotics resistance. Recent reports has confirmed the presence of multiple antibiotic-resistant bacteria in marine biota such as fish, mammals, seabirds, octopus and many more (Cabello et al. 2013; Rose et al. 2009). The transfer of antibiotic-resistant bacteria from marine ecosystem to terrestrial environment may follow three major routes: (a) transfer from marine to terrestrial environment through seabirds; (b) transfer through human consumption of seafood; (c) international dissemination through seafood export (Taylor et al. 2011; Ryu et al. 2012).

9.5 Alternative Methods of Disease Control in Aquaculture

There is a need to develop environmentally friendly alternatives for the control of pathogenic bacteria in aquaculture because of the harmful effects linked with the use of antibiotics in aquaculture (Sect. 9.4). In this regard, vaccination can be an ideal method for the control of infectious diseases but commercial availability of vaccines is limited for use in aquaculture. Use of probiotics and essential oils (EOs) can be effective and environmentally friendly alternatives for the control of bacterial diseases in aquaculture (Romero et al. 2012; Merrifield et al. 2010).

Probiotics are innocuous microorganisms which confers health benefits in host species, when administered in adequate amounts. In aquaculture, the application of probiotics offers several benefits such as (Irianto and Austin 2002; Romero and Navarrete 2006; Romero et al. 2012): (a) preventing bacterial multiplication in fish species as well as in their ambient environment; (b) improving the quality of water by reducing the level of pathogenic bacteria; (c) improving the efficacy of immune system in species; and (d) improving the digestion of food. Probiotics have mostly been applied at the larval stage of growth because the risk of a disease outbreak is more at this stage (Dierckens et al. 2009; Zhou et al. 2009; Tinh et al. 2008). Different microorganisms such as lactic acid bacteria, A*eromonas*, *Vibrio*, Yeasts and *Actinobacteria* have been reported to be effective for the prevention of diseases in prawns, bivalve molluscs, teleost fish and shrimps (Romero et al. 2012; Dimitroglou et al. 2011; Merrifield et al. 2010; Ninawe and Selvin 2009).

EOs extracted from different plant components such as flowers, seeds, fruits, leaves and roots have complex composition and can contain up to sixty different components. However, major constituents of EOs include terpenes, terpenoids, thymol and thyme (Bakkali et al. 2008; Marzoug et al. 2011). EOs have been reported to protect plants from pests and pathogens due to their antimicrobial properties. EOs provide protection against pathogens by (Chung et al. 2012; Ennajar et al. 2011; Romero et al. 2012): (a) disrupting the permeability of microbial cell membrane; (b) coagulating the contents of pathogenic cells; and (c) inhibiting the quorum sensing among bacteria. EOs have mostly been used to increase the shelf life of seafood (Lin et al. 2004; Mejlholm and Dalgaard 2002; Merrifield et al. 2010). For instance, in vivo application of EOs, namely oregano and thymol increased the shelf life of trout and carp up to a few weeks (Mahmoud et al. 2004; Pyrgotou et al. 2010). Although antimicrobial effects of EOs have not been investigated extensively in aquaculture, Randrianarivelo et al. (2010) observed that EO extracted from *Cinnamosma fragrans* reduced the concentration of heterotrophic bacteria in growing environment, thereby improving the survival rate of shrimp larvae. More research is needed to confirm the effectiveness of EOs for different aquatic species such as fish and clams.

9.6 Conclusion

Pharmaceutically active compounds (PhACs), human pharmaceuticals, veterinary medicines and illicit drugs have been detected in the marine ecosystem (both seawater and marine biota) across the globe. Reports published over the years have suggested that the concentration of PhACs detected can be harmful for marine lifeforms and human. Continued discharge of ineffectively treated wastewater into natural water bodies and indiscriminate use of pharmaceuticals in aquaculture may lead to elevated levels of PhACs in marine ecosystem in the near future. Due to the shared and unique environment in coastal waters and seawater, integrated mitigation measures are required to address the issue of PhACs. Moreover, establishment of a uniform method to monitor the presence and fate of PhACs and their impacts on marine ecosystem and human health is needed because current methods are not sensitive enough to effectively assess the toxicity of PhACs on coastal/marine ecosystem.

The exposure to PhACs can result in the reduction of lysosome membrane stability (LMS) as described in Sect. 9.4.1. Therefore, rapid screening of marine biota exposed to PhACs can be assessed by exploiting LMS. Another effective approach is to monitor the impacts of PhACs at the larval stage of marine biota due to their high sensitivity and rapid response. Moreover, assessment of biomarkers and endpoints in marine biota due to different mode of action may help in developing innovative biomonitoring programs for marine ecosystem. Notably, mode of action of PhACs in non-target spices is mostly unknown. Therefore, it is essential to further assess the unknown effects of PhACs on non-target species.

References

Aguirre-Martinez GV, del Valls TA, Martín-Díaz ML (2013a) Early responses measured in the brachyuran crab *Carcinus maenas* exposed to carbamazepine and novobiocin: application of a 2-tier approach. Ecotoxicol Environ Saf 97:47–58

Aguirre-Martinez GV, Buratti S, Fabbri E, del Valls TA, Martin-Diaz ML (2013b) Stability of lysosomal membrane in Carcinus maenas acts as a biomarker of exposure to pharmaceuticals. Environ Monit Assess 185:3783–3793

Aguirre-Martínez GV, Buratti S, Fabbri E, Delvalls AT, Martín-Díaz ML (2013c) Using lysosomal membrane stability of haemocytes in Ruditapes philippinarum as a biomarker of cellular stress to assess contamination by caffeine, ibuprofen, carbamazepine and novobiocin. J Environ Sci 25:1408–1418

Ahmad A, Khan MU, Patel I, Maharaj S, Pandey S, Dhingra S (2015). Knowledge, attitude and practice of B. Sc. Pharmacy students about antibiotics in Trinidad and Tobago. J Res Pharm Pract 4:37

Akinbowale OL, Peng H, Barton M (2006) Antimicrobial resistance in bacteria isolated from aquaculture sources in Australia. J Appl Microbiol 100:1103–1113

Almeida Â, Calisto V, Esteves VI, Schneider RJ, Soares AM, Figueira E, Freitas R (2014) Presence of the pharmaceutical drug carbamazepine in coastal systems: effects on bivalves. Aquat Toxicol 156:74–87

Alvarez-Muñoz D, Huerta B, Fernandez-Tejedor M, Rodríguez-Mozaz S, Barceló D (2015) Multi-residue method for the analysis of pharmaceuticals and some of their metabolites in bivalves. Talanta 136:174–182

Antunes S, Freitas R, Figueira E, Gonçalves F, Nunes B (2013) Biochemical effects of acetaminophen in aquatic species: edible clams *Venerupis decussata* and *Venerupis philippinarum*. Environ Sci Pollut Res 20:6658–6666

Arpin-Pont L, Bueno MJM, Gomez E, Fenet H (2016) Occurrence of PPCPs in the marine environment: a review. Environ Sci Pollut Res 23:4978–4991

Asif MB, Hai FI, Singh L, Price WE, Nghiem LD (2017a) Degradation of pharmaceuticals and personal care products by white-rot fungi—a critical review. Curr Pollut Rep 3:88–103

Asif MB, Hai FI, Hou J, Price WE, Nghiem LD (2017) Impact of wastewater derived dissolved interfering compounds on growth, enzymatic activity and trace organic contaminant removal of white rot fungie—a critical review. J Environ Manag 201:89–109

Austin B, Al-Zahrani A (1988) The effect of antimicrobial compounds on the gastrointestinal microflora of rainbow trout, *Salmo gairdneri* Richardson. J Fish Biol 33:1–14

Avendaño-Herrera R, Núñez S, Barja JL, Toranzo AE (2008) Evolution of drug resistance and minimum inhibitory concentration to enrofloxacin in Tenacibaculum maritimum strains isolated in fish farms. Aquacult Int 16:1–11

Bakkali F, Averbeck S, Averbeck D, Idaomar M (2008) Biological effects of essential oils—a review. Food Chem Toxicol 46:446–475

Balmer ME, Buser H-R, Müller MD, Poiger T (2005) Occurrence of some organic UV filters in wastewater, in surface waters, and in fish from Swiss lakes. Environ Sci Technol 39:953–962

Banni M, Sforzini S, Franzellitti S, Oliveri C, Viarengo A, Fabbri E (2015) Molecular and cellular effects induced in *Mytilus galloprovincialis* treated with oxytetracycline at different temperatures. PLoS ONE 10:e0128468

Barbier EB, Hacker SD, Kennedy C, Koch EW, Stier AC, Silliman BR (2011) The value of estuarine and coastal ecosystem services. Ecol Monogr 81:169–193

Bates JM, Mittge E, Kuhlman J, Baden KN, Cheesman SE, Guillemin K (2006) Distinct signals from the microbiota promote different aspects of zebrafish gut differentiation. Dev Biol 297:374–386

Bayen S, Zhang H, Desai MM, Ooi SK, Kelly BC (2013) Occurrence and distribution of pharmaceutically active and endocrine disrupting compounds in Singapore's marine environment: influence of hydrodynamics and physical–chemical properties. Environ Pollut 182:1–8

Benotti MJ, Brownawell BJ (2007) Distributions of pharmaceuticals in an urban estuary during both dry-and wet-weather conditions. Environ Sci Technol 41:5795–5802

Bertin A, Inostroza PA, Quiñones RA (2011) Estrogen pollution in a highly productive ecosystem off central-south Chile. Mar Pollut Bull 62:1530–1537

Boden AR, Reiner EJ (2004) Development of an isotope-dilution gas chromatographic-mass spectrometric method for the analysis of polycyclic aromatic compounds in environmental matrices. Polycyclic Aromat Compd 24:309–323

Brausch JM, Rand GM (2011) A review of personal care products in the aquatic environment: environmental concentrations and toxicity. Chemosphere 82:1518–1532

Brooks BW, Foran CM, Richards SM, Weston J, Turner PK, Stanley JK, Solomon KR, Slattery M, la Point TW (2003) Aquatic ecotoxicology of fluoxetine. Toxicol Lett 142:169–183

Broughton EI, Walker DG (2009) Prevalence of antibiotic-resistant *Salmonella* in fish in Guangdong, China. Foodborne Pathog Dis 6:519–521

Bueno MM, Boillot C, Fenet H, Chiron S, Casellas C, Gómez E (2013) Fast and easy extraction combined with high resolution-mass spectrometry for residue analysis of two anticonvulsants and their transformation products in marine mussels. J Chromatogr A 1305:27–34

Burridge L, Weis JS, Cabello F, Pizarro J, Bostick K (2010) Chemical use in salmon aquaculture: a review of current practices and possible environmental effects. Aquaculture 306:7–23

Caban M, Migowska N, Stepnowski P, Kwiatkowski M, Kumirska J (2012) Matrix effects and recovery calculations in analyses of pharmaceuticals based on the determination of β-blockers and β-agonists in environmental samples. J Chromatogr A 1258:117–127

Cabello FC (2006) Heavy use of prophylactic antibiotics in aquaculture: a growing problem for human and animal health and for the environment. Environ Microbiol 8:1137–1144

Cabello FC, Godfrey HP, Tomova A, Ivanova L, Dölz H, Millanao A, Buschmann AH (2013) Antimicrobial use in aquaculture re-examined: its relevance to antimicrobial resistance and to animal and human health. Environ Microbiol 15:1917–1942

Chen H, Liu S, Xu X-R, Liu S-S, Zhou G-J, Sun K-F, Zhao J-L, Ying G-G (2015) Antibiotics in typical marine aquaculture farms surrounding Hailing Island, South China: occurrence, bioaccumulation and human dietary exposure. Mar Pollut Bull 90:181–187

Christen V, Hickmann S, Rechenberg B, Fent K (2010) Highly active human pharmaceuticals in aquatic systems: a concept for their identification based on their mode of action. Aquat Toxicol 96:167–181

Chung I-M, Ahmad A, Kim S-J, Naik PM, Nagella P (2012) Composition of the essential oil constituents from leaves and stems of Korean *Coriandrum sativum* and their immunotoxicity activity on the Aedes aegypti L. Immunopharmacol Immunotoxicol 34:152–156

Comeau F, Surette C, Brun G, Losier R (2008) The occurrence of acidic drugs and caffeine in sewage effluents and receiving waters from three coastal watersheds in Atlantic Canada. Sci Total Environ 396:132–146

Commission TE (2010) Commission regulation on pharmacologically active substances and their classification regarding maximum residue limits in foodstuffs of animal origin. In: Union TE (ed) 37/2010 The European Union

Consortium S (2006) Sea lice resistance to chemotherapeutants: a handbook in resistance management. SEARCH, Sacramento, CA, USA

Corcoran E (2010) Sick water?: the central role of wastewater management in sustainable development: a rapid response assessment, London, UNEP/Earthprint. (ISBN: 9788277010755)

Cortez FS, Pereira CDS, Santos AR, Cesar A, Choueri RB, de Assis Martini G, Bohrer-Morel MB (2012) Biological effects of environmentally relevant concentrations of the pharmaceutical Triclosan in the marine mussel Perna perna (Linnaeus, 1758). Environ Pollut 168:145–150

Costanza R, D'Arge R, de Groot R, Farber S, Grasso M, Hannon B, Limburg K, Naeem S, O'Neill RV, Paruelo J (1998) The value of ecosystem services: putting the issues in perspective. Ecol Econ 25:67–72

Crain CM, Halpern BS, Beck MW, Kappel CV (2009) Understanding and managing human threats to the coastal marine environment. Ann N Y Acad Sci 1162:39–62

Das A, Saha D, Pal J (2009) Antimicrobial resistance and in vitro gene transfer in bacteria isolated from the ulcers of EUS-affected fish in India. Lett Appl Microbiol 49:497–502

Daughton CG, Ruhoy IS (2009) Environmental footprint of pharmaceuticals: the significance of factors beyond direct excretion to sewers. Environ Toxicol Chem 28:2495–2521

Daughton CG, Ternes TA (1999) Pharmaceuticals and personal care products in the environment: agents of subtle change? Environ Health Perspect 107:907

Defoirdt T, Sorgeloos P, Bossier P (2011) Alternatives to antibiotics for the control of bacterial disease in aquaculture. Curr Opin Microbiol 14:251–258

Díaz-Cruz MS, de Alda MJL, Barcelo D (2003) Environmental behavior and analysis of veterinary and human drugs in soils, sediments and sludge. TrAC Trends in Anal Chem 22:340–351

Dierckens K, Rekecki A, Laureau S, Sorgeloos P, Boon N, van den Broeck W, Bossier P (2009) Development of a bacterial challenge test for gnotobiotic sea bass (Dicentrarchus labrax) larvae. Environ Microbiol 11:526–533

Dimitroglou A, Merrifield DL, Carnevali O, Picchietti S, Avella M, Daniels C, Güroy D, Davies SJ (2011) Microbial manipulations to improve fish health and production—a Mediterranean perspective. Fish Shellfish Immunol 30:1–16

Dodder NG, Maruya KA, Ferguson PL, Grace R, Klosterhaus S, la Guardia MJ, Lauenstein GG, Ramirez J (2014) Occurrence of contaminants of emerging concern in mussels (Mytilus spp.) along the California coast and the influence of land use, storm water discharge, and treated wastewater effluent. Mar Pollut Bull 81:340–346

Du L, Liu W (2012) Occurrence, fate, and ecotoxicity of antibiotics in agro-ecosystems. Rev Agron Sustain Dev 32:309–327

Dung TT, Haesebrouck F, Tuan NA, Sorgeloos P, Baele M, Decostere A (2008) Antimicrobial susceptibility pattern of Edwardsiella ictaluri isolates from natural outbreaks of bacillary necrosis of *Pangasianodon hypophthalmus* in Vietnam. Microbial Drug Resist 14:311–316

Emke E, Evans S, Kasprzyk-Hordern B, de Voogt P (2014) Enantiomer profiling of high loads of amphetamine and MDMA in communal sewage: a Dutch perspective. Sci Total Environ 487:666–672

Emnet P, Gaw S, Northcott G, Storey B, Graham L (2015) Personal care products and steroid hormones in the Antarctic coastal environment associated with two Antarctic research stations, McMurdo Station and Scott Base. Environ Res 136:331–342

Ennajar M, Afloulous S, Romdhane M, Ibrahim H, Cazaux S, Abderraba M, Raies A, Bouajila J (2011) Influence of the process, season, and origin on volatile composition and antioxidant activity of Juniperus phoenicea L. leaves essential oils. J Food Sci 76:C224–C230

Evgenidou EN, Konstantinou IK, Lambropoulou DA (2015) Occurrence and removal of transformation products of PPCPs and illicit drugs in wastewaters: a review. Sci Total Environ 505:905–926

Fabbri E, Capuzzo A (2010) Cyclic AMP signaling in bivalve molluscs: an overview. J Exp Zool Part A: Ecol Genet Physiol 313:179–200

Fabbri E, Franzellitti S (2015) Human pharmaceuticals in the marine environment: focus on exposure and biological effects in animal species. Environ Toxicol Chem

Fabbri E, Franzellitti S (2016) Human pharmaceuticals in the marine environment: focus on exposure and biological effects in animal species. Environ Toxicol Chem 35:799–812

FAO (2014) The state of world fisheries and aquaculture opportunities and challenges. Food and Agriculture Organization of the United Nations, Rome. ISBN: 9789251082768 (Assessed from https://www.fao.org/3/a-i3720e.pdf)

Fedorova G, Nebesky V, Randak T, Grabic R (2014) Simultaneous determination of 32 antibiotics in aquaculture products using LC-MS/MS. Chem Pap 68:29–36

Fenet H, Arpin-Pont L, Vanhoutte-Brunier A, Munaron D, Fiandrino A, Bueno M-JM, Boillot C, Casellas C, Mathieu O, Gomez E (2014) Reducing PEC uncertainty in coastal zones: a case study on carbamazepine, oxcarbazepine and their metabolites. Environ Int 68:177–184

Fernández-Alarcón C, Miranda C, Singer R, Lopez Y, Rojas R, Bello H, Dominguez M, González-Rocha G (2010) Detection of the floR gene in a diversity of florfenicol resistant gram-negative bacilli from freshwater salmon farms in Chile. Zoonoses Pub Health 57:181–188

Fick J, Lindberg RH, Parkkonen J, Arvidsson B, Tysklind M, Larsson DGJ (2010) Therapeutic levels of levonorgestrel detected in blood plasma of fish: results from screening rainbow trout exposed to treated sewage effluents. Environ Sci Technol 44:2661–2666

Ford AT, Fong PP (2016) The effects of antidepressants appear to be rapid and at environmentally relevant concentrations. Environ Toxicol Chem 35:794–798

Franzellitti S, Buratti S, Capolupo M, Du B, Haddad SP, Chambliss CK, Brooks BW, Fabbri E (2014) An exploratory investigation of various modes of action and potential adverse outcomes of fluoxetine in marine mussels. Aquat Toxicol 151:14–26

Franzellitti S, Buratti S, Du B, Haddad SP, Chambliss CK, Brooks BW, Fabbri E (2015) A multibiomarker approach to explore interactive effects of propranolol and fluoxetine in marine mussels. Environ Pollut 205:60–69

Franzellitti S, Buratti S, Valbonesi P, Capuzzo A, Fabbri E (2011) The beta-blocker propranolol affects cAMP-dependent signaling and induces the stress response in Mediterranean mussels, *Mytilus galloprovincialis*. Aquat Toxicol 101:299–308

Franzellitti S, Buratti S, Valbonesi P, Fabbri E (2013) The mode of action (MOA) approach reveals interactive effects of environmental pharmaceuticals on *Mytilus galloprovincialis*. Aquat Toxicol 140–141:249–256

Fu W, Franco A, Trapp S (2009) Methods for estimating the bioconcentration factor of ionizable organic chemicals. Environ Toxicol Chem 28:1372–1379

Garcia SN, Foster M, Constantine LA, Huggett DB (2012) Field and laboratory fish tissue accumulation of the anti-convulsant drug carbamazepine. Ecotoxicol Environ Saf 84:207–211

Gaw S, Thomas KV, Hutchinson TH (2014) Sources, impacts and trends of pharmaceuticals in the marine and coastal environment. Phil Trans R Soc B 369:20130572

Gilroy ÈA, Balakrishnan VK, Solomon KR, Sverko E, Sibley PK (2012) Behaviour of pharmaceuticals in spiked lake sediments–effects and interactions with benthic invertebrates. Chemosphere 86:578–584

Glassmeyer ST, Koplin DW, Furlong ET, Focazio M (2008) Environmental presence and persistence of pharmaceuticals: An overview. In: Aga DS (ed) Fate of pharmaceuticals in the environment and in water treatment systems. CRC Press, Boca Raton. ISBN: 9781420052329

Gomez E, Bachelot M, Boillot C, Munaron D, Chiron S, Casellas C, Fenet H (2012) Bioconcentration of two pharmaceuticals (benzodiazepines) and two personal care products (UV filters) in marine mussels (*Mytilus galloprovincialis*) under controlled laboratory conditions. Environ Sci Pollut Res 19:2561–2569

Gómez MJ, Petrović M, Fernández-Alba AR, Barceló D (2006) Determination of pharmaceuticals of various therapeutic classes by solid-phase extraction and liquid chromatography–tandem mass spectrometry analysis in hospital effluent wastewaters. J Chromatogr A 1114:224–233

Gonzalez-Rey M, Bebianno MJ (2012) Does non-steroidal anti-inflammatory (NSAID) ibuprofen induce antioxidant stress and endocrine disruption in mussel *Mytilus galloprovincialis*? Environ Toxicol Pharmacol 33:361–371

Gonzalez-Rey M, Bebianno MJ (2014) Effects of non-steroidal anti-inflammatory drug (NSAID) diclofenac exposure in mussel *Mytilus galloprovincialis*. Aquat Toxicol 148:221–230

Grigorakis K, Rigos G (2011) Aquaculture effects on environmental and public welfare—the case of Mediterranean mariculture. Chemosphere 85:899–919

Guardabassi L, Courvalin P (2006) Modes of antimicrobial action and mechanisms of bacterial resistance. In: Aarestrup FM (ed) Antimicrobial resistance in bacteria of animal origin. American Society of Microbiology, Washington, DC. https://doi.org/10.1128/9781555817534.ch1

Gunnarsson L, Jauhiainen A, Kristiansson E, Nerman O, Larsson DJ (2008) Evolutionary conservation of human drug targets in organisms used for environmental risk assessments. Environ Sci Technol 42:5807–5813

Hai FI, Alturki A, Nguyen LN, Price WE, Nghiem LD (2016) Removal of trace organic contaminants by integrated membrane processes for water reuse applications. In: Ngo HH, Guo W, Surampalli RY, Zhang TC (eds) Green technologies for sustainable water management. American Society of Civil Engineers, New York. ISBN: 9780784414422

Hai FI, Nghiem LD, Khan SJ, Price WE, Yamamoto K (2014a) Wastewater reuse: removal of emerging trace organic contaminants (TrOC). In: Hai FI, Yamamoto K, Lee CH (eds) Membrane Biological Reactors: Theory, Modeling, Design, Management. IWA Publishing, London. ISBN: 9781780400655

Hai FI, Nguyen LN, Nghiem LD, Liao B-Q, Koyuncu I, Price WE (2014b) Trace organic contaminants removal by combined processes for wastewater reuse. In: Fatta-Kassinos D, Dionysiou DD, Kümmerer K (eds) Advanced treatment technologies for urban wastewater reuse. Springer, Gewerbestrasse. ISBN: 9783319238869

Hai FI, Tadkaew N, McDonald JA, Khan SJ, Nghiem LD (2011) Is halogen content the most important factor in the removal of halogenated trace organics by MBR treatment? Biores Technol 102:6299–6303

Hai FI, Yamamoto K, Lee CH (2014c) Membrane biological reactors: theory, modeling, design, management and applications to wastewater reuse. IWA Publishing, London. ISBN: 9781780400655

He X, Wang Z, Nie X, Yang Y, Pan D, Leung AO, Cheng Z, Yang Y, Li K, Chen K (2012) Residues of fluoroquinolones in marine aquaculture environment of the Pearl River Delta, South China. Environ Geochem Health 34:323–335

Heuer OE, Kruse H, Grave K, Collignon P, Karunasagar I, Angulo FJ (2009) Human health consequences of use of antimicrobial agents in aquaculture. Clin Infect Dis 49:1248–1253

Holmström K, Gräslund S, Wahlström A, Poungshompoo S, Bengtsson B-E, Kautsky N (2003) Antibiotic use in shrimp farming and implications for environmental impacts and human health. Int J Food Sci Technol 38:255–266

Howard PH, Muir DC (2011) Identifying new persistent and bioaccumulative organics among chemicals in commerce II: pharmaceuticals. Environ Sci Technol 45:6938–6946

Hughes SR, Kay P, Brown LE (2012) Global synthesis and critical evaluation of pharmaceutical data sets collected from river systems. Environ Sci Technol 47:661–677

Ingram T, Richter U, Mehling T, Smirnova I (2011) Modelling of pH dependent n-octanol/water partition coefficients of ionizable pharmaceuticals. Fluid Phase Equilib 305:197–203

Irianto A, Austin B (2002) Probiotics in aquaculture. J Fish Dis 25:633–642

Ishida Y, Ahmed AM, Mahfouz NB, Kimura T, El-Khodery SA, Moawad AA, Shimamoto T (2010) Molecular analysis of antimicrobial resistance in gram-negative bacteria isolated from fish farms in Egypt. J Vet Med Sci 72:727–734

Jia A, Hu J, Wu X, Peng H, Wu S, Dong Z (2011) Occurrence and source apportionment of sulfonamides and their metabolites in Liaodong Bay and the adjacent Liao River Basin, North China. Environ Toxicol Chem 30:1252–1260

Kemper N (2008) Veterinary antibiotics in the aquatic and terrestrial environment. Ecol Ind 8:1–13

Kim S-C, Carlson K (2006) Occurrence of ionophore antibiotics in water and sediments of a mixed-landscape watershed. Water Res 40:2549–2560

Klosterhaus SL, Grace R, Hamilton MC, Yee D (2013) Method validation and reconnaissance of pharmaceuticals, personal care products, and alkylphenols in surface waters, sediments, and mussels in an urban estuary. Environ Int 54:92–99

Knight J, Taylor GW, Wright P, Clare AS, Rowley AF (1999) Eicosanoid biosynthesis in an advanced deuterostomate invertebrate, the sea squirt (*Ciona intestinalis*). Biochim et Biophys Acta (BBA)-Mol Cell Biol Lipids 1436:467–478

Kookana RS, Williams M, Boxall AB, Larsson DJ, Gaw S, Choi K, Yamamoto H, Thatikonda S, Zhu Y-G, Carriquiriborde P (2014) Potential ecological footprints of active pharmaceutical ingredients: an examination of risk factors in low-, middle-and high-income countries. Phil Trans R Soc B 369:20130586

Kools SA, Boxall A, Moltmann JF, Bryning G, Koschorreck J, Knacker T (2008) A ranking of European veterinary medicines based on environmental risks. Integr Environ Assess Manag 4:399–408

Koutsogiannaki S, Evangelinos N, Koliakos G, Kaloyianni M (2006) Cytotoxic mechanisms of Zn^{2+} and Cd^{2+} involve Na^{+}/H^{+} exchanger (NHE) activation by ROS. Aquat Toxicol 78:315–324

Kumarasamy KK, Toleman MA, Walsh TR, Bagaria J, Butt F, Balakrishnan R, Chaudhary U, Doumith M, Giske CG, Irfan S (2010) Emergence of a new antibiotic resistance mechanism in India, Pakistan, and the UK: a molecular, biological, and epidemiological study. Lancet Infect Dis 10:597–602

Kümmerer K (2009) Antibiotics in the aquatic environment—a review—part I. Chemosphere 75:417–434

Lahti M, Brozinski J-M, Jylhä A, Kronberg L, Oikari A (2011) Uptake from water, biotransformation, and biliary excretion of pharmaceuticals by rainbow trout. Environ Toxicol Chem 30:1403–1411

Lahti M, Brozinski JM, Segner H, Kronberg L, Oikari A (2012) Bioavailability of pharmaceuticals in waters close to wastewater treatment plants: use of fish bile for exposure assessment. Environ Toxicol Chem 31:1831–1837

Lalumera GM, Calamari D, Galli P, Castiglioni S, Crosa G, Fanelli R (2004) Preliminary investigation on the environmental occurrence and effects of antibiotics used in aquaculture in Italy. Chemosphere 54:661–668

Langford K, Thomas KV (2011) Input of selected human pharmaceutical metabolites into the Norwegian aquatic environment. J Environ Monit 13:416–421

Lapworth D, Baran N, Stuart M, Ward R (2012) Emerging organic contaminants in groundwater: a review of sources, fate and occurrence. Environ Pollut 163:287–303

Lara-Martín PA, González-Mazo E, Petrovic M, Barceló D, Brownawell BJ (2014) Occurrence, distribution and partitioning of nonionic surfactants and pharmaceuticals in the urbanized Long Island Sound Estuary (NY). Mar Pollut Bull 85:710–719

Larsen J, Roney JM (2013) Farmed fish production overtakes beef. Earth Policy Institute, Washington, DC

Le TX, Munekage Y (2004) Residues of selected antibiotics in water and mud from shrimp ponds in mangrove areas in Vietnam. Mar Pollut Bull 49:922–929

Le TX, Munekage Y, Kato S-I (2005) Antibiotic resistance in bacteria from shrimp farming in mangrove areas. Sci Total Environ 349:95–105

Lee WY, Arnold C (1983) Chronic toxicity of ocean-dumped pharmaceutical wastes to the marine amphipod Amphithoe valida. Mar Pollut Bull 14:150–153

Li H (2003) Management of coastal mega-cities—a new challenge in the 21st century. Marine Policy 27:333–337

Li W, Shi Y, Gao L, Liu J, Cai Y (2012) Investigation of antibiotics in mollusks from coastal waters in the Bohai Sea of China. Environ Pollut 162:56–62

Lin Y, Labbe R, Shetty K (2004) Inhibition of Listeria monocytogenes in fish and meat systems by use of oregano and cranberry phytochemical synergies. Appl Environ Microbiol 70:5672–5678

Long ER, Dutch M, Weakland S, Chandramouli B, Benskin JP (2013) Quantification of pharmaceuticals, personal care products, and perfluoroalkyl substances in the marine sediments of Puget Sound, Washington, USA. Environ Toxicol Chem 32:1701–1710

Love DC, Rodman S, Neff RA, Nachman KE (2011) Veterinary drug residues in seafood inspected by the European Union, United States, Canada, and Japan from 2000 to 2009. Environ Sci Technol 45:7232–7240

Luo Y, Guo W, Ngo HH, Nghiem LD, Hai FI, Zhang J, Liang S, Wang XC (2014) A review on the occurrence of micropollutants in the aquatic environment and their fate and removal during wastewater treatment. Sci Total Environ 473–474:619–641

Mahmoud BS, Yamazaki K, Miyashita K, Il-Shik S, Dong-Suk C, Suzuki T (2004) Bacterial microflora of carp (*Cyprinus carpio*) and its shelf-life extension by essential oil compounds. Food Microbiol 21:657–666

Maranho L, Baena-Nogueras R, Lara-Martín P, Delvalls T, Martín-Díaz M (2014) Bioavailability, oxidative stress, neurotoxicity and genotoxicity of pharmaceuticals bound to marine sediments. The use of the polychaete Hediste diversicolor as bioindicator species. Environ Res 134:353–365

Maranho L, Garrido-Pérez M, Baena-Nogueras R, Lara-Martín P, Antón-Martín R, Delvalls T, Martín-Díaz M (2015) Are WWTPs effluents responsible for acute toxicity? Seasonal variations of sediment quality at the Bay of Cádiz (SW, Spain). Ecotoxicology 24:368–380

Martin-Diaz L, Franzellitti S, Buratti S, Valbonesi P, Capuzzo A, Fabbri E (2009) Effects of environmental concentrations of the antiepilectic drug carbamazepine on biomarkers and cAMP-mediated cell signaling in the mussel *Mytilus galloprovincialis*. Aquat Toxicol 94:177–185

Martínez M, Intralawan A, Vázquez G, Pérez-Maqueo O, Sutton P, Landgrave R (2007) The coasts of our world: ecological, economic and social importance. Ecol Econ 63:254–272

Maruya KA, Vidal-Dorsch DE, Bay SM, Kwon JW, Xia K, Armbrust KL (2012) Organic contaminants of emerging concern in sediments and flatfish collected near outfalls discharging treated wastewater effluent to the Southern California Bight. Environ Toxicol Chem 31:2683–2688

Marzoug HNB, Romdhane M, Lebrihi A, Mathieu F, Couderc F, Abderraba M, Khouja ML, Bouajila J (2011) Eucalyptus oleosa essential oils: chemical composition and antimicrobial and antioxidant activities of the oils from different plant parts (stems, leaves, flowers and fruits). Molecules 16:1695–1709

Matozzo V, Franchi N, Ballarin L (2014) In vitro effects of the nonsteroidal anti-inflammatory drug, ibuprofen, on the immune parameters of the colonial ascidian *Botryllus schlosseri*. Toxicol In Vitro 28:778–783

McEneff G, Barron L, Kelleher B, Paull B, Quinn B (2014) A year-long study of the spatial occurrence and relative distribution of pharmaceutical residues in sewage effluent, receiving marine waters and marine bivalves. Sci Total Environ 476:317–326

McIntosh D, Cunningham M, Ji B, Fekete FA, Parry EM, Clark SE, Zalinger ZB, Gilg IC, Danner GR, Johnson KA (2008) Transferable, multiple antibiotic and mercury resistance in Atlantic Canadian isolates of *Aeromonas salmonicida* subsp. salmonicida is associated with carriage of an IncA/C plasmid similar to the Salmonella enterica plasmid pSN254. J Antimicrob Chemother 61:1221–1228

Mehinto AC, Hill EM, Tyler CR (2010) Uptake and biological effects of environmentally relevant concentrations of the nonsteroidal anti-inflammatory pharmaceutical Diclofenac in Rainbow trout (*Oncorhynchus mykiss*). Environ Sci Technol 44:2176–2182

Mejlholm O, Dalgaard P (2002) Antimicrobial effect of essential oils on the seafood spoilage micro-organism *Photobacterium phosphoreum* in liquid media and fish products. Lett Appl Microbiol 34:27–31

Merrifield DL, Dimitroglou A, Foey A, Davies SJ, Baker RT, Bøgwald J, Castex M, Ringø E (2010) The current status and future focus of probiotic and prebiotic applications for salmonids. Aquaculture 302:1–18

Mesquita SR, Guilhermino L, Guimarães L (2011) Biochemical and locomotor responses of *Carcinus maenas* exposed to the serotonin reuptake inhibitor fluoxetine. Chemosphere 85:967–976

Milan M, Pauletto M, Patarnello T, Bargelloni L, Marin MG, Matozzo V (2013) Gene transcription and biomarker responses in the clam Ruditapes philippinarum after exposure to ibuprofen. Aquat Toxicol 126:17–29

Minh TB, Leung HW, Loi IH, Chan WH, So MK, Mao JQ, Choi D, Lam JC, Zheng G, Martin M (2009) Antibiotics in the Hong Kong metropolitan area: ubiquitous distribution and fate in Victoria Harbour. Mar Pollut Bull 58:1052–1062

Montezinho LP, Mørk A, Duarte CB, Penschuck S, Geraldes CF, Castro MMC (2007) Effects of mood stabilizers on the inhibition of adenylate cyclase via dopamine D2-like receptors. Bipolar Disord 9:290–297

Na G, Fang X, Cai Y, Ge L, Zong H, Yuan X, Yao Z, Zhang Z (2013) Occurrence, distribution, and bioaccumulation of antibiotics in coastal environment of Dalian, China. Mar Pollut Bull 69:233–237

Nash CE (2003) Interactions of Atlantic salmon in the Pacific Northwest: VI. A synopsis of the risk and uncertainty. Fish Res 62:339–347

Navarrete P, Mardones P, Opazo R, Espejo R, Romero J (2008) Oxytetracycline treatment reduces bacterial diversity of intestinal microbiota of Atlantic salmon. J Aquat Anim Health 20:177–183

Neumann B, Vafeidis AT, Zimmermann J, Nicholls RJ (2015) Future coastal population growth and exposure to sea-level rise and coastal flooding-a global assessment. PLoS ONE 10: e0118571

Nichols JW, Du B, Berninger JP, Connors KA, Chambliss CK, Erickson RJ, Hoffman AD, Brooks BW (2015) Observed and modeled effects of pH on bioconcentration of diphenhydramine, a weakly basic pharmaceutical, in fathead minnows. Environ Toxicol Chem 34:1425–1435

Nikaido H (2009) Multidrug resistance in bacteria. Annu Rev Biochem 78:119–146

Ninawe A, Selvin J (2009) Probiotics in shrimp aquaculture: avenues and challenges. Crit Rev Microbiol 35:43–66

Nödler K, Voutsa D, Licha T (2014) Polar organic micropollutants in the coastal environment of different marine systems. Mar Pollut Bull 85:50–59

OECD (2008) OECD guidelines for the testing of chemicals bioaccumulation in sediment-dwelling benthic oligochaetes. OECD Guideline

Organisation I (2003) International convention for the prevention of pollution from ships (MARPOL), Annex IV prevention of pollution by sewage from ships [Online]. Accessed 10 July 2016

Owen SF, Huggett DB, Hutchinson TH, Hetheridge MJ, Kinter LB, Ericson JF, Sumpter JP (2009) Uptake of propranolol, a cardiovascular pharmaceutical, from water into fish plasma and its effects on growth and organ biometry. Aquat Toxicol 93:217–224

Pal A, Gin KY-H, Lin AY-C, Reinhard M (2010) Impacts of emerging organic contaminants on freshwater resources: review of recent occurrences, sources, fate and effects. Sci Total Environ 408:6062–6069

Pal R, Megharaj M, Kirkbride KP, Naidu R (2013) Illicit drugs and the environment—a review. Sci Total Environ 463:1079–1092

Petrović M, Hernando MD, Díaz-Cruz MS, Barceló D (2005) Liquid chromatography–tandem mass spectrometry for the analysis of pharmaceutical residues in environmental samples: a review. J Chromatogr A 1067:1–14

Petrovic M, Perez S, Barcelo D (2013) Analysis, removal, effects and risk of pharmaceuticals in the water cycle: occurrence and transformation in the environment. Elsevier, Cambridge. ISBN: 9780444626578

Phan HV, Hai FI, McDonald JA, Khan SJ, van de Merwe JP, Leusch FDL, Zhang R, Price WE, Broeckmann A, Nghiem LD (2015) Impact of hazardous events on the removal of nutrients and trace organic contaminants by an anoxic–aerobic membrane bioreactor receiving real wastewater. Biores Technol 192:192–201

Pichon V (2000) Solid-phase extraction for multiresidue analysis of organic contaminants in water. J Chromatogr A 885:195–215

Pintado-Herrera MG, González-Mazo E, Lara-Martín PA (2013) Environmentally friendly analysis of emerging contaminants by pressurized hot water extraction—stir bar sorptive extraction—derivatization and gas chromatography—mass spectrometry. Anal Bioanal Chem 405:401–411

Pojana G, Gomiero A, Jonkers N, Marcomini A (2007) Natural and synthetic endocrine disrupting compounds (EDCs) in water, sediment and biota of a coastal lagoon. Environ Int 33:929–936

Pyrgotou N, Giatrakou V, Ntzimani A, Savvaidis IN (2010) Quality assessment of salted, modified atmosphere packaged rainbow trout under treatment with oregano essential oil. J Food Sci 75: M406–M411

Qi W, Müller B, Pernet-Coudrier B, Singer H, Liu H, Qu J, Berg M (2014) Organic micropollutants in the Yangtze River: seasonal occurrence and annual loads. Sci Total Environ 472:789–799

Rand-Weaver M, Margiotta-Casaluci L, Patel A, Panter GH, Owen SF, Sumpter JP (2013) The read-across hypothesis and environmental risk assessment of pharmaceuticals. Environ Sci Technol 47:11384–11395

Randrianarivelo R, Danthu P, Benoit C, Ruez P, Raherimandimby M, Sarter S (2010) Novel alternative to antibiotics in shrimp hatchery: effects of the essential oil of *Cinnamosma fragrans* on survival and bacterial concentration of *Penaeus monodon* larvae. J Appl Microbiol 109:642–650

Rawn DF, Krakalovich T, Forsyth DS, Roscoe V (2009) Analysis of fin and non-fin fish products for azamethiphos and dichlorvos residues from the Canadian retail market. Int J Food Sci Technol 44:1510–1516

Richardson BJ, Lam PK, Martin M (2005) Emerging chemicals of concern: pharmaceuticals and personal care products (PPCPs) in Asia, with particular reference to Southern China. Mar Pollut Bull 50:913–920

Rico A, van den Brink PJ (2014) Probabilistic risk assessment of veterinary medicines applied to four major aquaculture species produced in Asia. Sci Total Environ 468:630–641

Robinson BJ, Hui JP, Soo EC, Hellou J (2009) Estrogenic compounds in seawater and sediment from Halifax Harbour, Nova Scotia, Canada. Environ Toxicol Chem 28:18–25

Rodríguez-Navas C, Björklund E, Bak SA, Hansen M, Krogh KA, Maya F, Forteza R, Cerdà V (2013) Pollution pathways of pharmaceutical residues in the aquatic environment on the island of Mallorca, Spain. Arch Environ Contam Toxicol 65:56–66

Romero J, Feijoó CG, Navarrete P (2012) Antibiotics in aquaculture-use, abuse and alternatives. In: Carvalho ED, David GS, Silva RJ (eds) Health and environment in aquaculture. INTECH Open Access Publisher, Rijeka. ISBN: 9789535104971

Romero J, Navarrete P (2006) 16S rDNA-based analysis of dominant bacterial populations associated with early life stages of coho salmon (*Oncorhynchus kisutch*). Microb Ecol 51:422–430

Rose JM, Gast RJ, Bogomolni A, Ellis JC, Lentell BJ, Touhey K, Moore M (2009) Occurrence and patterns of antibiotic resistance in vertebrates off the Northeastern United States coast. FEMS Microbiol Ecol 67:421–431

Rosi-Marshall E, Snow D, Bartelt-Hunt SL, Paspalof A, Tank J (2015) A review of ecological effects and environmental fate of illicit drugs in aquatic ecosystems. J Hazard Mater 282:18–25

Roth M, Richards R, Sommerville C (1993) Current practices in the chemotherapeutic control of sea lice infestations in aquaculture: a review. J Fish Dis 16:1–26

Roy W, Sutherland I, Rodger H, Varma K (2000) Tolerance of Atlantic salmon, *Salmo salar* L., and rainbow trout, *Oncorhynchus mykiss* (Walbaum), to emamectin benzoate, a new orally administered treatment for sea lice. Aquaculture 184:19–29

Ryu S-H, Park S-G, Choi S-M, Hwang Y-O, Ham H-J, Kim S-U, Lee Y-K, Kim M-S, Park G-Y, Kim K-S (2012) Antimicrobial resistance and resistance genes in *Escherichia coli* strains isolated from commercial fish and seafood. Int J Food Microbiol 152:14–18

Santos LH, Araújo AN, Fachini A, Pena A, Delerue-Matos C, Montenegro M (2010) Ecotoxicological aspects related to the presence of pharmaceuticals in the aquatic environment. J Hazard Mater 175:45–95

Santos N, Medina W, Martins N, Rodrigues MC, Curti C, Santos A (2008) Involvement of oxidative stress in the hepatotoxicity induced by aromatic antiepileptic drugs. Toxicol In Vitro 22:1820–1824

Sarmah AK, Meyer MT, Boxall AB (2006) A global perspective on the use, sales, exposure pathways, occurrence, fate and effects of veterinary antibiotics (VAs) in the environment. Chemosphere 65:725–759

Schmitt H, Boucard T, Garric J, Jensen J, Parrott J, Pery A, Rombke J, Straub JO, Hutchinson TH, Sanchez-Arguello P, Wennmalm A, Duis K (2010) Recommendations on the environmental risk assessment of pharmaceuticals: effect characterization. Integr Environ Assess Manag 6 (Suppl):588–602

Schulze AD, Alabi AO, Tattersall-Sheldrake AR, Miller KM (2006) Bacterial diversity in a marine hatchery: balance between pathogenic and potentially probiotic bacterial strains. Aquaculture 256:50–73

Semblante GU, Hai FI, Huang X, Ball AS, Price WE, Nghiem LD (2015) Trace organic contaminants in biosolids: impact of conventional wastewater and sludge processing technologies and emerging alternatives. J Hazard Mater 300:1–17

Shenker M, Harush D, Ben-Ari J, Chefetz B (2011) Uptake of carbamazepine by cucumber plants —a case study related to irrigation with reclaimed wastewater. Chemosphere 82:905–910

Shpakov A, Shipilov V, Bondareva V (2005) Sensitivity of adenylyl cyclase signaling system of the mollusk *Anodonta cygnea* ganglions to serotonin and adrenergic agonists. Ann N Y Acad Sci 1040:466–468

Small C, Nicholls RJ (2003) A global analysis of human settlement in coastal zones. J Coast Res, 584–599

Smith P (2008) Antimicrobial resistance in aquaculture. Rev Sci Tech 27:243–264

Smith P, Christofilogiannis P (2007) Application of normalised resistance interpretation to the detection of multiple low-level resistance in strains of *Vibrio anguillarum* obtained from Greek fish farms. Aquaculture 272:223–230

Solé M, Shaw JP, Frickers PE, Readman JW, Hutchinson TH (2010) Effects on feeding rate and biomarker responses of marine mussels experimentally exposed to propranolol and acetaminophen. Anal Bioanal Chem 396:649–656

Son S, Wang M, Shon J-K (2011) Satellite observations of optical and biological properties in the Korean dump site of the Yellow Sea. Remote Sens Environ 115:562–572

Spongberg AL, Witter JD, Acuna J, Vargas J, Murillo M, Umana G, Gomez E, Perez G (2011) Reconnaissance of selected PPCP compounds in Costa Rican surface waters. Water Res 45:6709–6717

Stewart M, Olsen G, Hickey CW, Ferreira B, Jelić A, Petrović M, Barcelo D (2014) A survey of emerging contaminants in the estuarine receiving environment around Auckland, New Zealand. Sci Total Environ 468:202–210

Taylor NG, Verner-Jeffreys DW, Baker-Austin C (2011) Aquatic systems: maintaining, mixing and mobilising antimicrobial resistance? Trends Ecol Evol 26:278–284

Teo JW, Suwanto A, Poh CL (2000) Novel β-lactamase genes from two environmental isolates of Vibrio harveyi. Antimicrob Agents Chemother 44:1309–1314

Tinh NTN, Dierckens K, Sorgeloos P, Bossier P (2008) A review of the functionality of probiotics in the larviculture food chain. Mar Biotechnol 10:1–12

Todar K (2002) Antimicrobial agents used in the treatment of infectious disease. Department of Bacteriology, University of Wisconsin-Madison, Madison

Trautwein C, Berset J-D, Wolschke H, Kümmerer K (2014) Occurrence of the antidiabetic drug Metformin and its ultimate transformation product Guanylurea in several compartments of the aquatic cycle. Environ Int 70:203–212

Vanderford BJ, Snyder SA (2006) Analysis of pharmaceuticals in water by isotope dilution liquid chromatography/tandem mass spectrometry. Environ Sci Technol 40:7312–7320

Vanneste J (2013) Recent progress on detecting, understanding and controlling Pseudomonas syringae pv. actinidiae: a short review. NZ Plant Prot 66:170–177

Verlicchi P, Al Aukidy M, Zambello E (2012) Occurrence of pharmaceutical compounds in urban wastewater: removal, mass load and environmental risk after a secondary treatment—a review. Sci Total Environ 429:123–155

Viarengo A, Lowe D, Bolognesi C, Fabbri E, Koehler A (2007) The use of biomarkers in biomonitoring: a 2-tier approach assessing the level of pollutant-induced stress syndrome in sentinel organisms. Comp Biochem Physiol C: Toxicol Pharmacol 146:281–300

Waddy S, Burridge L, Hamilton M, Mercer S, Aiken D, Haya K (2002) Rapid communication/communication rapide emamectin benzoate induces molting in American lobster, *Homarus americanus*. Can J Fish Aquat Sci 59:1096–1099

Waddy S, Merritt V, Hamilton-Gibson M, Aiken D, Burridge L (2007) Relationship between dose of emamectin benzoate and molting response of ovigerous American lobsters (*Homarus americanus*). Ecotoxicol Environ Saf 67:95–99

Wang G, Ma P, Zhang Q, Lewis J, Lacey M, Furukawa Y, O'Reilly SE, Meaux S, McLachlan J, Zhang S (2012) Endocrine disrupting chemicals in New Orleans surface waters and Mississippi Sound sediments. J Environ Monit 14:1353–1364

Wang J, Gardinali PR (2013) Uptake and depuration of pharmaceuticals in reclaimed water by mosquito fish (*Gambusia holbrooki*): a worst-case, multiple-exposure scenario. Environ Toxicol Chem 32:1752–1758

Wijekoon KC, Hai FI, Kang J, Price WE, Guo W, Ngo HH, Nghiem LD (2013) The fate of pharmaceuticals, steroid hormones, phytoestrogens, UV-filters and pesticides during MBR treatment. Biores Technol 144:247–254

Xu Y, Nguyen TV, Reinhard M, Gin KY-H (2011) Photodegradation kinetics of p-tert-octylphenol, 4-tert-octylphenoxy-acetic acid and ibuprofen under simulated solar conditions in surface water. Chemosphere 85:790–796

Yang S, Hai FI, Nghiem LD, Nguyen LN, Roddick F, Price WE (2013a) Removal of bisphenol A and diclofenac by a novel fungal membrane bioreactor operated under non-sterile conditions. Int Biodeterior Biodegradation 85:483–490

Yang S, Hai FI, Nghiem LD, Price WE, Roddick F, Moreira MT, Magram SF (2013b) Understanding the factors controlling the removal of trace organic contaminants by white-rot fungi and their lignin modifying enzymes: a critical review. Biores Technol 141:97–108

Yu Y, Wu L (2012) Analysis of endocrine disrupting compounds, pharmaceuticals and personal care products in sewage sludge by gas chromatography—mass spectrometry. Talanta 89:258–263

Zenker A, Cicero MR, Prestinaci F, Bottoni P, Carere M (2014) Bioaccumulation and biomagnification potential of pharmaceuticals with a focus to the aquatic environment. J Environ Manage 133:378–387

Zhang X, Gao Y, Li Q, Li G, Guo Q, Yan C (2011) Estrogenic compounds and estrogenicity in surface water, sediments, and organisms from Yundang Lagoon in Xiamen, China. Arch Environ Contam Toxicol 61:93–100

Zheng Q, Zhang R, Wang Y, Pan X, Tang J, Zhang G (2012) Occurrence and distribution of antibiotics in the Beibu Gulf, China: impacts of river discharge and aquaculture activities. Mar Environ Res 78:26–33

Zheng S, Qiu X, Chen B, Yu X, Liu Z, Zhong G, Li H, Chen M, Sun G, Huang H (2011) Antibiotics pollution in Jiulong River estuary: source, distribution and bacterial resistance. Chemosphere 84:1677–1685

Zhou X-X, Wang Y-B, Li W-F (2009) Effect of probiotic on larvae shrimp (*Penaeus vannamei*) based on water quality, survival rate and digestive enzyme activities. Aquaculture 287:349–353

Chapter 10
Waste Treatment in Recirculating Shrimp Culture Systems

Raj Boopathy

Abstract Disposal of effluent from conventional shrimp farms containing high concentration of organic matter and nutrients into different environmental systems can have detrimental impacts on their adjacent ecosystems (e.g. coastal ecosystem). To minimize the environmental impacts of shrimp farming effluents, recirculating raceway system has recently been introduced, producing high density shrimp yields. Although it is a zero water exchange system, a certain portion of water from the recirculating system needs to be treated or disposed of on regular basis due to an increase in the concentration of nitrate and nitrites owing to the protein enriched diet of shrimps. Hence, an effective approach for the treatment of nitrogen enriched wastewater produced by the recirculating raceway system is needed. Biological wastewater treatment capable of nitrification and denitrification is simple and environmentally friendly approach. In this regard, removal of ammonia and nitrates was assessed in aerobic sequencing batch reactor (SBR), providing almost complete removal of organic impurities (above 99%) as well as of ammonia. On the other hand, anaerobic SBR achieved efficient denitrification and also provided above 99% total nitrogen removal. Notably, the addition of Bacillus consortium in SBR can be helpful to control the growth of shrimp pathogen, *Vibrio harveyi* in the wastewater. This chapter discusses global shrimp production, biosecurity, recirculating raceway system, and the use of SBR in treating shrimp wastewater.

Keywords Denitrification · Nitrification · Sequencing batch reactor
Recirculating raceway system · Zero water exchange · Probiotic

---

R. Boopathy (✉)
Department of Biological Sciences, Nicholls State University, Thibodaux, LA 70310, USA
e-mail: Ramaraj.Boopathy@nicholls.edu

F. I. Hai et al. (eds.), *Sustainable Aquaculture*, Applied Environmental Science and Engineering for a Sustainable Future,
https://doi.org/10.1007/978-3-319-73257-2_10

10.1 Introduction

Over the last two decades there has been an increase in consumer demand for shrimp (Weirich et al. 2002). In 1998, 80% of shrimp consumed in the United States was imported to meet consumer demand (Browdy 1998), which has led to a 3.2 billion dollar trade deficit (Aquaculture Outlook 1999). This deficit increased, due mostly to a strong United States domestic economy, resulting in an increase in both restaurant sales and home consumption of shrimp (Jory 2000). In order to reduce this trade deficit, the United States Department of Agriculture developed the United States Marine Shrimp Farming Program (USMSFP) to increase shrimp production in the United States.

In order for the United States to compete within the global shrimp market and decrease the trade deficit, new research and development must create technology to increase domestic shrimp farm production. Traditional pond systems, which are used to produce the bulk of farm raised shrimp in the United States, have three main limiting factors: length of growing season, land limitations, and high rates of water exchange (Browdy and Moss 2005). Shrimp have a growing season based on temperature. Therefore, ponds can only produce shrimp during the time of year when the temperature is conducive for growth. Land limitations refer to the fact that pond aquaculture is typically located close to the coast for easy water exchange with estuaries and other bodies of water. Coastal land is often expensive and limited in availability due to coastal development, making farming activities difficult (Landesman 1994). Finally, effluents from conventional shrimp farming ponds having a wide range of organic and inorganic impurities generally contains high level of organic carbon and nutrients. The potential environmental problems caused by aquaculture effluents include, but are not limited to, oxygen depletion, degradation of benthic communities, and exacerbation of toxic algae bloom (Goldburg and Triplett 1997). Because of the potential detrimental effects of aquaculture effluents on adjacent water bodies and ecosystems criteria and standards for the disposal of aquaculture effluent have been developed and enforced by different government agencies (Kinne et al. 2001). For instance, the United Stated Environmental Protection Agency enforced the code of a federal regulation (part 451) under the Clean Water Act in 2004, establishing a narrative of technology based limitations as well as establishing environmental quality standards for the disposal of effluent from aquatic animal production facilities.

The USMSFP developed recirculating raceway system for high intensive shrimp farming in the U.S. In a recirculating aquaculture system, 95–99% of water is reused. This means that the system is required to have some type of water treatment process, such as a biological bead filter or sedimentation cones, to control the accumulation of suspended solids (Summerfelt et al. 2001). Because recirculating systems do not require water exchange with the environment, these systems can operate with minimal water exchange. Raceways are culture units in which the water flow is sufficient to induce a current to which aquatic organisms respond and in which detrital material can be transported. The advantages of this system include increased production of shrimp per unit space and the ease of harvesting and feeding (Mazik and Parker 2001). When recirculating and raceway systems are combined the system is called a recirculating raceway system. The recirculating raceway system combines advantages from both the recirculating and raceway systems. The recirculating raceway system can grow shrimp in high densities, increase the ease of feeding and harvesting, and operate with minimal water exchange.

Recirculating raceway systems can be very intensive system that produce large yields of shrimp per unit volume. However, the waste produced due to the high shrimp density, can cause substantial environmental impacts if discharged into the environment (Kinne et al. 2001). This waste consists of uneaten fish feed, fecal matter, and urine. The feed and fecal wastes produced in the raceway system is made almost entirely of organic matter (Goldburg and Triplett 1997), and is characterized by high amounts of COD, BOD, total suspended solids (TSS), dissolved particulate matter, volatile suspended solids (VSS) and nutrients (Kinne et al. 2001; Cohen et al. 2005).

Waste-treatment methods for aquaculture are largely adapted from municipal wastewater (sewage) treatment (Goldburg and Triplett 1997). In many shrimp farms, an industrial bead filter constantly filters the water. The bead filter works by passing the water through a filter bed without the addition of chemicals. The granular material inside the filter will remove suspended solids, through a series of complex processes involving one or more removal mechanisms. These mechanisms include straining, interception, impaction, sedimentation, and adsorption. Once the filtration process is complete, the filter must be backwashed to remove the suspended solids that have accumulated in the filter. This is achieved by reversing the flow of water through the filter. The granular material is fluidized and the suspended solids are washed out of the filter (Cerra and Maisel 1979). Bacterial processes, such as nitrification, also occur within the bead filter. In some shrimp farms sedimentation cones are used to remove suspended solids. Water from the raceway systems are pumped into the sedimentation cones, where the suspended solids are allowed to settle out into the bottom of the cones, and then the water on top of the cone is pumped back into the recirculating raceway system. Valves and gravity are used to empty the solids from the bottom of the cones. Both sedimentation and filtration methods produce large amounts of wastewater. This wastewater gets pumped out of the recirculating raceway systems via emptying the sedimentation cones or backwashing the bead filter, and has to be disposed. Disposal of this wastewater is complicated by the wastewater's saline properties, and reuse of this water will cause problems due to the toxic concentrations of ammonia and nitrite. Most of the studies focusing on the treatment of wastewater from aquaculture have been carried out in small scale systems under controlled environmental conditions. For instance, Boopathy et al. (2005) achieved significant removal of organic impurities as well as nutrients following the treatment of shrimp farming wastewater in a sequencing batch reactor (SBR). A SBR consisting of a single reactor utilizes activated sludge for the treatment of wastewater and can be operated in aerobic as well as anaerobic mode (Boopathy et al. 2005; Fontenot et al. 2007; Morgenroth and Wilder 1998). Despite the excellent performance of SBR, it is

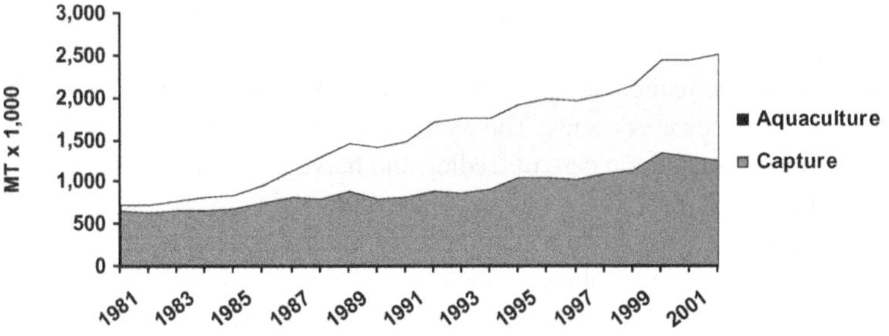

Fig. 10.1 Global Peneaid shrimp supply for the period of 1981–2001 (Ostrowski et al. 2005)

imperative to assess its operational feasibility as well as treatment efficiency at large scale for its practical applications (Brune et al. 2003). This chapter discusses global shrimp production, biosecurity, recirculating raceway system, and the use of SBR in treating shrimp wastewater.

10.1.1 Global Shrimp Production

Total global shrimp production consists of (i) planned shrimp farming activities; and (ii) wild harvesting. According to an estimate by Keithly et al. (2005), total global production of shrimps increased significantly with an approximate growth rate of 220 million pounds per annum from 1980 to 2001 (Fig. 10.1) 220 million pounds is a significant number in the fact that the southeast (North Carolina to Texas) United States shrimp harvest usually falls within the 220–280 million pounds per year (Keithly et al. 2005). As the population increases in the United States, the southeast shrimp harvest cannot meet the United States consumer demand for marine shrimp. In 1980, 258 million pounds of headless shell-on shrimp were imported into the United States and accounted for 55% of the total United States shrimp supply. By 2001, imports into the United States advanced to 1.18 billion pounds at which point they represented 85% of total United States shrimp supply (Keithly et al. 2005). In 1998, approximately 80% of shrimp consumed in the United States were imported and of this amount over 50% of the shrimp came from shrimp farms located in Asia (Browdy 1998). To reduce the trade deficit, shrimp farming must increase, because the wild-caught harvest can be variable due to weather variation, increasing oil prices, and fishing pressure on shrimp populations. In 1980, global production of farmed shrimp equaled about 160 million pounds, and by 2001, global production of farmed shrimp increased to 2.8 billion pounds, or more than 35% of total shrimp output. Overall farmed shrimp production increased by approximately 130 million pounds per year during 1980–2001 (Keithly et al. 2005). The increased farmed shrimp production allowed more shrimp product to enter the global trade market. This made shrimp more affordable for consumers dropping the price of shrimp per pound from $5.82 in 1980 to $2.87 in 2001 (Keithly et al. 2005). Csavas (1994) stated that farm-raised shrimp are of greater importance to global trade than wild-caught product because farm-raised product has greater freshness, can be grown year round, sizes can be controlled better in a farm-based system, and the farming system will help meet consumer demands. Although the United States demand for marine shrimp is high, the United States' contribution to world farm-raised shrimp production is insignificant. In the United States there are only a few facilities that focus on the production of the pacific white shrimp, *Litopenaeus vannamei*. These facilities are primarily located in Florida, Hawaii, South Carolina, and Texas, and of these facilities some produce only larval shrimp, which are sold and distributed to foreign buyers (Weirich et al. 2002).

10.1.1.1 Litopenaeus Vannamei

Litopenaeus vannamei, or the pacific white shrimp, is the most widely produced shrimp in recirculating raceway systems (Williams et al. 1996). In 1996, this shrimp constituted 30% of the farmed penaeid shrimp worldwide (Williams et al. 1996). *L. vannamei* grows naturally in salinities ranging from 1 to 40 g/L (Menz and Blake 1980), by using different osmotic regulation mechanisms such as hyper-osmotic regulation in low salinity and hypo-osmotic regulation in high salinity (Castille and Lawrence 1998). However, most marine shrimps are grown in water with salinities higher than 15 g/L.

L. vannamei is a rapidly growing species, which is disease tolerant, and has a good survival rate in a high density system (Williams et al. 1996). Even though *L. vannamei* is the most widely produced species in aquaculture, there are still problems associated with mass production of the species. *L. vannamei* can be an invasive species if released into areas that are non-native.

10.1.2 Biosecurity

Biosecurity has been defined as the sum of all procedures in place to protect living organisms from contracting, carrying, and spreading diseases and other non-desirable health conditions (Pruder 2004). Biosecurity is comprised of a series of activities that include preventive medicine, adequate diagnosis, containing outbreaks that occur, and the eradication or disinfection of the pathogen (Pruder 2004). A biosecure facility should also be designed around the concept of biosecurity. Materials used in ponds and raceways should be easy to disinfect, the facility should not have unauthorized vehicles or visitors, and also the facility should prevent the escape of the organisms while preventing the entry of others (Pruder 2004).

The main objective of biosecurity for shrimp farms is to control pathogens such as Taura Syndrome Virus (TSV), White Spot Virus (WSV), the Hepatopancreatic Parvovirus (HPV), and other major shrimp pathogens. The main vector for these pathogens appears to be shrimp larvae, both wild-caught shrimp larvae and shrimp larvae that have been raised in a hatchery (Pruder 2004; Otoshi et al. 2003). In order to reduce pathogens related to shrimp larvae, researchers have developed high health shrimp or specific pathogen-free (SPF) shrimp. These shrimp go through rigorous quarantine and screening efforts. Once the SPF shrimp leave the hatchery the SPF shrimp are considered high health shrimp (Pruder 2004; Browdy and Moss 2005; Otoshi et al. 2003).

The second vector that is responsible for introducing disease and pathogens into aquaculture systems is water exchange (Pruder 2004; Cohen et al. 2005; Browdy and Moss 2005). The use of untreated water can put the entire aquaculture facility at risk of disease. With some shrimp farming methods, water is discharged daily into a receiving body of water, and then replaced with untreated water (Cohen et al. 2005; Otoshi et al. 2003). This technique helped to ensure water quality in the aquaculture system, but is no longer practiced due to conflicts with other potential users, pathogen induction, and government regulations on effluent. The discharge issue has generated interest in the shrimp aquaculture industry to achieve zero-water exchange using recirculating systems (Cohen et al. 2005). The zero-water exchange method simply continues to use the same water, helping to guarantee biosecurity. For the recirculating raceway system to be successful when using the zero-water exchange method, proper feed management, adequate aeration and circulation, and nitrogen cycling processes must be managed carefully (Cohen et al. 2005).

Finally, the third major vector for pathogen induction is through excess feed. Elevated concentrations of nitrogen and phosphorous from the excess feed can stimulate growth or blooms of phytoplankton (algae), a process termed eutrophication (Goldburg and Triplett 1997). Due to the lack of water-exchange in aquaculture systems, nutrients can easily build up and cause eutrophication. When algae die in large numbers, there is a large influx of organic matter into the water column. BOD is used to measure the concentration of organic matter available for degradation by microorganisms. When BOD is high, microorganisms are using oxygen in the water to decompose organic matter (Goldburg and Triplett 1997). So, eutrophication due to excess feed can cause a population increase in microbial, algal, and other microscopic communities. The increase of microscopic communities can impact the carrying capacity of the culture system by disrupting water quality parameters such as dissolved oxygen (Cohen et al. 2005).

10.1.3 Recirculating Raceway Systems

Recirculating raceway systems have many advantages over traditional pond aquaculture. First, this system allows for biosecurity measures to be implemented very easily and prevents losses due to infectious diseases (Bratvold and Browdy 1999; Browdy and Moss 2005). Also, recirculating raceway systems can achieve year round production if kept indoors (Browdy and Moss 2005). This is because recirculating raceway systems can be enclosed in greenhouses that can be heated using heat-exchange units located on the bottom of the raceway or cooled by fans built into the greenhouses in order to maintain the optimal growth temperature for the shrimp (Weirich et al. 2002). With the ability to grow shrimp year round farmers can time production to market conditions rather than the seasons of the year (Goldburg and Triplett 1997). An additional benefit to the recirculating raceway being indoors is the ability to reduce losses of shrimp to predators, which also helps to reduce pathogen induction from wild organisms. Finally, environmental impacts due to effluent discharges are reduced or eliminated because of the minimal water exchange between the recirculating raceway system and the environment. Additionally, the recirculating raceway systems also use less land than traditional pond systems (Browdy and Moss 2005). Greater environmental control means that recirculating raceway systems offer better control of contaminants, product quality, predators, and introduction of pathogens. There are disadvantages to the recirculating raceway system, most notably the fact that the raceway has to treat and circulate large volumes of water and typically require larger capital investments. Additionally, recirculating raceway systems have higher operating costs due to energy, labor, and supplies such as supplemental oxygen.

When managing recirculating raceway systems, heterotrophic and autotrophic bacteria are important, and an adequate understanding of the role bacteria play is essential in maintaining recirculating raceway water quality. Heterotrophic bacteria obtain carbon and energy for growth from organic compounds, survive during periods of stress (such as limited food sources and low oxygen) by forming spores, and are important dietary components of detritivores such as shrimp (McGraw 2002). Autotrophic bacteria obtain energy from light (photoautotroph) or the oxidation of inorganic compounds such as ammonia (chemoautotrophs), survive during periods of stress through inactivity, and are a poor source of nutrition for detritivores (McGraw 2002). However, autotrophic bacteria are more efficient at nitrification than heterotrophic bacteria. Most recirculating raceway systems are heterotrophic systems. Heterotrophic systems are managed by a large input of feed, which maintains the C:N ratio in the system. With the excessive carbon, bacteria populations increase, and become a feed source for the shrimp. Significant nitrification occurs within the raceway system, although the overall rate of nitrification is limited by the faster growing heterotrophic microbial population (Brune et al. 2003). Nitrification limitation in the recirculating raceway system is due to the availability of biodegradable organic matter, which supports the growth of heterotrophic bacteria. These heterotrophs compete with the autotrophic nitrifiers for oxygen, nutrients, and space (Sharma and Ahlert 1977; Zhu and Chen 2001). Heterotrophic recirculating raceway systems are also designed to maximize aeration to mix suspended solids to improve nitrification (Avinmelech et al. 1986, 1999). Excessive carbon matter, typically in the form of molasses, can be added to the recirculating raceway system in order to stimulate heterotrophic bacteria growth and increase nitrogen uptake.

10.1.4 Sequencing Batch Reactor

An activated sludge based wastewater treatment process was studied for the first time in 1914 and was operated in batch mode. This batch-fed wastewater treatment system was comprised of following steps: (i) addition of wastewater; (ii) aeration for biological degradation of pollutants; and (iii) sedimentation to allow the settlement of activated sludge. After the settlement of activated sludge, the supernatant containing low levels of organic and nutrients was collected for discharge and above steps were repeated for the treatment of next wastewater batch (Fang et al. 1993). However, this batch operation was discontinued in favor of continuous operation, because at the time batch operation did not seem as effective at treating wastewater, compared to other wastewater treatments. Eventually, interest in batch-fed reactors were resuscitated again during 1950–1960 mainly to test newly developed treatment process as well analytical equipment (Fang et al. 1993). The fill-and-draw type of batch operation was re-examined and renamed Sequencing Batch Reactor (SBR). Owing to the development of improved control devices as well as aeration equipment, SBR can compete with continuous flow conventional activated sludge treatment system such as the plug flow reactor, continuous stir tank reactor, and an arbitrary flow reactor, providing comparable reduction in pollutant concentrations (Fang et al. 1993).

The SBR is used in shrimp industry (Boopathy et al. 2005, 2007; Fontenot et al. 2007) because of its ability to accomplish equalization, aeration, and clarification in a timed sequence in a single reactor basin (Fig. 10.2). Other activated sludge systems use multiple structures to achieve equalization, aeration, and clarification, which require extensive plant space and pumping and piping systems. The SBR promises to reduce operating costs and plant space (Jang et al. 2004). The sequencing series for wastewater treatment using the SBR consists of the following process stages: fill, react, settle, decant and idle (Boopathy et al. 2005; Kargi and

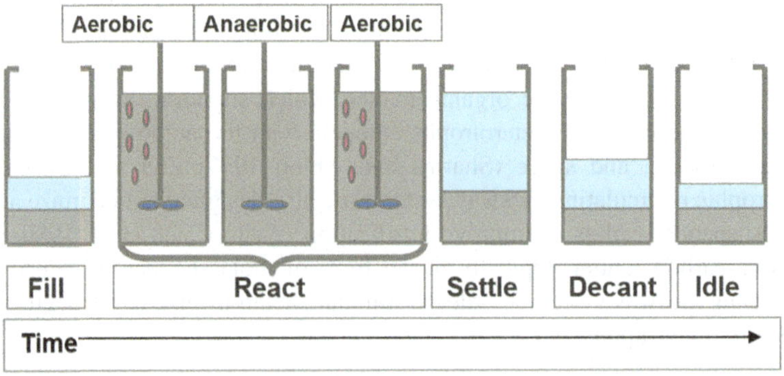

Fig. 10.2 A schematic of a sequencing batch reactor showing different processes/stages of wastewater treatment. All stages of treatment are carried out in a single reactor at different time intervals (Boopathy et al. 2005)

Uygur 2005; Morgenroth and Wilder 1998). The SBR has been well studied in terms of the potential for simultaneous removal of carbon and nitrogen (Murat et al. 2002), and has been successfully used to treat both municipal and industrial wastewater (Peters et al. 2004). Murat et al. (2002) used a SBR to successfully remove high levels of COD and nitrogen from tannery wastewater. The SBR is also extensively used in the treatment of swine wastewater (Juteau et al. 2005; Deng et al. 2006). Peters et al. (2004) showed that SBR can be used to effectively remove high levels of COD and nitrogen from sewage effluent.

10.1.5 Bacteriology of Nitrification and Denitrification

Nitrogen is an essential nutrient for all organisms, as part of important molecules such as proteins, nucleic acids, adenosine phosphates, pyridine nucleotides, and pigments (Hagopian and Riley 1998). Shrimp expel nitrogen through urination and excretion. Uneaten feed and decomposing deceased shrimp also contributes to nitrogenous waste in aquaculture systems (Hagopian and Riley 1998; Cripps and Bergheim 2000). Both un-ionized ammonia and nitrite are toxic to shrimp at low concentrations. *L. vannamei*, exhibited a 96-h LC50 (median lethal concentration) of 24.39 mg/L ammonia with a salinity at 15 ppt, 8.05 pH, and a temperature of 23 °C (Lin and Chen 2001). The 96-h LC50 for nitrite in *L. vannamei* is 76.5 mg/L at 15 ppt salinity, with a water temperature at 18 °C, and the pH at 8.02 (Lin and Chen 2003). Therefore, nitrification and denitrification are very important processes in the treatment of shrimp aquaculture wastewater, so that ammonia and nitrite do not accumulate in recirculating raceway systems. Ammonia and nitrite become mineralized through nitrification into nitrate compound and then nitrate becomes volatilized through denitrification and into nitrogen gas.

10.1.5.1 Nitrification

Autotrophic as well as heterotrophic bacteria, under aerobic conditions, can be involved in a biological nitrification process (Zhu and Chen 2001). Notably, two groups of bacteria with different phylogenetic evolution, namely ammonia oxidizing bacteria and nitrite-oxidizing bacteria, are known to perform nitrification. Ammonia-oxidizing bacteria such as *Nitrosomonas*, *Nitrosovibrio* and *Nitrosospira* perform nitrification via catabolic conversion of ammonia to nitrite (Reaction 1), while nitrite-oxidizing bacteria such as *Nitrobacter* and *Nitrospira* (Reaction 2) are responsible for the conversion of nitrite to nitrate (Hagopian and Riley 1998; Zhu and Chen 2001). Compared to the conversion of nitrite to nitrate, ammonia conversion to nitrite generates more energy (Hagopian and Riley 1998; Remde and Conrad 1990; Rijn 1995) as shown:

$$NH_3 + 1.5O_2 \leftrightarrow NO_2^- + H_2O + H^+ + 84\ \text{kcal mol}^{-1} \quad \text{(Reaction1)}$$

$$NO_2^- + 0.5O_2 \leftrightarrow NO_3^- + 17.8\ \text{kcal mol}^{-1} \quad \text{(Reaction2)}$$

As mentioned above, *Nitrosomonas*, *Nitrosovibrio* and *Nitrosospira* are involved in Reaction 1 i.e. conversion of ammonia to nitrate, while Reaction 2 i.e. conversion of nitrite to nitrate is carried out by *Nitrobacter* and *Nitrospira*, indicating that nitrifying bacteria belongs to Nitrobacteraceae family that is a group of chemoautotrophic gram negative bacteria (Tanaka et al. 1983). Chemoautotrophic bacteria utilize inorganic chemical substrates (NH_3, NO_2) as an electron source to immobilize inorganic carbon (CO_2), thereby allowing carbon fixation to occur (Hagopian and Riley 1998; Ritchie and Nicholas 1972; Sundermeyer-Klinger et al. 1984).

10.1.5.2 Denitrification

Nitrification oxidizes toxic ammonia and nitrite, to the relatively non-toxic form of nitrate; however, over a long period of time nitrate concentrations could become toxic to aquatic organisms (Sauthier et al. 1998; Ng et al. 1993). This is even more prevalent in recirculating raceway systems where the water is continuously reused. Denitrification is the process, where heterotrophic bacteria, such as *Pseudomonas denitrificans* and *Escherichia coli*, under anaerobic conditions convert nitrate to nitrogen gas, and completely volatize nitrogen from the system (Sauthier et al. 1998). The bacteria oxidize organic matter using the following electron acceptors in

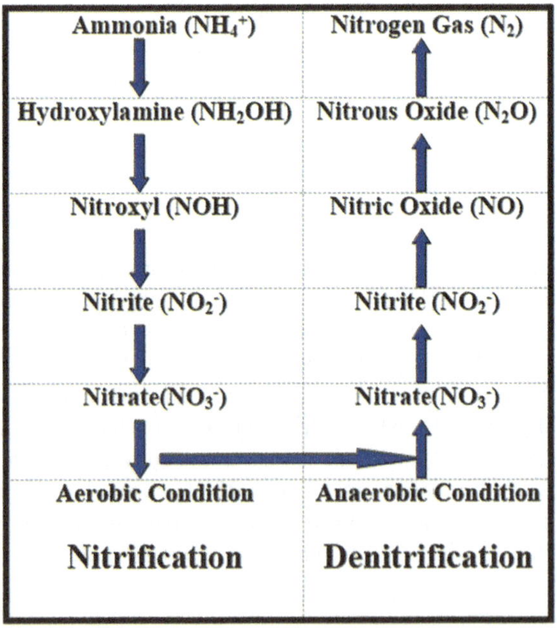

Fig. 10.3 The biochemical transition of nitrogen from ammonia to nitrogen gas by nitrification and denitrification

the following order: dissolved oxygen, nitrate, and sulfates. If there are poor reducing conditions then partial denitrification can occur, and if there are drastic reducing conditions sulfates can be reduced to toxic sulfides (Sauthier et al. 1998). Figure 10.3 shows how ammonia is converted into nitrate under aerobic conditions (nitrification), and then how nitrate is converted into nitrogen gas under anaerobic conditions (denitrification).

10.2 Case Study of Shrimp Wastewater Treatment

Performance of a sequencing batch reactor (SBR) for the treatment of shrimp farming effluent/sludge was studied to improve the quality of water in recirculating raceway system. Timing to remove wastewater and sludge is same in recirculating raceway shrimp farming system. This study was conducted with the aim to assess: (i) the performance of SBR for the removal of carbon and nitrogen from shrimp farming effluent; and (ii) the suitability of treated effluent for reuse in shrimp ponds.

10.2.1 Shrimp Waste Sludge

Bead filter backwash was collected in 3 L sealed containers from a recirculating raceway shrimp farming system located at Waddell Mariculture Center, South Carolina. Effluent/waste sludge samples were stored at 4 °C until use.

10.2.2 Sequencing Batch Reactor (SBR)

For the treatment of shrimp farming effluent, four laboratory scale SBRs each having a working volume of 19 L were erected. Initially, each SBR was fed with 4 L shrimp farming effluent and was operated under aerobic mode by providing aeration using air stones. A stirring motor (Model RW 20/RW 20DZM; Tekmar Company, Cincinnati, OH) was installed in each reactor and operated at 100 rpm to keep the contents of reactors well mixed. After a certain period of time, SBRs were operated in anaerobic mode by turning off aeration as well as mixing. Different time sequences for aerobic and anaerobic modes were tested until the end of the experiment to optimize aerobic and anaerobic time sequences for the removal of carbon and nitrogen from shrimp farming effluent. Aerobic and anaerobic mode of operation in SBR are vital because organic carbon oxidation and nitrification occur in aerobic mode, while denitrification occurs in anaerobic/anoxic mode. Without anaerobic mode of operation, total nitrogen removal can deteriorate due to poor denitrification. Since all four SBRs were operated simultaneously under identical operating condition, carbon and nitrogen removal represents the average of results obtained from four SBRs.

10.2.3 Pilot Scale SBR

After optimizing the operating conditions in laboratory scale SBRs for the treatment of shrimp farming sludge, two identical pilot scale SBRs (500 L) were installed at the recirculating raceway farming system, Waddell Mariculture Center, South Carolina. Arrangement of a SBR as well as water flow directions with respect to culture system and bead filter is shown in Fig. 10.4. SBRs were operated in aerobic and anaerobic mode for 3 and 6 days, respectively. Carbon and nitrogen removal was reported as the average of duplicate SBRs.

10.2.4 Chemical Analysis

30 mL samples from SBRs were collected at regular intervals to quantify COD, ammonia, nitrate and nitrite concentrations. Supernatant of each sample centrifuged at 5000 rpm for 10 min was used for chemical analyses. COD was analyzed as per the method described in standard methods for the examination of water and wastewater (APHA 1998). Ammonia, nitrate and nitrite were quantified via colorimetric method using HACH water analysis kit (Hach 1999). The dissolved oxygen (DO) and salinity were measured using an YSI DO and salinity probe (Model No. 85-10FT, Yellow Spring, OH), respectively. The pH was measured using a pH probe (Model UB 10, Denver Instruments, Boulder, CO).

10.2.5 Data Analysis

Data of removal efficiencies was subjected to a number of statistical analyses. Paired t-test ($p \leq 0.05$; SAS Institute 2003) was used to analyze total concentration of COD. Analysis of variance (ANOVA) followed by tukey "post hoc" analysis (SAS Institute 2003) were used for all results.

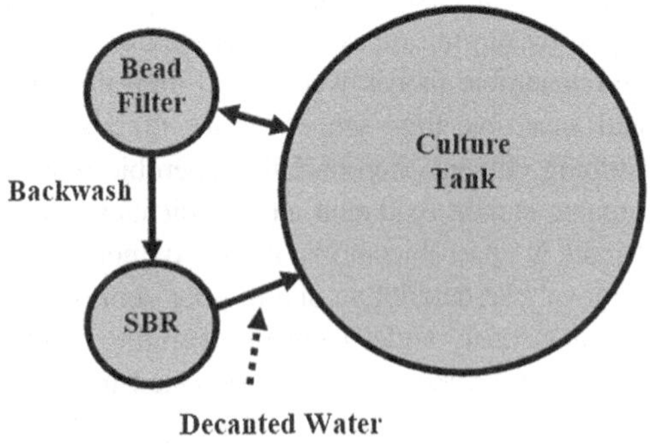

Fig. 10.4 A recirculating raceway shrimp farming system and location of bead filter and SBR. Solid arrows are showing the direction of water flow

Table 10.1 Characteristics of shrimp farming effluent

Parameters	Units	n	Concentration
pH	NA	4	8.1 ± 0.1
Salinity	ppt	4	28.6 ± 0.4
Total solids	mg/L	4	33.1 ± 3.9
Total chemical oxygen demand	mg/L	4	1593 ± 36
Ammonia	mg/L	4	83.7 ± 6.1
Nitrate	mg/L	4	31.3 ± 1.4
Nitrite	mg/L	4	250 ± 22.7

n No. of samples
NA Not applicable

10.3 Results of Case Study

10.3.1 Characterization of Shrimp Farming Effluent

Wastewater characterization is vital to: (i) understand the physical and chemical composition of wastewater; and (ii) select or design an effective wastewater treatment system. Based on the characteristics of shrimp farming effluent (Table 10.1), an activated sludge based treatment system may be efficient because of high levels of COD and nitrogen (nitrate, nitrite and ammonia). However, adequate nitrification and denitrification may be required for efficient removal of total nitrogen.

10.3.2 Performance of Laboratory SBR

Four laboratory scale SBRs (19 L each) were operated simultaneously under identical operating conditions in aerobic (0–3 days) and anaerobic (4–9 days) modes. SBRs achieved above 99% ammonia removal under aerobic conditions (Fig. 10.5). On the other hand, concentration of nitrates increased from 47 and 93 mg/L (Fig. 10.5) in SBRs at the end of aerobic operating mode, indicating that an effective nitrification occurred. When SBRs were operated in anaerobic operating conditions from 4–9 day, nitrate concentration decreased gradually. Concentration of nitrates reduced from 93 mg/L (day 3) to 2 mg/L (day 8). Similarly, concentration of nitrite increased from 235 to 401 mg/L under aerobic conditions (0–3 days) and gradually reduced to 5 mg/L at the end of anaerobic operating mode (Fig. 10.5). Reduction in nitrite and nitrate concentration under anaerobic operating mode suggests that an effective denitrification process occurred in SBRs.

Influent concentration of organic matter quantified in terms of COD was on the higher side (1596 mg/L) probably because shrimp effluent mainly contains shrimp food contents. COD concentration reduced from 1596 to 400 mg/L in SBR

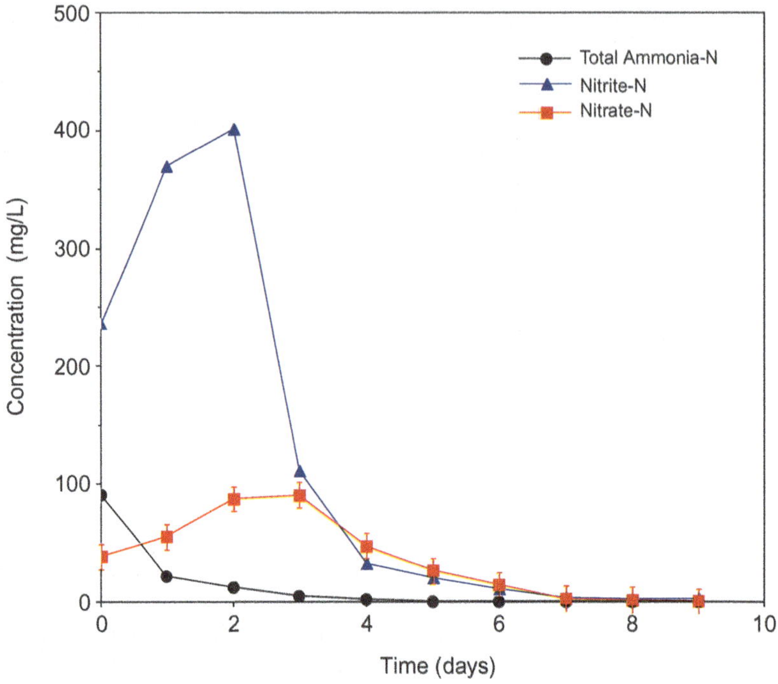

Fig. 10.5 Variations in the concentration of ammonia, nitrate and nitrite in laboratory scale SBRs during the treatment of shrimp farming effluent. SBRs were operated in aerobic and anaerobic modes from 0 to 3 and 4 to 9 days, respectively. Results are presented as average ± standard-deviation achieved in four SBRs

operated in aerobic mode for 3 days (Fig. 10.6). Similarly, SBR operated in anaerobic operating mode further reduced the concentration of COD from 4 to 6 days (Fig. 10.6) but COD removal did not improve significantly during the last three days of anaerobic operating condition i.e. 7–9 days. Operation of laboratory scale SBRs for 9 days achieved significant COD, ammonia, nitrate and nitrite removal. Nitrogen removal (Fig. 10.5) from shrimp effluent also indicate that level of nitrifying and denitrifying bacteria was adequate in sludge. Moreover, there was no need to add specific microbes to carry out nitrification and/or denitrification.

10.3.3 Performance of Pilot Scale SBR

Performance of pilot SBR plants (500 L) was assessed for the treatment of shrimp farming effluent after confirming the efficient COD and nitrogen removal in laboratory scale SBRs (Figs. 10.5 and 10.6). Two identical pilot scale SBRs were installed at Waddell Mariculture Center, SC as shown in Fig. 10.4 and the

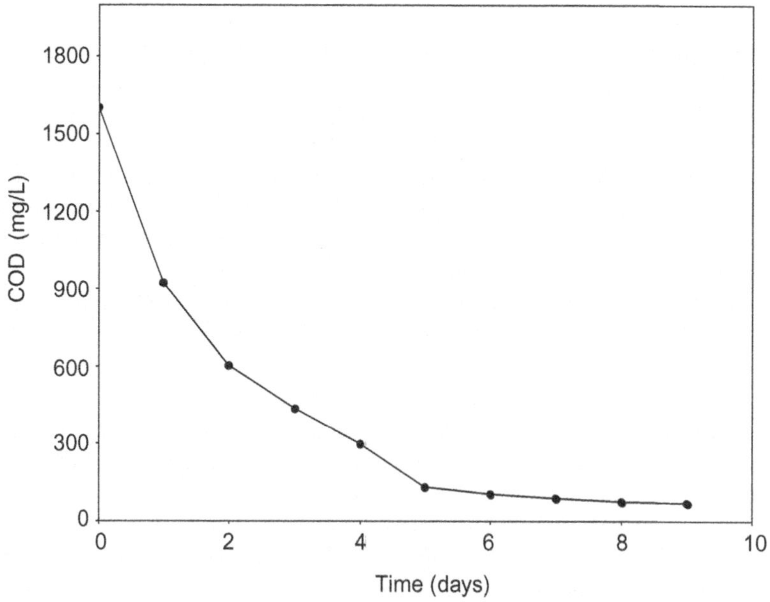

Fig. 10.6 COD concentration in the laboratory scale SBR. The results are average with S.D. for four reactors

Table 10.2 Performance[a] of pilot scale SBRs for the treatment of shrimp farming effluent/sludge

Time	Condition	NH_4	NO_2	NO_3	COD
0	Aerobic	93.7 ± 54.9	266 ± 74	21.3 ± 20.5	1593 ± 811
1		55.7 ± 42.2	661 ± 298	27.8 ± 14.8	1177 ± 669
2		19.4 ± 25.9	94 ± 70	19.2 ± 8.9	190 ± 7.8
3		9.8 ± 4.7	58.1 ± 19.3	65.0	–
4	Anaerobic	3.6 ± 8.6	46.3 ± 12.5	20.5 ± 4.1	–
5		–	20	16.8 ± 20.1	–
6		–	–	–	–
7		–	18	0	–

[a]Time (days) of each aerobic and anaerobic period and mean (N = 2) total ammonia-nitrogen (NH_4, mg/L), nitrite-nitrogen (NO_2, mg/L), nitrate-nitrogen (NO_3, mg/L), and chemical oxygen demand (COD, mg/L)

backwash of bead filter was used as an influent to pilot SBR plants. The objective of the treatment of shrimp farming effluent was to recycle the treated effluent back to shrimp culture. Pilot scale SBRs achieved 100% removal of COD, ammonia, nitrite and nitrate within a week of their operation (Table 10.2).

SBRs (laboratory and pilot scale) successfully achieved almost complete removal of organic matter as well as ammonia, nitrate and nitrites from shrimp farming effluent or sludge. SBR have a number of advantages over continues flow activated sludge treatment system including simple reactor design as well as simplicity of operation. SBR have been studied for the treatment of different type of wastewater such as slaughterhouse effluents, dairy effluent, swine manure and sewage (Irwine and Ketchum 1989; Masse and Masse 2000; Fernandes et al. 1991; Lo et al. 1991). Although performance of SBRs was excellent for shrimp farming effluents, a number of treatment systems such as conventional activated sludge process as well as the use of filtration systems have also been studied for the treatment of shrimp farming effluents. Moreover, apparatus for the removal of sludge and foam fractions from shrimp farming effluents has also been designed and investigated (Arbiv and Rijn 1995; Browdy et al. 1995; Hopkins 1994; Holloway 2002). However, these systems can be expensive and may have higher operating expenditures. Design of SBR is simple and all processes occur in a single tank compared to the requirement of multiple tanks in continuous flow activated sludge process. This case study showed that shrimp farming effluents or sludge can be efficiently treated in SBRs. Simple operation of SBR in aerobic (0–3 days) and anaerobic mode (4–9 days) led to almost complete removal (above 99%) of COD and total nitrogen. Although nitrifying bacteria bacteria are slow growing microbes, efficient ammonia removal was achieved, meaning that shrimp farming effluent already contained these microbes. On the other hand, denitrification *i.e.* nitrate and nitrite removal was achieved only during the anaerobic operating mode (Fig. 10.5). At the end of the operation the sludge can be dewatered and the water can be recycled back into shrimp production. The application of SBR technology for intensive shrimp production is an attractive alternative to various methods currently used in shrimp aquaculture.

The practice of introducing microbes has become a common practice in commercial aquaculture activities, particularly in shrimp farming, around the globe. Purpose of bacterial amendments is to improve: (i) the digestion of food in aquaculture; (ii) the immune system of cultures against pathogens; and (iii) the quality of water as well as ponds bottom conditions. These products containing certain strains of bacteria used to improve the environmental quality of the aquaculture ponds are marketed as "probiotics". Use of probiotics in aquaculture industry has grown rapidly, suggesting a positive perception towards these products. On the other hand, there are many products in the market which range in price and quality. Some have followed meticulous quality control systems and are based on years of scientific research. Unfortunately, the market also has many low quality products, which offer the grower little to no value. These types of low quality products cast doubt on the use of microbial amendments in aquaculture.

10.3.4 Use of Probiotics in Shrimp Aquaculture

Effects of probiotics in aquaculture is not completely understood because available studies are not enough to develop a definitive opinion. Probiotics have mostly been studied with the view point of their impacts on the environmental quality of aquaculture ponds as well as on the diversity of pathogens. Bacterial amendments are not only used in aquaculture but are also used in wastewater treatment and/or bioremediation. Bacterial amendments in bioremediation are known to provide favorable conditions for the growth of certain bacterial communities to: (i) competitively exclude other microbial communities; and (ii) provide direct bioremediation or biodegradation. On the other hand, impacts of bacterial amendments in aquaculture have not been demonstrated clearly mainly because environmental conditions are highly variable in aquaculture ponds, thereby making it difficult to either maintain effective controls or to replicate environmental conditions. A better understanding of comparative efficacy and mode of action of various probiotic products will depend upon the development of more controlled laboratory testing models.

In order to study the efficacy of any probiotic, focus should be on the in vitro microbiological assessment of probiotics. FDA Bam and AOAC procedures were used to verify bacterial populations in the product in the order of a billion per gram. Disc diffusion methods and broth inhibition assays were used to confirm inhibition rates against pathogens of interest including *Vibrio harveii.*

A commercially available probiotic, namely MeraBac W (Novus corporation, MO, USA) was tested in a small scale assay system under controlled environmental conditions with the objective to: (i) develop as small scale assay for testing the effects of probiotic in shrimp ponds; and (ii) elucidate the effects of selected probiotic on the water quality of shrimp ponds as well as on the abundance of a shrimp pathogen (*Vibrio harveyii*). In this study, 250 mL of shrimp farming effluent containing frozen aquaculture sludge collected from a shrimp ponds was incubated separately in two benchtop reactor (500 mL each) and was aerated. Moreover, autoclaved shrimp farming effluent was also incubated separately in two benchtop reactors mainly to assess the impact of the probiotic on shrimp pathogen. COD removal in bench top reactors from non-autoclaved shrimp effluent was measured with and without the addition of the probiotic. Initial COD concentration of shrimp effluent was ranged from 2800 to 3300 mg/L.

Notably, addition of the probiotic significantly improved the removal of COD in compared to that obtained without the addition of the probiotic (Fig. 10.7). Moreover, it was observed that improvement in the rate of COD removal depends on the concentration of probiotics. Addition of the probiotic at a concentration of 1 g/L achieved above 99% removal of COD within 4 days. On the other hand, addition of the probiotic at 0.0001 and 0.1 g/L required 6 and 8 days, respectively, to achieve above 99% COD removal (Fig. 10.7). The complete removal of waste COD within 8 days at recommended concentrations for pond application is a very important finding, particularly in consideration of the fact that most growers treat pond weekly assuring continual enhancement of waste digestion. Of interest was the lack of significant change in digestion rates with the addition of glucose. This suggests that despite the high nitrogen content of the shrimp waste tested here, addition of labile carbon did not change sludge digestion rates. Further research

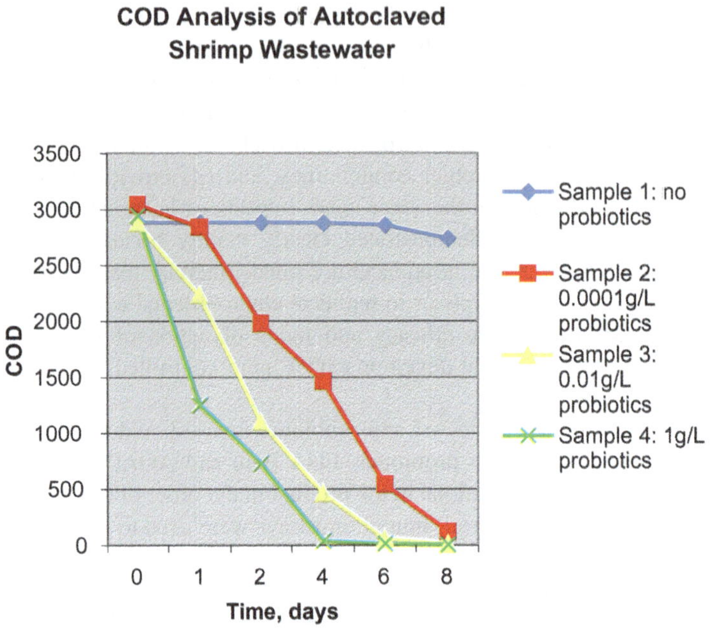

Fig. 10.7 Chemical oxygen demand (COD) on shrimp culture wastewater in presence and absence (control) of the probiotic MERA bac W™ (Novus Int.) of analysis in the reactor with various probiotic concentrations

using this model system could help determine if and when use of supplementary molasses can improve probiotic activity in pond systems.

Impact of probiotic addition on the abundance of *Vibrio harveyii* (a shrimp pathogen) was also assessed in this study by adding a known concentration of *V. harveyii* i.e. 10^9 cell/mL in benchtop reactors at the beginning of the experiment and its concentration was monitored until the end of the experiment. A significant reduction in the concentration of *V. harveyii* was observed in the presence of a probiotic (Fig. 10.8). Initial concentration of *V. harveyii* i.e. 10^9 cell/mL was reduced to 10^2 cell/mL (corresponding to a 5 log removal) within 8 days. On the other hand, concentration of *V. harveyii* in the absence of the probiotic remained at 10^9 cell/mL till the end of the experiment (Fig. 10.8). Interestingly, concentration of *V. harveyii* increased from 10^9 to 10^{12} cell/mL in the reactor fed with autoclaved shrimp farming effluent within 8 days. This study clearly showed that the supplemental heterotrophic bacterial amendments have a competitive edge over *V. harveyii* and thus decreased the population of *Vibrio*. It is interesting to note that *Vibrio* proliferated in the absence of other heterotrophic bacteria in autoclaved waste water. Only the selected bacterial strains in the Mera Bac W were effective in significantly reducing the *Vibrio* concentration in the wastewater.

This study suggests that the application of bacterial amendments contributed significantly to digestion of organic material and suppressed *Vibrio* populations in

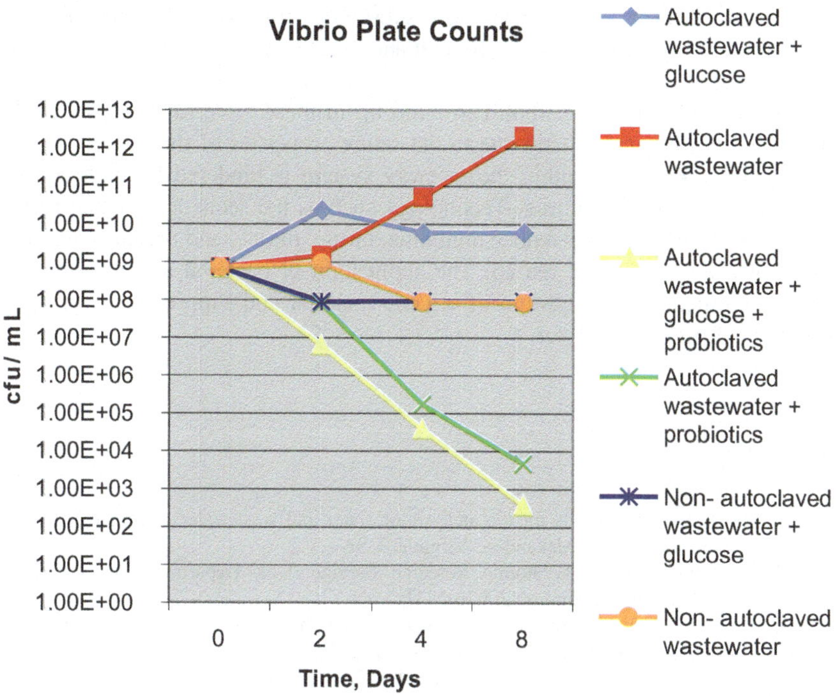

Fig. 10.8 *Vibrio harveyii* bacterial counts in the bioreactors with low concentration of probiotic and native bacteria bioreactor with and without glucose

shrimp wastewater. These controlled and replicated trials can shed light on mode of action and relative efficiency of probiotic activity. The data from these types of comparisons can provide a more robust statistical analysis than pond-based trials. Nevertheless, these laboratory scale trials have important practical implications for use in shrimp production systems. One of the most significant criticisms of probiotic technologies relate to the application rates relative to pond volumes. This study shows at the recommended application rates, the bacteria had significant effects on water quality.

10.4 Conclusion

Around 85% of shrimp consumed in the United States are imported. This leads to a trade deficit of over 3 billion dollar. In order for the United States to reduce this deficit and compete on a global scale in the shrimp market, new technologies, such as recirculating raceway systems, must be explored. Recirculating raceway systems produce large amounts of shrimp per unit volume. The waste produced due to the high density of these systems can cause substantial environmental problems, and disposing of the waste can be very difficult and expensive. Because the recirculating raceway system is a zero-water exchange system, nutrients such as nitrogen, phosphorous, and organic carbon can build up in these systems. Disposal of this wastewater can be problematic due to the saline properties of the wastewater and reuse of the wastewater within the raceway system is hindered by high concentrations of ammonia and nitrite. Technology such as the sequencing batch reactor (SBR) is being used to help reduce ammonia, nitrite, nitrate, and organic carbon, so that the wastewater can be reused. The application of bacterial amendments contributed significantly to digestion of organic material and suppressed *Vibrio* populations in shrimp wastewater.

References

APHA (1998) Standard methods for the analysis of water and wastewater, 20th edn. American Public health Association, Alexandria, Virginia, USA

Aquaculture Outlook (1999) Economic Research Service, U.S. Department of Agriculture, Washington, DC. Report number, 20036-5831

Arbiv R, Rijn VJ (1995) Performance of a treatment system for inorganic nitrogen removal in intensive aquaculture systems. Aquac Eng 14:189–203

Avnimelech Y, Weber B, Hepher B, Mitstein A, Zorn M (1986) Studies in circulated fish ponds; organic matter recycling and nitrogen transformation. Aquac Fish Manage 17:231–242

Avnimelech Y, Kochva M, Hargreaves JA (1999) Sedimentation and resuspension in earthen fish ponds. J World Aquaculture Soc 30:401–409

Boopathy R, Fontenot Q, Kilgen MB (2005) Biological treatment of sludge form a recirculating aquaculture system using a sequencing batch reactor. J World Aquaculture Soc 36:542–545

Boopathy R, Bonvillain C, Fontenot Q, Kilgen MB (2007) Biological treatment of low-salinity shrimp aquaculture wastewater using sequencing batch reactor. Int Biodeterior Biodegradation 59:16–19

Bratvold D, Browdy CL (1999) Disinfection, community establishment and production in a proto-type biosecure shrimp pond. J World Aquaculture Soc 30:422–432

Browdy CL (1998) Recent developments in penaeid broodstock and seed production technologies: improving the outlook for superior captive stocks. Aquaculture 164:3–21

Browdy CL, Moss SM (2005) Shrimp culture in urban, super-intensive closed systems. Urban Aquac 6:173–185

Browdy CL, Stokes AD, Hopkins JS, Sandifer PA (1995) Improving sustainability of shrimp pond water resource utilization. In: 3rd Ecuadorian aquaculture conference, Guayaquil, Ecuador

Brune DE, Schwartz G, Eversole AG, Collier JA, Schwedler TE (2003) Intensification of pond aquaculture and high rate photosynthetic systems. Aquac Eng 28:65–86

Castille FL, Lawerence AL (1998) The effect of salinity on osmotic, sodium, and chloride concentrations in the hemolymph of euryhaline shrimp of the genus *Penaeus*. Comp Biochem Physiol 68A:75–80

Cerra FJ, Maisel JW (1979) Wastewater engineering treatment, disposal, reuse. McGraw-Hill Inc., New York, p 228

Cohen JM, Samocha TM, Fox JM, Gandy RL, Lawrence AL (2005) Characterization of water quality factors during intensive raceway production of juvenile *Litopenaeus vannamei* using limited discharge and biosecure management tools. Aquac Eng 32:425–442

Cripps SJ, Bergheim A (2000) Solids management and removal for intensive land-based aquaculture production systems. Aquac Eng 22:33–56

Csavas L (1994) Important factors in the success of shrimp farming. World Aquaculture 25:34–56

Deng LW, Zheng P, Chen ZA (2006) Anaerobic digestion and post-treatment of swine wastewater using IC-SBR process with bypass of raw wastewater. Water Sci Technol 41:965–969

Fang HP, Yeong CY, Book KM, Chiu CM (1993) Removal of COD and nitrogen in wastewater using a sequencing batch reactor with fibrous packing. Water Sci Technol 28:125–131

Federal Register (2004) Effluent limitations guidelines and new source performance standards for the concentrated aquatic animal production point source category; final rule. 40 CFR Part 451. Monday, 23 Aug 2004. http://epa.gov/guide/aquaculture

Fernandes L, McKyes E, Warith M, Barrington S (1991) Treatment of liquid swine manure in the sequencing batch reactor under aerobic and anaerobic conditions. Can Agr Eng 33:373–379

Fontenot Q, Bonvillain C, Kilgen MB, Boopathy R (2007) Effects of temperature, salinity, and carbon: nitrogen ratio on sequencing batch reactor treating shrimp aquaculture wastewater. Biores Technol 98:1700–1703

Goldburg R, Triplett T (1997) Murky waters: environmental effects of aquaculture in the US. http://www.environmentaldefense.org/documents/490_AQUA.pdf

Hach D (1999) Hach DR/2000 spectrophotometer handbook. Loveland, Colorado, USA

Hagopian DS, Riley JG (1998) A closer look at the bacteriology of nitrification. Aquac Eng 18:223–244

Holloway M (2002) Blue revolution. Discover 23:56–63

Hopkins JS (1994) An apparatus for continuous removal of sludge and foam fractions in intensive shrimp ponds. Progressive Fish Culturist 56:135–139

Irwine RL, Ketchum LH (1989) Sequencing batch reactors for biological wastewater treatment. CRC Crit Rev Environ Control 18:255–294

Jang JD, Bradford JP, Rennebery R (2004) Application of biochemical oxygen demand (BOD) biosensor for optimization of biological carbon and nitrogen removal from synthetic wastewater in a sequencing batch reactor. Biosens Bioelectron 19:805–812

Jory DE (2000) Shrimp—potential for development of marine shrimp farming in Florida. Aquac Mag 26:72–80

Juteau P, Tremblay D, Villemur R, Bisaillon JG, Beaudet R (2005) Analysis of the bacterial community inhabiting an aerobic thermophilic sequencing batch reactor (AT-SBR) treating swine waste. Appl Microbiol Biotechnol 66:122–155

Kargi F, Uygur A (2005) Improved nutrient removal from saline wastewater in an SBR by *Halobacter* supplemented activated sludge. Environ Eng Sci 22:170–171

Kinne PN, Samocha TM, Jones ER, Browdy CL (2001) Characterization of intensive shrimp effluent and preliminary studies on biofiltration. North Am J Aquac 63:25–33

Keithly WR, Diop H, Kazmierczak RF, Travis MD (2005) An economic analysis of the southeast US shrimp processing industry responses to an increasing import base. Proc Gulf Caribb Fish Inst 56:133–149

Landesman L (1994) Negative impacts of coastal aquaculture development. World Aquaculture 25:12–17

Lin Y, Chen J (2001) Acute toxicity of nitrite on *Litopenaeus vannamei* (Boone) juveniles at different salinity levels. Aquaculture 244:193–201

Lin Y, Chen J (2003) Acute toxicity of ammonia on *Litopenaeus vannamei* boone juveniles at different salinity levels. J Exp Mar Biol Ecol 259:109–119

Lo KV, Liao PH, Van Kleeck RJ (1991) A full scale sequencing batch reactor treatment of dilute swine wastewater. Can Agric Eng 33:193–195

Masse DI, Masse L (2000) Treatment of slaughterhouse wastewater in anaerobic sequencing batch reactors. Can Agric Eng 42:131–137

Mazik PM, Parker NC (2001) Semi-controlled systems. In: Wedemeyer GA (ed) Fish hatchery management, 2nd edn. American Fisheries Society, Bethesda, Maryland, pp 241–284

McGraw WJ (2002) Utilization of heterotrophic and autotrophic bacteria in aquaculture. Global Aquac Advocate 12:82–83

Menz A, Blake BF (1980) Experiments on the growth of *Penaeus vannamei* boon. J Exp Mar Biol Ecol 48:99–111
Morgenroth E, Wilder PA (1998) Sequencing batch reactor technology: concepts, design and experiences. Water Environ Manag 12:314–321
Murat S, Genceli EA, Tasli R, Artan N, Orhon D (2002) Sequencing batch reactor treatment of tannery wastewater for carbon and nitrogen. Water Sci Technol 46:219–227
Ng WJ, Kho K, Ong SL, Sin TS, Ho JM, Tay SH (1993) Preliminary estimation of tropical ornamental fish metabolite production rates. Aquaculture 110:263–269
Ostrowski A, Warren A, Atwood H, Browdy C, Boopathy R, Lotz JM, Lawrence A, Lightner D, Moss S (2005) Strategies and focus of the U.S. marine shrimp farming program in a changing world market. Paper presented at the aquaculture America 2005 conference, New Orleans, 17–20 Jan 2005
Otoshi CA, Arce SM, Moss SM (2003) Growth and reproductive performance of broodstock shrimp reared in a biosecure recirculating aquaculture system versus a flow-through pond. Aquac Eng 29:93–107
Peters M, Newland M, Seviour T, Broom T, Bridle T (2004) Demonstration of enhanced nutrient removal at two full-scale SBR plants. Water Sci Technol 50:115–120
Pruder GD (2004) Biosecurity: application in aquaculture. Aquac Eng 32:3–10
Remde A, Conrad R (1990) Production of nitric oxide in *Nitrosomonas europea* by reduction of nitrite. Arch Microbiol 154:187–191
Rijn VJ (1995) The potential for integrated biological treatment systems in recirculation fish culture—a review. Aquaculture 139:181–201
Ritchie GA, Nicholas DJ (1972) Identification of the sources of nitrous oxide produced by oxidative and reductive processes in *Nitrosomonas europaea*. Biochem J 126:1181–1191
Statistical Analysis System (SAS) Institute (2003) Version 9.1. SAS Institute, Cary, North Carolina
Sauthier N, Grasmick A, Blancheton JP (1998) Biological denitrification applied to a marine closed aquaculture system. Water Res 32:1932–1938
Sharma B, Ahlert RC (1977) Nitrification and nitrogen removal. Water Res 11:897–925
Summerfelt S, Bebak-Williams J, Tsukuda S (2001) Controlled systems: water reuse and recirculation. In: Wedemeyer GA (ed) Fish hatchery management, 2nd edn. American Fisheries Society, Bethesda, Maryland, pp 285–397
Sundermeyer-Klinger H, Meyer W, Warninghoff B, Brock E (1984) Memebrane-bound nitrite oxidoreductase of *Nitrobacter*: evidence for a nitrate reductase system. Arch Microbiol 140:153–158
Tanaka Y, Fukumori Y, Yamanaka T (1983) Purification of cytochrome a_1c_1 from *Nitrobacter agilis* and characterization of nitrite oxidation system of the bacterium. Arch Microbiol 135:265–271
Weirich CR, Browdy CL, Bratvold D, McAbee BJ, Stokes AD (2002) Preliminary characterization of a prototype minimal exchange super-intensive shrimp production system. In: Proceedings of the IVth international conference on recirculating aquaculture, Virginia Tech University, Blacksburg, July 2002. pp 255–270
Williams AS, Davis DA, Arnold CR (1996) Density-dependent growth and survival of *Penaeus setiferus* and *Penaeus vannamei* in a semi-closed recirculating system. J World Aquac Soc 27:107–112
Zhu S, Chen S (2001) Effects of organic carbon on nitrification rate in fixed film biofilters. Aquac Eng 25:1–11

Index

F. I. Hai et al. (eds.), *Sustainable Aquaculture*, Applied Environmental Science and Engineering for a Sustainable Future,
https://doi.org/10.1007/978-3-319-73257-2

The manufacturer's authorised representative in the EU is Springer Nature Customer Service Centre GmbH, Europaplatz 3, 69115 Heidelberg, Germany. If you have any concerns regarding our products, please contact ProductSafety@springernature.com

Printed and bound by CPI Group (UK) Ltd, Croydon, CR0 4YY
30/09/2026
02225658-0011